CTS®-I
Certified Technology Specialist-Installation

EXAM GUIDE

ABOUT THE AUTHOR

Andy Ciddor is a technical writer, consultant, educator, and systems technician. His unhealthy interest in how things work began when he terminally dismembered an army field telephone at the age of eight. Andy's interests eventually expanded into wireless, audio, lighting, electronics, AV, IT, software, control systems, communications, technical production, cosmology, rocket science, neurobiology, etc., as the opportunities arose. He has contributed on a wide range of subjects to publications across the English-speaking world and was the founding editor of *AV Technology* magazine. His favorite place is at the bottom of a steep learning curve.

About the Technical Editor

Greg Bronson, CTS-D, is a technical advisor for AVIXA. He has worked as a technician, service manager, and system designer/project manager over a 30+ year career in AV. Bronson is a longtime volunteer and SME for AVIXA, including leadership roles within its standards, education, committee, and council programs.

CTS®-I
Certified Technology Specialist-Installation
EXAM GUIDE

Second Edition

Andy Ciddor

New York   Chicago   San Francisco
Athens   London   Madrid   Mexico City
Milan   New Delhi   Singapore   Sydney   Toronto

McGraw Hill books are available at special quantity discounts to use as premiums and sales promotions, or for use in corporate training programs. To contact a representative, please visit the Contact Us pages at www.mhprofessional.com.

CTS®-I Certified Technology Specialist-Installation Exam Guide, Second Edition

1 2 3 4 5 6 7 8 9   LCR   25 24 23 22 21

Library of Congress Control Number: 2021931147

ISBN   978-1-260-13609-8
MHID   1-260-13609-4

Sponsoring Editor	**Technical Editor**	**Production Supervisor**
Tim Green	Greg Bronson	Thomas Somers
Editorial Supervisor	**Copy Editor**	**Composition**
Patty Mon	Lisa McCoy	KnowledgeWorks Global Ltd.
Project Managers	**Proofreaders**	**Illustration**
Garima Poddar, Neelu Sahu	Paul Tyler, Richard Camp	KnowledgeWorks Global Ltd.
KnowledgeWorks Global Ltd.	**Indexer**	**Art Director, Cover**
Acquisitions Coordinator	Ted Laux	Jeff Weeks
Emily Walters		

Image Credits for Table 6-4:
Page 105, DisplayPort: SooniosPro / Getty Images
Page 105, Mini DisplayPort: Beeldbewerking / Getty Images
Page 107, USB-type A: phanasitti / Getty Images
Page 107, USB-type B: phanasitti / Getty Images

CONTENTS AT A GLANCE

CONTENTS

FOREWORD

What does it take to create an exceptional audiovisual experience? What creates a spectacle that wows people or the type of integrated collaboration space that makes companies more productive? It takes the right combination of content, space, and technology, but it also takes you. It takes trained professionals—integrators, designers, manufacturers, distributors, and technology managers.

AVIXA™, the Audiovisual and Integrated Experience Association, believes in the value of creating experiences through technology—experiences that lead to better outcomes through enhanced communication, entertainment, and education. To deliver these experiences and outcomes, our industry must continually master its craft. That means staying well-trained, using and understanding standards, and committing to professional certifications. Building integrated audiovisual experiences is guided by the science of sight and sound, activated by technology, and realized by dedicated specialists.

For more than 30 years, AVIXA has administered the Certified Technology Specialist™ (CTS') program, which is recognized as the leading AV professional credential. There are three CTS credentials: general (CTS), design (CTS-D), and installation (CTS-I). There are currently more than 12,000 CTS holders, and more than 2,000 of them are CTS-D or CTS-I holders. The CTS program is accredited by the American National Standards Institute (ANSI) to meet the International Organization of Standardization (ISO) and International Electrotechnical Commission (IEC) ISO/IEC 17024:2012 certifications of personnel standards.

As you will learn in this guide, a CTS-I installs and maintains AV systems through a series of industry-accepted skills and best practices. Those who hold the CTS-I credential, like its counterpart for the AV design community, the Certified Technology Specialist– Design (CTS-D), are members of a special group of AV professionals who have gone beyond foundational experience and dedicated themselves to quality work, focused expertise, and the confidence of the people they work with.

So, congratulations on your decision to pursue your CTS-I certification. We hope this guide helps you reach your goals and continue to advance this exciting industry.

What's inside these pages? For starters, everything you might expect.

Despite changes in AV technology, many skills remain the same. CTS-I professionals are proficient in building racks, terminating cables, setting audio gain, and verifying the performance of installed AV systems. All of those subjects are covered here.

Today's CTS-I can also configure and troubleshoot modern AV protocols, such as extended display identification data (EDID) and high-bandwidth digital content protection (HDCP)—two important attributes of today's high-definition video systems. Both, and many more, are addressed in this guide.

Finally, a CTS-I knows that the so-called convergence of AV and IT technologies is yesterday's news. Convergence has happened. As a result, AV installers must know how to integrate and control AV gear on a network. They must understand Internet protocols, network security, and a host of other IT concerns that help ensure networked AV systems operate as promised without affecting other services. The *CTS-I Certified Technology Specialist–Installation Exam Guide, Second Edition* includes a networking chapter devoted to such AV/IT skills.

All told, the document in your hands encapsulates the AVIXA knowledge base for aspiring CTS-I holders. Even if you have never taken the CTS-I exam, this guide represents a handy reference for those in the field. It reflects the necessities of being a modern, expert AV installer; as such, it also reflects that the CTS-I exam as it has changed over time. For more on the exam itself, see Chapter 2.

The *CTS-I Certified Technology Specialist-Installation Exam Guide, Second Edition* will prepare you for the CTS-I exam, but it is not required reading, and for good reason.

The CTS-I credential is accredited by ANSI under the ISO/IEC 17024 General Requirements for Bodies Operating Certification Schemes of Persons program. For many, an ANSI/ISO/IEC certification is an additional mark of distinction. There are roughly 1 million ANSI-certified professionals across different industries. In accordance with globally recognized principles, no single publication or class will necessarily prepare you for the CTS-I exam, nor are you obligated to enroll in AVIXA courses to take the exam. If, however, you are interested in other ways that AVIXA can help establish your professional qualifications, visit us at avixa.org/certification-section/cts-i.

This is an exciting time to be part of the AV industry. By certifying your skills, you've shown already that you are committed to your own success and to the success of AV professionals everywhere. Now you're ready to go a step further and be a leader in this industry. We thank you for your commitment and wish you luck in your certification journey.

David Labuskes, CTS, CAE, RCDD
Chief Executive Officer
AVIXA

ACKNOWLEDGMENTS

Thank you to the countless audiovisual professionals who have contributed to the knowledge base that resides at AVIXA. These professionals have also recognized the need for advanced study and experience in the field of AV systems integration, resulting in the Certified Technology Specialist-Installation (CTS-I) exam and certification.

The repository of knowledge at AVIXA continues to grow as inventors, manufacturers, engineers, system designers, system integrators, subject matter experts, AV technicians, and instructors share their expertise and develop new solutions. This edition of the *CTS-I Certified Technology Specialist-Installation Exam Guide* draws heavily on that accumulated knowledge and expertise, together with help from the volunteer subject matter experts whose efforts assisted in steering this book down the ever-changing technology path that is twenty-first-century AV.

I would like to thank AVIXA for giving me the opportunity to work on this exciting project, with special thanks to the project supervisor, Bob Higginbotham, CTS-I, CTS-D; our technical editor, Greg Bronson, CTS-D; and Charles Heureaux and Will Murillo, who stepped out of their normal roles to prepare the artwork for this edition. At McGraw Hill, my profound thanks go to Tim Green, Emily Walters, Claire Yee, and Patty Mon for once more patiently guiding me through the process of creation. At KnowledgeWorks Global Ltd. my thanks are to Garima Poddar and Neelu Sahu and their team for wrangling the manuscript through the production process.

Most of all, I would like to offer my heartfelt thanks to the countless legions of technicians, directors, designers, engineers, trainers, educators, clients, and innocent bystanders who have patiently (and sometimes not so patiently) answered my questions during a lifetime of inquiring as to what it is, how it works, and why they did it that way.

Like every twenty-first-century technical writer, I must acknowledge the contributions of Vannevar Bush and Ted Nelson, whose concept of hypertext, and its child, the World Wide Web, made researching this work possible. I must also tip my hat to Jimmy Wales and his imperfectly amazing Wikipedia project, which serves as such a valuable launching point for journeys into the unknown.

This guide happened only because of my partner Val's encouragement, support, and tolerance, together with the cheering on from the sidelines from my adult offspring, Rivka, Lachlan, and Rhian.

Congratulations for extending your skills to the CTS-I level, and best wishes for success in the exam.

—Andy Ciddor, 2021

Special thanks go to the AVIXA volunteer task group members Mike Tomei, CTS-D, CTS-I, Tomei AV Consulting; Mark Madison, CTS-D, CTS-I, senior director, Diversified, and Jason Antinori, CTS-D, senior design specialist, TELUS, for sharing both time and expertise throughout the technical edit process.

2021 AVIXA Board of Directors

Leadership Search Committee Chair: Jeff Day, President, North of IO Advisors

Chairman: Jon Sidwick, President, Collabtech Group

Vice Chair: Samantha Phenix, CEO, Phenix Consulting

Secretary-Treasurer: Martin Saul, CEO, ICAP Global

Directors: Ian Harris, CTS-D, President and Principal Consultant, ihD Ltd

Alexis La Broi, CTS, Director of Sales, Mid-Atlantic and New England, ihD Ltd

Cathryn Lai, Senior Vice President and General Manager of the US Scientific Games, Digital

Tobias Lang, CEO, Lang AG

Alexandra Rosen, Director of Communications and Thought Leadership, GoDaddy

Kay Sargent, Senior Principal/Director of Workplace, HOK

Jatan Shah, Chief Operating and Technology Officer, QSC

Brad Sousa, Chief Technology Officer, AVI Systems

AVIXA Staff

David Labuskes, CTS, CAE, RCDD, CEO

Dan Goldstein, Chief Marketing Officer

Pamela M. Taggart, CTS, Vice President, Content Creation

Bob Higginbotham, CTS-D, CTS-I, Director, On-Demand Training and Standards

Jodi Hughes, Director of Content Delivery

Nicole R. Verardi, Senior Director Marketing

Kelly Smith, Manager Marketing

Zachary Fisher, CTS, Senior Developer On-Demand Content

Chuck Espinoza, CTS-D, CTS-I, Senior Staff Instructor

Leslie Rivera, Developer On-Demand Content

Charles Heureaux, Instructional Graphic Designer

William Murillo, Senior Designer

Michelle Bollen, Standards Developer

Loanna Overcash, Standards Developer

PART I

The CTS-Installation and CTS-I Exam

Getting CTS-I Certification

In this chapter, you will learn about
- What types of work an AV installer does
- The benefits of earning a CTS-I credential
- Eligibility criteria for taking the CTS-I exam
- AVIXA certifications and work on establishing standards

As a Certified Technology Specialist (CTS), you already know what audiovisual (AV) systems integration firms do for their clients. You may have performed a number of the basic tasks routinely in your career. At this point, you are ready to make a greater commitment to your professional life in AV systems installation and integration by preparing for the Certified Technology Specialist-Installation (CTS-I) exam. Congratulations on your decision, and welcome to this self-study guide.

You may have noticed that some of your colleagues and supervisors with advanced skills are engaged in specific aspects of complex and innovative AV projects. For many decades, the commercial AV industry has been at the forefront of deploying new digital audio, video, and AV networking technologies that increase workplace efficiencies and enhance the end-user experience. AVIXA has already completed 80 years of service and support to the professional AV community.

The industry employs thousands of technicians, installers, and designers who are responsible for creating some of the most exciting AV experiences at museums, trade shows, hotels, arenas, entertainment complexes, parks, shopping centers, and theaters, as well as installing dependable AV communication and collaboration systems at corporations, universities, government, schools, and other nonprofit institutions. Commercial-grade AV systems are utilized in just about every industry, and as new technologies emerge, AV installers are involved in replacing old technologies and installing new systems.

What Does a CTS-I Certified Installer Do?

AV projects involve a multitude of skills and knowledge that draw from several disciplines, including engineering, science, and the arts. When you prepared for your CTS exam, you studied the basics of how audio and video systems work. Your study for certification and work experience would have given you a strong foundation of the basic skills and general principles that come into play when working on an AV project.

As you prepare for the CTS-I exam, you will need to hone some of those basic skills and move ahead with mastering more specific ones. An installer performs specialized tasks to set up and maintain AV systems by doing the following:

- Following specifications, schematics, codes, and safety protocols
- Administering installation process logistics
- Troubleshooting and problem-solving systems
- Maintaining tools and equipment
- Communicating with clients, designers, other trades, other installers, and staff to provide the best AV solutions for client needs, on time and within budget

These are the overall or general activities performed by an AV installer. AVIXA has developed a job task analysis (JTA), which is a comprehensive list of the key responsibilities (referred to as *duties*) and the tasks that an installer should demonstrate proficiency in. Based on the JTA, AVIXA's independent certification committee has created a CTS-I exam content outline. Both the JTA and outline are available at the organization's website and are included in the free *CTS-I Candidate Handbook* (available in print and online at www.avixa.org/training-certification/certification/cts-i).

The content and practice exercises in this book do not follow the CTS-I exam content outline perfectly, or the order in which the exam questions are presented. Instead, the content follows the course of installation from pre-installation activities to completion and sign-off.

AVIXA Certifications

Certification shows your commitment to being the best in the AV field. The continuing education that accompanies certification will help keep you up to date with advancing technologies. Pursuing advanced certification is an excellent decision for your company and career.

Currently, three CTS certifications are available:

- Certified Technology Specialist (CTS)
- Certified Technology Specialist-Design (CTS-D)
- Certified Technology Specialist-Installation (CTS-I)

All three of AVIXA's certifications have achieved accreditation through the International Organization for Standardization (ISO) and the International Electrotechnical Commission (IEC) as administered by the American National Standards Institute (ANSI) in the United States. They have been accredited by ANSI to the ISO/IEC 17024 personnel standard—the AV industry's only third-party accredited personnel certification program. These are the only certifications in the AV industry to achieve ANSI accreditation.

The certification programs are administered independently by AVIXA's certification committee. You can learn more about how the exams are developed and administered, as well as how to maintain your certification, by visiting the certification website at www.avixa.org/certification.

Why Earn Your CTS-I Credential?

Earning your CTS-I credential demonstrates to your employer that you take your job in the AV industry seriously. A career in the installation world is a commitment. You are dedicating your professional life to a higher level of excellence that can be achieved only through education and expertise in the AV field.

Although certification is not a guarantee of performance by certified individuals, Certified Technology Specialist holders at all levels of certification (CTS, CTS-D, CTS-I) have demonstrated AV knowledge and skills. Certified individuals adhere to the CTS Code of Ethics and Conduct and maintain their status through continued education. Certification demonstrates commitment to professional growth in the AV industry and is strongly supported by AVIXA.

The CTS credential is recognized worldwide as the leading AV professional credential. Holding a CTS shows your professionalism and technical proficiency and increases your credibility and your customers' confidence. So why take the next step toward specialized certification?

The general CTS is exactly that: general. It demonstrates a broad base of knowledge and experience in the AV industry. The CTS-I credential demonstrates mastery of a particular sector of the industry: systems integration. It is narrower than the CTS, but it is also more technical, more in-depth, and more advanced. That is why more than half of independent design consultants require the CTS-I in their bid specifications, according to the results of the *CTS Survey of Consultants*, published by AVIXA.

Employers, technology managers, and design consultants want to know that the people installing their systems can do the following:

- Perform all the common AV cable termination types
- Spot an improperly connected termination on sight
- Set up systems to meet exacting design specifications
- Configure cutting-edge technologies, including networked devices, digital signal processors, and distributed digital video
- Verify that the system performs exactly according to need, intent, and design
- Perform in-depth troubleshooting
- Shoulder the responsibility of maintaining a system over the long term

All that takes experience and know-how that a general CTS cannot necessarily be assumed to have. When a company hires a CTS-I, however, it can feel confident of getting a seasoned professional who is ready to execute all these duties independently from day one. A CTS is a valuable team member. A CTS-I leads the team.

When are you ready to take the next step from general CTS certification to specialized CTS-I certification? If you are ready to take on the most challenging technical aspects of systems integration, if you are ready to take responsibility for the overall quality of the systems you install, or if you are ready to lead, then you are ready for the CTS-I.

Are You Eligible for the CTS-I Exam?

To be considered eligible to sit for the CTS-I certification examination, you must meet the following prerequisites:

- Hold a current CTS certification
- Be in good standing with the certification committee (in other words, have no ethics cases or sanctions)
- Have two years of AV industry experience

There are several other process requirements, such as proof of identity and an application fee. The *CTS-I Candidate Handbook* contains all the information on the eligibility requirements, as well as the application form.

AVIXA Standards

AVIXA is an ANSI-accredited standards developer. When working in the AV industry, you will need to know the standards set by AVIXA for guidance in your work or running your AV business. Questions on the CTS-I exam may also be based on these standards. So, in addition to what you will be learning in preparation for the CTS-I exam, you should take a few minutes to become familiar with these standards.

ANSI/AVIXA-approved standards include the following:

- Audio Coverage Uniformity in Enclosed Listener Areas
- Standard Guide for Audiovisual Systems Design and Coordination Processes
- Cable Labeling for Audiovisual Systems
- Image System Contrast Ratio
- Rack Design for Audiovisual Systems
- Audiovisual Systems Energy Management
- Rack Building for Audiovisual Systems
- Audiovisual Systems Performance Verification

Want to learn more about standards? AVIXA has many more standards currently in development. Visit www.avixa.org/standards to learn more. AVIXA Elite-level members can download the standards for free from AVIXA's website. All completed standards can be purchased from the AVIXA web store at https://store.avixa.org/.

As technologies evolve and innovations spawn new ones, the need to develop and establish standards for new devices, software, and system solutions is ongoing. You can count on AVIXA to address the need for standardization in the future.

Getting Started

When preparing for a computer-based exam like the CTS-I exam, your ability to read, analyze, and respond to multiple-choice questions quickly will make a big difference on your exam scores. There are many strategies for locating the correct answer in a multiple-choice question.

VIDEO   Watch the videos on multiple-choice questions and on demystifying AVIXA certification. The tips provided in the multiple-choice questions video could save you time when separating the right answer from the distracters, either by reinforcing the correct answer or by eliminating one or two wrong answers. Check Appendix D for a link to these and other videos.

The next chapter will suggest more ways in which you can prepare for the CTS-I exam.

The CTS-I Exam

In this chapter, you will learn about
- The scope of the CTS-I exam
- Exam preparation strategies
- Math strategies
- Sample exam questions
- The CTS-I exam application

Now that you understand what a professional audiovisual installer does, the benefits of earning Certified Technology Specialist-Installation (CTS-I) certification, and the eligibility requirements, you will want to start preparing for the exam. Your preparation for the CTS-I exam should include courses recognized for CTS-I eligibility and a study of the material in this exam guide. What you learn on the job will also help reinforce some of the knowledge you gain from classroom and online courses, as well as self-study.

This chapter describes the content that the CTS-I exam questions are based on and the strategies you will need to focus on to accurately answer the questions. It also provides math strategies to solve equations by correctly using the order of operations with the TI-30XS MultiView calculator, as well as a primer on making calculations based on Ohm's law.

AVIXA regularly updates the CTS-I exam content and procedures for taking the exam, so be sure to visit AVIXA's website to obtain the latest information and requirements.

The Scope of the CTS-I Exam

The CTS-I exam tests the knowledge and skills required by an audiovisual professional to earn the CTS-I certification. To create the CTS-I exam, a group of volunteer audiovisual subject matter experts (SMEs), guided by professional test development experts, participated in an audiovisual (AV) system installation job task analysis (JTA) study. The results of this study form the basis of a valid, reliable, fair, and realistic assessment of the skills, knowledge, and abilities required for competent job performance by AV professionals who specialize in AV system installation and integration.

In creating the JTA, the group of volunteer SMEs identified major categories or duties, as well as topics within each duty, based on the tasks that a certified individual might perform on an AV installation job. The exam development team examined the importance, criticality, and frequency of AV installation tasks on typical projects and used the data to determine the number of CTS-I exam questions related to each duty and task.

Based on the JTA, the CTS-I exam content outline categorizes the subject matter in five duties and lists the tasks in each of the areas that will be addressed in the exam. It includes the percentage each topic represents and the number of questions dedicated to each on the exam. Table 2-1 lists the duties and tasks, as well as the number of questions on the exam for each.

CTS-I Duties and Tasks	% of Exam	# of Items
Duty A: Conducting Pre-Installation Activities	**20%**	**20**
Task 1: Review Audiovisual Project Documentation	4%	4
Task 2: Conduct Technical Site Survey	4%	4
Task 3: Prepare for Audiovisual Installation	4%	4
Task 4: Evaluate Overall Facility Readiness	4%	4
Task 5: Conduct On-Site Preparations for Installation	4%	4
Duty B: Conducting Site Rough-In/First-Fix	**10%**	**10**
Task 1: Deinstallation of Existing Equipment and Cabling	2%	2
Task 2: Mount Substructure	4%	4
Task 3: Pull Cable	4%	4
Duty C: Installing Audiovisual Systems	**38%**	**38**
Task 1: Assemble AV Rack	4%	4
Task 2: Wire the AV Equipment Rack	5%	5
Task 3: Distribute AV Equipment	4%	4
Task 4: Mount AV Equipment	4%	4
Task 5: Terminate Cables	5%	5
Task 6: Configure Network Properties of Equipment	4%	4
Task 7: Load Configuration and Control Programs	4%	4
Task 8: Test the AV System	4%	4
Task 9: Calibrate the AV System	4%	4
Duty D: Performing Systems Close Out	**11%**	**11**
Task 1: Demonstrate to Client or Client's Representative that System Performs to Specifications	4%	4
Task 2: Provide Training on System Operation	4%	4
Task 3: Obtain Project Completion Sign Off from Client or Client's Representative	3%	3

Table 2-1 CTS-I Exam Content Outline *(continued)*

CTS-I Duties and Tasks	% of Exam	# of Items
Duty E: Conducting Ongoing Project Responsibilities	**21%**	**21**
Task 1: Complete Progress Reports	3%	3
Task 2: Coordinate with Other Contractors	4%	4
Task 3: Address Needed Field Modifications	4%	4
Task 4: Repair AV Systems	3%	3
Task 5: Maintain AV Systems	3%	3
Task 6: Maintain Tools and Equipment	4%	4
Total	**100%**	**100**

Table 2-1   CTS-I Exam Content Outline

Exam Preparation Strategies

You can prepare for the CTS-I exam in a number of ways, including studying this book. One place to start is to perform a self-assessment of your AV installation knowledge in order to identify your strengths and weaknesses. You will find free sample questions that are similar to the questions presented on the CTS-I at the link listed in Appendix E.

NOTE   Because the CTS-I exam is designed to comply with ANSI standards, the CTS-I sample exam is not allowed to include actual exam questions, and practice questions may not be informed by the exam itself. Any practice question you find here or elsewhere is written to be *similar* to an actual CTS-I exam question.

It is also a good idea to thoroughly review the exam content outline and the JTA. When reviewing the exam outline, focus on the major content areas and the tasks listed for each duty. The greater the number of exam questions for a task, the more emphasis you might place on studying that content and mastering how to perform that task. For example, because 37 percent of the exam covers installing AV systems (duty C), you will want to spend more of your study time focused on that content and the ten tasks within that area that the questions will be based on.

The JTA provides a comprehensive list of topics and the skills in each area. Your work experience may give you confidence in performing certain tasks and knowing the answers to related questions. For example, you may have been involved in conducting a technical site survey (duty A, task 2) and know how to calculate conduit capacities.

But if you were not involved in conducting field engineering, you might not know how to make installation decisions in response to site assessment (duty E, task 4). Hence, you may want to concentrate your study on topics in the JTA that you are not familiar with. You may find it useful to download the JTA (from the free *CTS-I Candidate Handbook* at www.avixa.org/ctsi) to your mobile device for easy and frequent review, as well as highlight and annotate the topics and skills you need to work on.

The content relating to the duties and tasks in the JTA are covered in this book in five main sections:

I Pre-installation activities such as interpreting drawings and other documentation

II System fabrication (building AV racks)

III Audio and video systems installation and integration

IV System verification

V Conducting system closeout, including cleanup and customer training

The content and sample questions in this book are based on the same JTA and exam content outline as the CTS-I exam. However, because the exam questions are secret, there can be no guarantee that this book will cover every question on the exam or that the exam will address every topic in this book. This book prepares you for a career as a certified AV installer—not just for the credentialing exam.

While working in the AV industry, you may have come across numerous acronyms and technical terms. Study the glossary in this book to get familiar with the definition of terms.

CTS-I Exam Sample Questions

The CTS-I exam is composed of 100 multiple-choice questions that address each of the duties and tasks listed in Table 2-1. The questions focus primarily on issues that an AV professional may encounter when working on a specific job or task, rather than on general AV technology knowledge. Ten of the questions are pilot questions; these questions will not be scored, but you will not know which of the questions are the pilot ones.

Here are five examples of the types of questions that may appear on the CTS-I exam. For each question, the duty and task from which the question is drawn are identified within brackets preceding the question. The correct answer to each question is provided after these five questions.

1. [Conducting Pre-installation Activities/Evaluate Overall Facility Conditions]

   During a technical site survey for an installation with a display, information gathered should include:

   A. Color of the walls

   B. Receptionist name and phone number

   C. Ceiling height and construction method

   D. Rules governing use of the freight elevator

2. [Conducting Site Rough-In/First-Fix/Mount Substructure]

   When mounting a 34kg (75lb) projector with a 7kg (15lb) mount, the substructure should be able to support a minimum weight of:

   A. 82kg (180lb)

   B. 41kg (90lb)

    C. 204kg (450lb)

    D. 122kg (270lb)

3. [Installing Audiovisual Systems/Prepare Audiovisual Rack]

Where should you place the wireless microphone receiver antenna in a rack assembly?

    A. Line of sight to the transmitter

    B. Inside the transmitter room

    C. In the front of the rack

    D. At the rear of the rack

4. [Performing Systems Closeout/Provide Training on Equipment Operation]

When training a customer on the operation of a touch-panel interface, an example that the GUI is a good design that meets the customer's needs is:

    A. The customer can easily use the touch panel with little explanation

    B. The customer requires intensive training on how to use the touch panel

    C. The customer thinks the panel looks cool

    D. The customer wants only a few changes

5. [Installing Audiovisual Systems/Terminate Cables]

The following image illustrates:

    A. A properly stripped cable for a linear compression BNC connector

    B. A poorly prepared termination point for a DB-9 connection

    C. A poorly stripped cable for use with a G type connector

    D. A properly prepared connection for an RJ-45 termination

Answers to Sample CTS-I Exam Questions

The following are the correct answers to the preceding sample questions:

1. **C**. Information on ceiling height and construction method should be gathered during a technical site survey for an installation with a display.

2. **C**. 204kg (450lb).

3. **C**. In the front of the rack.

4. **A**. The customer can easily step in and use the touch panel with little explanation.

5. **A**. The image shows a properly stripped cable for a linear compression BNC connector.

In addition to multiple-choice questions, the CTS-I exam may include advanced item types, wherein you are presented with an image and asked to click the correct item/zone.

Mathematical Strategies

Candidates for the CTS-I exam might not have had to solve a word problem or might not have used a mathematical formula in years. Some of the questions on the CTS-I exam can be answered only by solving mathematical equations. Use the order of operations formula and Ohm's law presented here for self-study. If you are familiar with both, they will serve as a refresher in these mathematical principles and will build your confidence with AV mathematics applications.

Mathematical symbols are shorthand marks that represent mathematical concepts. Although most of the symbols are used universally, you may be accustomed to using only the ones you have learned. The symbol for addition is a plus sign (+) and for subtraction is a minus sign (–), but there are two symbols each for multiplication and division.

- Multiplication symbols × or *
- Division symbols ÷ or /

The * and / symbols have been adopted because standard computer keyboards do not include keys for ÷ or ×. You will want to memorize these symbols if you are not accustomed to using them.

When preparing for the CTS-I exam, you should use a Texas Instruments TI-30XS MultiView calculator to practice the relevant calculations. A virtual version of the TI-30XS is provided on the display screen at your Pearson VUE testing station when you take this exam. You will not be provided with a physical calculator during the exam, but when preparing for the exam, it is a good idea to use a physical TI-30XS MultiView calculator (see Figure 2-1) or an on-screen emulator to familiarize yourself with its functions. A search of the Texas Instruments Education website will locate a free, downloadable TI-30XS emulator for either Windows or Macintosh computers.

Figure 2-1
The TI-30XS
MultiView
calculator

Your use of AV calculations should continue long after you have earned your CTS-I certification. In your job, applying mathematics gives your work more credibility. Your supervisors and your co-workers can look at your work to see whether you made decisions based on a trusted method. They can test and verify the quality of your work. The focus on testing and verifying is the mark of a professional.

The Exam Calculator

A scientific calculator is designed to perform complex operations (for example, exponent, square root, logarithm, and so on) that are not common in everyday arithmetic. Many AV installation tasks call for complex calculations, so you must learn how to use a scientific calculator. While scientific calculator applications (apps) are available for most smart personal devices and computers, you are not permitted to take such equipment into the CTS-I testing facility. You should therefore learn to use the TI-30XS MultiView calculator, shown in Figure 2-1, because you will need to use it to answer some questions on the CTS-I exam.

As you familiarize yourself with your TI-30XS MultiView calculator, make sure you can find the following buttons:

- Log
- Antilog
- Reciprocal
- Exponent
- Square Root

The more you use the TI-30XS before you take the exam, the faster you will be able to locate the buttons during the exam.

Order of Operations

Many AV mathematical formulas use addition, subtraction, multiplication, division, exponents, and logarithms. These formulas require a solid foundation in the order of operations, which will help you correctly calculate the desired result by prioritizing which part of the formula to calculate first. It is a way to rank the order in which you work your way through a formula.

The priority of operations is

1. Any numbers within parentheses or brackets

2. Any exponents, indices, or orders

3. Any multiplication or division

4. Any addition or subtraction

If there are multiple operations with the same priority, proceed from left to right: parentheses, exponents, multiplication, division, addition, and subtraction.

You can remember the order of operations by using the acronyms BODMAS (brackets, orders, division, multiplication, addition, subtraction) or PEMDAS (parentheses, exponents, multiplication, division, addition, subtraction).

 TIP If the priority of operations is the same, begin solving from left to right.

Practice Exercise 1

Solve the following equation using the order of operations:

$$2 + (5 \div 8^2 \times 9)$$

Step 1 Anything inside parentheses is processed first. Inside the parentheses, calculate the exponent first.

$$2 + (5 \div 8^2 \times 9)$$
$$2 + (5 \div 64 \times 9)$$

Step 2 Inside the parentheses are now two operations that have the same priority: multiply and divide.

Because they are the same priority, begin solving them from left to right.

$$2 + (5 \div 64 \times 9)$$
$$2 + (0.078125 \times 9)$$
$$2 + (0.703125)$$

This step is typically where a mistake might be made. If the formula is at a stage where the operations are of the same priority, continue solving from *left to right*. If this step is not completed, you will arrive at the wrong answer.

Examine the following *incorrect* processing shown here:

$$2 + (5 \div 64 \times 9)$$
$$2 + (5 \div 576)$$
$$2 + (0.009)$$

In this example of incorrect processing, multiplication was performed first. The incorrect processing of right to left within the parentheses means that the final answer will be wrong.

Step 3 Now the only remaining operation is addition.

$$2 + (0.703125) = 2.703125$$

Answer Rounded to one decimal place, the result is 2.7.

Use the next practice exercise to reinforce your calculation skills.

Practice Exercise 2

Solve the following equation using the order of operations:

$$6^2 + 4 \div (3 \times 8) - 8$$

Step 1 Anything inside parentheses is processed first. Inside the parentheses, multiply first.

$$6^2 + 4 \div (24) - 8$$

Step 2 There are no more operations in parentheses. Calculate the exponent next.

$$36 + 4 \div 24 - 8$$

Step 3 The next operator is divide. Calculate the division.

$$36 + 0.16667 - 8$$

Step 4   The only two operations remaining are addition and subtraction. Both addition and subtraction are the same priority. Solve the formula from left to right. Remember to process from *left to right*.

$$36.16667 - 8 = 28.16667$$

Answer   Rounded to one decimal place, the result is 28.2.

Electrical Calculations

Now that you have practiced basic mathematical principles, you can apply them to a fundamental concept in AV installation: Ohm's law, which you will have studied to attain your CTS certification.

Electrical Basics

From your prior electrical knowledge, you should be able to recall concepts and terms such as Ohm's law, voltage, current, resistance, impedance, and power, which are summarized in Table 2-2.

If you are not comfortable with these terms, please consider revising your electrical fundamentals before proceeding with the CTS-I exam.

Ohm's Law

Voltage, current, and resistance/impedance are all related. The current in an electrical circuit is proportional to the applied voltage. An increase in voltage means an increase in current, provided resistance/impedance stays the same. The relationship between current and resistance/impedance is inversely proportional, meaning that as one increases, the other decreases. An increase in resistance/impedance produces a decrease in current, provided voltage stays the same.

Parameter	Symbol	Definition	Unit of Measurement	Abbreviation
Voltage or Electromotive Force	V or E	Potential difference (electrical pressure)	Volt	V
Current	I	Electron flow	Ampere (Amp)	A
Resistance	R	Opposition to electron flow due to the materials in the circuit	Ohm	Ω
Impedance	Z	Total opposition to electron flow in a circuit	Ohm	Ω
Power	P	Rate at which work is done	Watt	W

Table 2-2   Characteristics of Voltage, Current, Resistance, Impedance, and Power

This relationship between voltage, current, and resistance is defined in mathematical form by Ohm's law.

$$I = V \div R$$
Current = Voltage ÷ Resistance

So, 1 amp is equal to the steady current produced by 1 volt applied across a resistance of 1 ohm.

The relationship between power, current, and voltage is defined in the power equation:

$$P = I \times V$$
Power = Current × Voltage

So, 1 watt is the power produced by a current of 1 amp driven by a potential difference of 1 volt.

Figure 2-2 shows Ohm's law and the power law equation and gives the formulas for calculating any one parameter if the other two are known. If you know the values of two of these variables, you can easily calculate the third.

- If the two known variables are on top of each other, divide the top variable by the bottom.

- If the two known variables are next to each other, multiply them.

Combining Ohm's law and the power law equation allows us to make other calculations such as these:

$$P = I^2 \times R$$
$$R = P/I^2$$
$$P = V^2/R$$
$$I = \sqrt{(P/R)}$$

You will not be provided with any kind of formula sheet to reference during the CTS-I exam. You should rely on these simple diagrams during the exam because they are much easier to memorize.

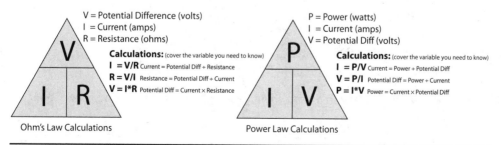

Figure 2-2   Ohm's law and power law formulas

For example, if you want to determine the power (P) resulting when there are 3 amps of current (I) at 12 volts (V), simply cover the *P* in the power law diagram, which leaves *V* and *I* next to each other in a multiply relationship. Multiply the values for *V* and *I* (12 × 3), and the result (36) is the number of watts.

Apply the math operation shown here:

$$P = I \times V$$
$$P = 12 \times 3$$
$$P = 36$$

There are 36 watts of power.

 VIDEO Watch AV Math Online: *Ohm's Law and the Power Formula* is a short video tutorial that explains how to use the Ohm's law and power law relationships. Appendix D provides a link to the video.

The best way to master these calculations is to keep practicing. If you find the following practice exercises simple to do, look for more challenging equations to solve at your place of work or in an online course.

Practice Exercise 1: Ohm's Law Calculation

Perform the following calculation using Ohm's law:

Calculate the current in a circuit where the voltage is 2 volts and the resistance is 8 ohms.

Step 1 In this example, voltage and resistance are known, and you are solving for current. The correct formula to use is $I = V \div R$.

Step 2 Divide voltage by resistance.

$$I = 2 \div 8$$
$$I = 0.25$$

Answer

$$I = 0.25A \ (250mA)$$

Practice Exercise 2: Ohm's Law Calculation

Perform the following calculation using Ohm's law:

Calculate the voltage in a circuit where the current is 4 amps and the resistance is 25 ohms.

Step 1 In this example, current and resistance are known, and you are solving for voltage. The correct formula to use is $V = I \times R$.

Multiply current times resistance.

$$V = 4 \times 25$$

Step 2 Calculate.

$$V = 100$$

Answer

100v

Practice Exercise 3: Ohm's Law and Power Formula Calculation

Perform the following calculation using Ohm's law and the power formula:

Calculate the resistance in a circuit where the voltage is 4 volts and power is 2 watts.

Step 1 In this case, you have only one known Ohm's law variable: voltage. However, you have two known power equation variables: voltage and power. You can use the power formula to derive another Ohm's law variable, current, and then solve for resistance.

First, use the power equation to solve for current.

$$I = P \div V$$
$$I = 2 \div 4$$
$$I = 0.5$$

The current is 0.5A (500mA).
Now that you know the current, use Ohm's law to solve for resistance.

$$R = V \div I$$
$$R = 4 \div 0.5$$
$$R = 8$$

Answer The resistance is 8 ohms.

In fact, using the two relationship diagrams, you can easily derive any two variables from any other two, unless your only two given variables are power and resistance. Just in case, it might be a good idea to memorize $V = \sqrt{(P \times R)}$ for the CTS-I exam.

Practice Exercise 4: Ohm's Law and Power Formula Calculation

Perform the following calculation using Ohm's law and the power formula:

Calculate the voltage in a circuit where the resistance is 16 ohms and power is 4 watts.

Step 1   In this example, resistance and power are known, and you are solving for voltage. Because you already have values for power and resistance, the correct formula to use is the following:

$$V = \sqrt{(P \times R)}$$
Enter the values from the question.
$$V = \sqrt{(16 \times 4)}$$

Step 2   Calculate the value in parentheses first.

$$V = \sqrt{(64)}$$

Step 3   Calculate the square root.

$$V = 8$$

Answer   The voltage is 8 volts.

 NOTE   If you encounter a question or topic in this book that is unfamiliar to you, write it down for further study. That way, you know to focus on these topics when you develop your personal study plan.

The CTS-I Exam Application and Processes

When you are ready to take the CTS-I exam, you will want to review the application process. All the information you need is posted on AVIXA's website at www.avixa.org/ctsi and in the *CTS-I Candidate Handbook*.

These are the important steps to bear in mind:

1. You must meet all eligibility requirements as of the date of the application.

2. You may apply for the exam by using the application in the *CTS-I Candidate Handbook* or by downloading the most recent application. You may also apply online at www.avixa.org/training-certification/certification/cts-i. Mail, fax, or e-mail the application to the AVIXA certification office for review and subsequent approval.

3. Applications will not be processed unless all required information on the application is completed and the application fee is received.

4. You must provide phone and e-mail contact information to facilitate e-mail confirmation of receipt of the application and any necessary phone contact prior to or following the exam.

5. AVIXA will review and respond to your application within approximately ten business days. For applications that are incomplete or lack documentation and/or payment, AVIXA will contact the applicant regarding the missing requirements.

Once you have been approved for eligibility, AVIXA will notify you within one day of notifying Pearson VUE of your eligibility. You may then contact Pearson VUE after a 24-hour period to make your testing appointment. You can find the list of available testing locations at www.pearsonvue.com/avixa.

6. Your application is approved for a period of 120 days from the date of the eligibility approval notice, and you must arrange for and be tested during that 120-day period. The exam application fee must be paid using a major credit card or by check, money order, or wire transfer at the time the application is submitted.

Getting to the Testing Center

On the scheduled day of the CTS-I exam, you should report to the exam center as instructed in your appointment confirmation letter. Plan to arrive at least 30 minutes prior to the scheduled start time. It is not necessary (although it is preferred) to bring your e-mail or letter of confirmation with you. However, you must have proper identification (as described shortly). The name and address on the ID must match the information on file with AVIXA and the vendor responsible for presenting the exam.

Helpful Tips

If you live more than an hour from the exam center, you might consider staying at a nearby accommodation the night before so you can get a good night's rest and make sure you arrive on time. It may also be a good idea to visit the testing center prior to the exam to ensure you know exactly where to go and how to get there. On the day of the exam, make sure to allow extra time for unforeseen events, such as traffic delays. These measures can help reduce unnecessary stress on exam day. If you arrive after your assigned exam time, you will be considered a "no-show" and will not be admitted. To take the exam, you will need to reapply by contacting AVIXA and paying a reinstatement fee.

Identification Requirements

Candidates must check in at the testing center with two forms of valid ID, one of which must be a government-issued photo ID with signature (driver's license, government ID, or passport). (See the "On the Day of the Exam" section in the *CTS-I Candidate Handbook*.) For testing center identification purposes, you must bring both a valid government-issued ID and a secondary ID that has a matching signature to the name on the government ID. *The first and last names on the ID must match exactly with the name submitted on the application, or you will be denied admission.* Candidates can make changes to their names by contacting AVIXA (certification@avixa.org) prior to scheduling their exam appointment. Candidates will also be required to provide a digital signature and have a digital photo taken when checking in. This information is retained in a secure database for no more than five years from the last exam date. The candidate's electronic signature is not linked to the candidate's personal identification information, such as address or credit card information.

For certain countries, including China, Hong Kong, and Taiwan, identification requirements may differ. Please see the *CTS-I Candidate Handbook* for more details.

Items Restricted from the Exam Room

You are not allowed to bring anything into the exam room. Secure lockers are provided to store personal items while taking the exam. The following are examples of items that are *not* permitted in the exam room or testing center:

- Slide rules, papers, dictionaries, or other reference materials
- Phones and signaling devices such as pagers
- Alarms
- Recording/playback devices of any kind
- Calculators (a calculator will be displayed on the test computer screen)
- Photographic or image-copying devices
- Electronic devices of any kind
- Jewelry or watches (time will be displayed on the computer screen and wall clocks in each testing center)
- Caps or hats (except for religious reasons)

Note that the previous list is not exhaustive. Do not expect to be allowed to bring any item into the testing room.

About the Exam

The exam will be presented via computer. The exam uses two different question types: multiple choice and hotspot. For multiple-choice questions, the computer will display each question, with four possible answers (A, B, C, and D). One of the answers represents the single correct response, and credit is granted only if you select that response. For hotspot questions, the computer will display an image and a question. You will be required to mark the correct area of the image with your cursor in order to answer the question. There is only one correct image area, and credit is granted only if you mark the correct area.

As noted earlier, candidates are currently provided 150 minutes for 110 questions, 10 of which are nonscoring pilot questions. There is a brief on-screen computer-based tutorial just prior to starting the exam and a brief online survey at the end of the exam. The time allotted to complete the tutorial and survey is in addition to the 150-minute exam time.

To get familiar with the testing interface, use the online tutorial sample exam from Pearson VUE's AVIXA Testing page at www.pearsonvue.com/avixa. The questions are not AV-related; the tutorial is simply intended to familiarize you with the testing interface. You can use this free resource any time before you take the CTS-I exam.

During the Exam

Bear in mind the following points on how things are handled during the exam:

- Candidates should listen carefully to the instructions given by the exam supervisor and read all directions thoroughly.

- Questions concerning the content of the exam will not be answered during the exam.

- The exam center supervisor will keep the official time and ensure that the proper amount of time is provided for the exam.

- Restroom breaks are permitted, but the clock will not stop during the 150 minutes allotted for the actual exam.

- During the exam, candidates will be reminded when logging in to the testing center computer screen and prior to being allowed to take the exam that they have agreed to follow the CTS Code of Ethics and Conduct and nondisclosure agreements presented earlier in the application process.

- Candidates will have access to a computer-based calculator and a wipe-off note board provided by the testing center.

- Candidates will have the capability to provide comments for any question, as well as mark questions and return to them for review.

- There will be an on-screen reminder when only five minutes remain to complete the exam.

- No exam materials, notes, documents, or memoranda of any kind may be taken from the exam room.

For best results, pace yourself by periodically checking your progress. This will allow you to make adjustments to the speed at which you answer the questions, if necessary. Remember that the more questions you answer, the better your chance of achieving a passing score. If you are unsure of a response, eliminate as many options as possible and choose from the answers that remain. You will also be allowed to mark questions for review prior to the end of the exam.

Be sure to record an answer for each question, even if you are not sure the answer is correct. Again, you can note which questions you want to review and return to them later. There is no penalty for guessing.

Dismissal or Removal from the Exam

During the exam, the exam supervisor may dismiss a candidate from the exam for any of the following reasons:

- The candidate's admission to the exam is unauthorized

- A candidate creates a disturbance or gives or receives help

- A candidate attempts to remove exam materials or notes from the testing room
- A candidate possesses items that are not permitted, as specified in the "Items Restricted from the Exam Room" section earlier in this chapter
- A candidate exhibits behavior consistent with attempting to memorize or copy exam items

Any individual who removes or attempts to remove exam materials or is observed cheating in any manner while taking the exam will be subject to disciplinary or legal action. Sanctions could result in removing the credential or denying the candidate's application for any AVIXA credential.

Any unauthorized individual found in possession of exam materials will be subject to disciplinary procedures in addition to possible legal action. Candidates in violation of AVIXA testing policies are subject to forfeiture of the exam fee.

Hazardous Weather or Local Emergencies

In the event of hazardous weather conditions or any other unforeseen local emergencies occurring on the day of the exam, the exam presentation vendor will determine whether circumstances require cancellation. Every attempt will be made to administer all exams as scheduled.

When an exam center must be closed, the vendor will contact all affected candidates to reschedule the exam date and time. Under those circumstances, candidates will be contacted through every means available: e-mail and all phone numbers on record. This is an important reason for candidates to provide and maintain up-to-date contact information with AVIXA and the exam vendor.

Special Accommodations for Exams

AVIXA complies with the Americans with Disabilities Act (or country equivalent) and is interested in ensuring that no individual is deprived of the opportunity to take the exam solely by reason of a disability as defined under the Americans with Disabilities Act (or equivalent). Two forms must be submitted to receive special accommodations:

- Request for AVIXA (CTS, CTS-D, CTS-I) Exam Special Accommodations
- AVIXA (CTS, CTS-D, CTS-I) Exam, Healthcare Documentation of Disability Related Needs

Applicants must complete both forms and submit them with their application information to the AVIXA certification office no later than 45 days prior to the desired exam date.

Requests for special testing accommodations require documentation of a formally diagnosed or qualified disability by a qualified professional who has provided evaluation or treatment for the candidate.

You can find these forms, along with more information about the process, at the AVIXA CTS-I website and in the *CTS-I Candidate Handbook*.

Exam Scoring

Candidates receive their results immediately upon test completion. The final passing score for each examination is established by a panel of SMEs using a criterion-referenced process. This process defines the minimally acceptable level of competence and takes into consideration the difficulty of the questions used on each examination.

Candidates who do not pass the exam receive their score and the percentages of questions they answered correctly in each duty. AVIXA provides these percentages in order to help candidates identify their strengths and weaknesses, which may assist them in studying for a retest. It is not possible to arrive at your total exam score by averaging these percentages because there are different numbers of exam items in each duty on the exam.

Retesting

Candidates who do not pass the CTS-I exam may retest two more times following their original exam date. There is a minimum 30-day waiting period between each retest. After the retest application has been approved, the candidate has 120 days from the date of their newly reissued eligibility notice to retake the exam.

After two retests, if you still have not passed the exam, you must wait 90 days before restarting the application process. This period allows the applicant time to adequately prepare and prevents overexposure to the exam.

Currently certified CTS-I individuals may not retake the CTS-I exam, except as specified by AVIXA's CTS-I renewal policy.

Candidates must meet all eligibility requirements in effect at the time of any subsequent application. You can find the CTS-I Exam Retest Application form and current retest fees at the AVIXA website.

Chapter Review

Upon completion of this chapter, you should have a clear understanding of the following:

- What audiovisual content areas the CTS-I exam questions are based on
- How to study and prepare for the CTS-I exam
- How to solve equations using the order of operations correctly with the TI-30XS MultiView calculator
- How to make calculations using Ohm's law formulas with no errors
- How to apply for the exam
- What to expect on the day of the exam, including how the exam is conducted and other relevant factors

With all this information at your fingertips, you will want to move ahead with your study of the content, skills, and responsibilities involved in AV systems installation and integration.

PART II

The Basics of AV Installation

Managing an AV Project

3

In this chapter, you will learn about
- Scope of work
- Developing a project plan
- Managing resources and activities
- Project management documentation
- Types of AV projects
- Key stages of AV installation

Every successful project involves planning and management, but audiovisual (AV) installations demand a high level of attention to detail at every stage of the project. This is because AV projects involve many stakeholders, a wide range of trades and contractors, and a long list of manufacturers and products. A change in a single component could impact the timeline and budget. That is why every aspect of the project requires proper planning and tracking. Planning and management are not just for project managers—everyone involved in the project needs to have the essential skills to plan and monitor the activities of their portion of the job and coordinate with other stakeholders. A CTS-I–certified AV professional is expected to be a leader in the project management process, not just a participant.

Planning is a pre-installation activity. It is a way to take stock of and prioritize all the elements of a project. During the planning stages, you will need to have a clear understanding of the client's objectives and purpose for the AV installation. Typically, at the start of a simple project, your company and client will meet to discuss what needs to be done. On larger and more complex projects, the planning meetings will also include other stakeholders and vendors, such as the AV design consultant, *general contractor* (GC, also known as the *building contractor, prime contractor,* or *main contractor*) and electrical contractor. A huge amount of planning time is spent on communication about resources, activities, scheduling, cost, and ongoing communication about the project.

As a professional on the AV team, you need to know what steps to follow to reach your specific goal efficiently. You need to know where you are in the project; who to talk to for clarification; who is in charge; and your project deadlines, milestones, and schedule.

You must know how to read and modify spreadsheets and other planning documents to track the activities and components you are responsible for to ensure that the project will be completed on time and within budget. Good communication skills and flexibility to make changes are desired qualities during the planning process. Understand that some things will happen a little sooner than you planned and others will be delayed. But along the way, you will need to complete certain milestones based on the agreement your company has with the client.

Scope of Work

It is important to read the scope statement before developing a project plan. The scope statement is a contractual agreement between the client, the project sponsor, the key stakeholders, and the AV project team. It defines the boundaries of the project. Any deviation from the agreed scope must be formally negotiated and documented through the change processes described in the "Change Documentation" section.

A scope statement has two distinct components:

- **Deliverable scope** The features and functions that characterize a product, process, or service to be delivered. It states what will be delivered.

- **Project scope** All of the work that must be accomplished to create and deliver a product or service with the specified features and functions. It states how the product or service should be delivered.

Because the scope statement is written to provide the basis for future project decisions, it should include items that are both in and out of scope. Separately listing both in-scope and out-of-scope items will ensure that you include all activities in the scope statement.

Based on the client's needs and expectations, the scope statement should clearly list what your company will, and will not, be doing and what it will, and will not, be delivering. For example, an in-scope task would be the proper termination of cables, and an out-of-scope task on some projects might be the ordering of program source streams because the client wants to research available offerings and make a decision at a future date.

Project Planning

The importance of proper planning at the start of a project should not be minimized. Even experienced installers should not be tempted to skip this first step because it is vital to the success of a project. Sometimes, a project may appear simple to implement because you have done hundreds of similar installations. But if you skip the planning step, you could end up with numerous problems. As a professional installer, you will need to justify the actions you take to solve the problem.

Let's look at an example. Say a customer hires you to mount a projector from the ceiling. When you get to the site, you find that a sprinkler system has been installed right where the projector needs to be mounted. What do you do? You cannot rip the sprinkler system out of the ceiling to install the projector. If you install the projector to the right, left, back, or forward, it will not be at the appropriate throw distance or angle to the screen.

Headaches such as this can be easily avoided by having a documented plan, which includes a site survey. If you had conducted a site survey, you might have identified the problem and communicated it to the client. If the problem arose after your initial survey because of the work of other trades, you will communicate the problem through established channels and record it on your documents. Since not every installation job is the same, you will face fewer problems and delays by establishing a project plan.

Developing a Plan

Thinking through the key stages and the roles of other teams on the project is critical. When major construction is involved, you do not want to wait to install a motorized screen or projector lift into the ceiling until after the new room is painted and the electrical wiring has been completed. Effectively communicating with all the people involved in the project is essential for success.

There should be two plans at the start of an AV project. Plan 1 involves discovering what the project requires through a needs analysis and concept design/program report, while Plan 2 focuses on the work that is actually going to be done to accomplish Plan 1.

Since Plan 1 was accomplished in the design phase, you need to focus your efforts on Plan 2. Your job as an installer is to take information from the needs analysis and concept design/program report and work with the project manager to build a timeline for when the systems will be built and installed. If the plan already exists, then your task is not to discover but to understand and confirm the details and to raise any issues you find through the correct channels.

The project manager tracks the key elements such as scope, time, cost, quality, and risk, as well as the progress of the project by using project management software.

While most system integration firms use commercially-available project management software, some of them utilize custom software because no single program is a perfect fit for AV installation jobs. It is common for a company to use different software for the various components: job costing, bill of materials (BOM), scheduling, and tracking customer information. You will need to become confident with whatever software your company is using.

Managing Project Resources and Activities

Project management skills are critical to the success of any AV project. These skills are regarded as a significant asset to a systems integration firm. That is why many AV professionals, in addition to earning their CTS credentials, pursue studies for their Project Management Professional (PMP) or Certified Associate in Project Management (CAPM) certification from the Project Management Institute (PMI). The growing demand for more sophisticated solutions such as interactive visualization systems and command-and-control centers, together with the convergence with digital information technologies, is creating a need for AV professionals with the skills to shepherd large and complex projects from inception to completion, on budget and on schedule.

The five key elements of a project are

- **Scope**   Covers deliverables and activities that must be accomplished
- **Time**   Includes estimation and actual duration of activities
- **Cost**   Estimates human and material resources
- **Quality**   Makes sure performance meets expectations
- **Risk**   Includes threats, opportunities, and response strategies

All of these elements have to be driven by the client's purpose or objectives. The AV project manager will coordinate the activities of your company's staff with stakeholders from the client side of the project, as well as with other contractors. Project managers are also responsible for effectively managing resources and producing the documentation required for the project. Whether or not project management is explicitly part of your role, it is important for you to identify relevant project issues and know how to address them.

TIP   The key elements of a project are driven by the client's need. The AV system's ability to address their need is how the client will ultimately measure the quality of the system.

Complex AV systems for fixed installation or live events require detailed project records and analysis of all activities and system components, including the schedule for coordinating with other vendors performing work at the site. The AV team will typically use project management software to plan the work and ensure task coordination.

Team members responsible for various aspects of the project must be able to perform a wide range of typical project management tasks, including the following:

- Create a work breakdown structure
- Create a project schedule
- Create a logic network for coordination of resources
- Manage the project budget
- Monitor work at the client site

Project management software facilitates the monitoring of resources and activities. There are three basic ways to track resources (both labor and materials) and the schedule using project management software.

These are the charts used in AV project management:

- **Work breakdown structure**   A work breakdown structure (WBS) presents deliverables on products or services.
- **Gantt chart**   A Gantt chart provides a timeline for all activities that are scheduled.
- **Logic network diagram**   A logic network diagram shows which tasks have to be completed before you can begin the next task.

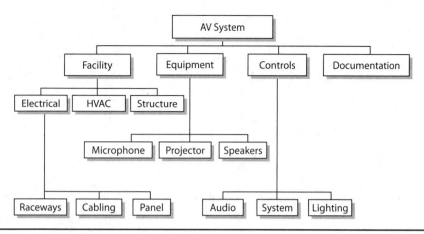

Figure 3-1   A general work breakdown structure

These charts provide a view of the project at different stages. Let's look at what is involved with each.

Work Breakdown Structure

To begin scoping the labor needed for a project, you need to create a work breakdown structure. Figure 3-1 shows an example of a WBS. The purpose of this document is to make certain that all those in a group working together on a project know exactly what results are required by the rest of the project team and the client. The WBS defines aspects of the project as products or services. These are always stated as nouns and do not specify who will accomplish something, when it will be done, or how much of a specific product or service will be delivered.

The WBS divides the project's work into clearly understandable chunks. It also shows the relationship between these elements and to the entire project or system. Every lower level of the WBS shows a more specific component of the total statement and traces that specific component's relationship to a more general task.

The WBS can serve as an outline for estimates, and it also makes it easier to assign clearly defined responsibilities. Additionally, the WBS can create a stable basis for recognizing and managing changes in the project plan, and even for evaluating performance.

Gantt Chart for Project Schedules

A general project schedule is typically established during the design and bid phases. The owner's project manager finalizes it before work begins and will bring it to planning meetings. This way, every team project manager can see their project deliverables. Each trade or discipline must have a corresponding schedule of work (developed by each team's project manager), which is incorporated into the main project schedule. The AV project team needs to be represented on this construction schedule for the entire project. This responsibility often falls to the AV designer or consultant.

University Audiovisual Project

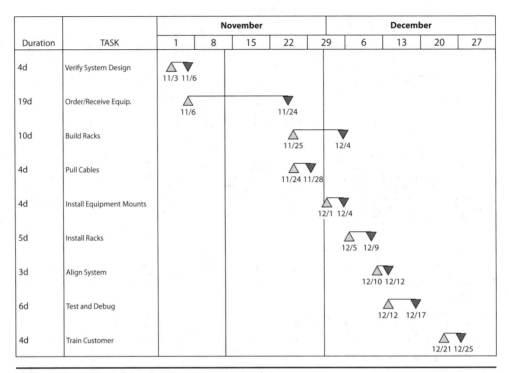

Duration	TASK	November					December			
		1	8	15	22	29	6	13	20	27
4d	Verify System Design	11/3 11/6								
19d	Order/Receive Equip.	11/6			11/24					
10d	Build Racks				11/25		12/4			
4d	Pull Cables				11/24 11/28					
4d	Install Equipment Mounts					12/1 12/4				
5d	Install Racks						12/5 12/9			
3d	Align System							12/10 12/12		
6d	Test and Debug							12/12 12/17		
4d	Train Customer									12/21 12/25

Figure 3-2   A Gantt chart showing a project schedule

The most common way to present a project schedule is a type of bar chart known as a *Gantt chart*, which depicts the timeline for tasks and subtasks as horizontal bars or lines. It shows the sequence in which tasks should be performed and any project milestones, such as the completion of major categories of tasks. Project management software is often used to create Gantt charts, like the example shown in Figure 3-2.

Because some tasks must be completed before others may start, the Gantt chart identifies these types of dependencies in a manner that clearly shows their sequence. As you know, walls within a room must be finished and painted prior to mounting some AV components, so the start dates of these AV tasks will be identified as dependent on the end dates of the room-preparation tasks. In Figure 3-2, the second task (ordering and receiving equipment) is dependent on the first task (verifying the system design).

Logic Network Diagram for Coordinating Resources

Resources such as tools, test equipment, vehicles, and staff time are always limited, even in large organizations. Since your company will simultaneously be working on more than one project, it is essential to plan for the allocation of resources for each project. This will help avoid conflicts and shortages when projects have overlapping needs. An effective resource plan involves creating a clear schedule, showing what resources will be required and when they will be needed, and highlighting possible shortages and related problems. Once these challenges are identified, you can initiate actions to assure the availability of more people, tools, or other resources needed to properly complete the job on schedule.

After the WBS has been completed, specific activities and milestones can be linked to the project's deliverables. A *deliverable* is an object or a service that is required to be delivered to a customer as a component of a complete project. The deliverable could be a single device, a complex functional system, a document, a piece of software, or a training session. These deliverables and their requirements form the basis for a *logic network diagram,* such as the one shown in Figure 3-3.

A logic network diagram helps organize and finally schedule a project's activities and milestones. The uses of a logic network diagram include

- Listing activities and their sequential relationships
- Identifying dependencies and their impacts
- Introducing activities into the network according to their dependencies, not on the basis of time constraints
- Identifying successors
- Identifying predecessors

The logic network diagram includes the activities and tasks that are required to create the deliverable, as they consume time and resources. The functions of a logic network are expressed as verbs, such as *build, test, fabricate,* and *develop.* Each activity should have an associated deliverable.

In Figure 3-3, *effort* refers to the amount of time it will take to complete an *activity* or task in its entirety. Many companies calculate an effort value based on best-case scenarios. This optimistic practice usually increases risk and can lead to underbidding the labor

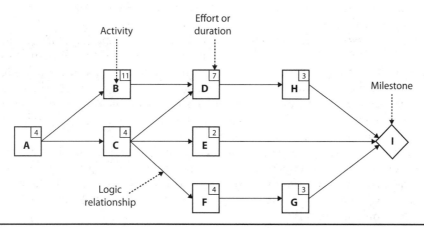

Figure 3-3   A logic network diagram

component of an installation. For example, it may take only a few minutes to replace a faulty component, but if a technician must travel to a distant location to get the component, those few minutes can turn into hours.

Mature organizations look at resource requirements across all of their current projects before establishing timelines. They prefer to estimate *duration,* which refers to the amount of time it will actually take to complete a task, not just the time devoted to the task. Duration factors in several components, such as the number of work periods, resource availability based on other projects, and personal and organizational calendars (weekends, holidays).

Milestones are key events in the project, such as the completion of a major deliverable or the occurrence of an important event. They can be associated with scheduled payments, client approvals, and similar events. Examples include *materials delivered, racks programmed,* and *projectors calibrated.* As shown in Figure 3-3, milestones are typically shown as diamonds or triangles in the logic network diagram.

In addition to project management documentation, you should be adept at reading technical diagrams and drawings that will show where on the job site the AV equipment needs to be installed, and how to install it. Chapter 4 discusses architectural drawings and AV documentation.

Managing the Project Budget

The project budget is based on the estimated costs of the labor, supplies, and equipment that are necessary to complete the agreed-upon work. As part of the project team, installers are expected to provide input during budget planning, as well as help keep under control any unforeseen expenditures during installation. That is why it is important to know how the budget can be affected. Table 3-1 shows an example of an AV project budget estimate.

The fee that a company charges a client is based on the cost to the company to provide the needed labor, supplies, and equipment necessary to complete the agreed-upon contract, plus a markup that is intended to cover overhead costs (such as office rental, insurance, taxes, and staff benefits) and profit. The fee an AV company charges a client should never equal cost.

Based on the AV team's estimates for labor and materials, the project manager allocates these resources and monitors how they are expended to ensure that the actual costs do not exceed the amount that has been allocated to complete the project. If the project is taking more of these resources, the project manager needs to determine the cause and make adjustments to get the project budget back on track. The project manager should also keep track of staff activities and make sure the work is performed according to the standards defined within the contract.

Modifications or additions because of changes in site conditions, client requests, or other factors will have an impact on the project budget. The costs incurred in making changes, including the resulting additions to the workload, should be agreed to by both the AV company and the client in a formal document (see "Providing Project Documentation" later in this chapter).

Estimated Budget

Item	Estimated Unit Cost	Estimated Quantity	Estimated Total Cost
Videowall panels	$3,000	8	$24,000
Videowall controllers	$3,500	2	$7,000
Audio/video switching systems	$3,500	1	$3,500
Surround-sound audio system	$3,500	1	$3,500
HDMI switching system	$3,500	1	$3,500
Ceiling loudspeakers	$150	8	$1,200
Videoconferencing endpoint equipment	$8,000	1	$8,000
Videoconference cameras (PTZ)	$4,000	2	$8,000
Cable television receiver	$500	1	$500
Document cameras	$1,750	2	$3,500
Wireless mouse	$350	1	$350
Presentation monitors	$3500	2	$7,000
Computer video interfaces	$500	2	$1,000
Half-height equipment racks (18RU)	$500	8	$4,000
UHF wireless microphone	$700	1	$700
Audio reinforcement system	$11,000	1	$11,000
Ceiling array microphone	$1,000	1	$1,000
Audioconference interface	$2,000	1	$2,000
Audiovisual control system	$16,000	1	$16,000
Auxiliary audiovisual connections	$250	5	$1,250
Instructor workstation	$3,000	1	$3,000
Wall boxes	$500	4	$2,000
Miscellaneous hardware, cable, and connectors	$3,000	1	$3,000
Audiovisual contractor labor	$12,000	1	$12,000
Total Estimated AV Cost			$127,000

Table 3-1 A Project Budget Estimate

The project manager determines how labor and material resources are expended to ensure that the actual costs do not exceed the amount allocated for completion of the project. Sometimes, modifications or additions are necessary because of changes in site conditions or client requests, and these changes affect the project budget. The additional work and costs must be clearly communicated to the client and explicitly agreed upon by your company and the client.

Monitoring Work at the Client Site

The scope and complexity of a project will determine how much onsite management is necessary. An AV project manager will have several objectives to meet simultaneously. The work must be performed with adherence to safety procedures, conform to relevant regulations and codes, and meet the client's and AV company's standards for quality. Troubleshooting is a critical part of an installer's job and involves addressing problems or concerns that could affect the schedule or budget.

Other Project Management Activities

A crucial part of AV work is coordination with other trades. The AV project manager and installer must be aware of the different aspects of a construction project that can affect AV installation. For example, you may need to provide specifications for cabinetry, millwork, or carpentry that integrates with your AV systems.

Coordination with other trades is typically accomplished through weekly construction meetings. During these meetings, each team's project manager provides a progress report on the status of their portion of the job. Important information is exchanged, schedules are updated, and coordination issues are resolved. This information may result in changes to the original design. It is important to identify and note how any change will affect the AV installation. These and other factors that must be considered are addressed in Chapter 5 on pre-installation activities.

Coordinating project activities with other stakeholders is important both on construction projects and on live-event AV projects. Stakeholders for live-event projects may include groups that are not typical in fixed installation or construction projects. These may include the event promoters, producers, designers, rigging providers, equipment rental companies, specialist technical labor hire companies, pyrotechnics and effects suppliers, and staging vendors.

NOTE Some of the activities mentioned in this chapter are conducted before the AV installation. Other pre-installation tasks will be covered in subsequent chapters.

Providing Project Documentation

Consistent and accurate communication is crucial to the success of a project. It is important to establish your lines of reporting at the outset of the project. You should clearly identify who you need to be communicating with and make certain that they receive the appropriate information at the expected time. It is also important to gain an understanding of the less formal communication structures that may exist within a particular project's hierarchy and to establish the interpersonal relationships that will assist in the smooth flow of the project.

In addition to discussions at regular project meetings and onsite verbal understandings, every important change or clarification of work should be confirmed in writing.

Where necessary, consider reviewing meeting notes, or taking your own notes of the discussions, and circulating them to the appropriate stakeholders to keep them aware of agreed decisions and action items. This is especially important with any communication from the client, or the client's representative, regarding changes in the project scope or specification.

Changes should always be formally documented, agreed to, and signed by all major parties (for example, owner or owner's rep, general contractor/building contractor, design consultant), as well as the appointed member (sales executive, project manager) of the AV systems integration firm. This may seem to be an excessive precaution, but it is critical to have all significant changes in writing to avoid costly misunderstandings or delays later in the project. These updated understandings often become part of the formal agreement or contract as change orders or another related addendum.

Members of the design and installation teams rely on project documents to understand project goals, envision how the space is constructed, and implement system requirements. Knowing which documentation to use when you need to communicate something will help you speed along the installation process.

Successful project documentation will achieve the following:

- Clearly and accurately communicate facility details, project requirements, and system functions
- Define the scope of project participants' roles
- Indicate who is responsible for completing each aspect of the project
- Help all project team members coordinate effectively and efficiently

The forms used to document decisions and changes related to a project include both information-related ones, such as letters of transmittal, and others to request and receive approval on items that will be changed, such as change orders. These and similar documents should all contain project-specific information, including the owner's name and address, specific installation locations, contractor names, and project numbers, as well as other relevant contact and contract information.

The ANSI/AVIXA 2M-2010 *Standard Guide for Audiovisual Systems Design and Coordination Processes (2010)* lays out a procedure for determining what project communication (documentation, meetings, and so on) needs to take place on a given project. The following is from the standard's abstract, available at www.avixa.org:

> A successful professional audiovisual system installation depends on the clear definition and coordination of processes, resources, and responsibilities of the design and installation project teams. A properly documented audiovisual system provides the information necessary to understand and implement the system goals and project requirements in a logical and efficient manner. The documentation should complement and coordinate related architectural, engineering, and construction documentation. This standard outlines a consistent set of the standard tasks, responsibilities, and deliverables required for professional audiovisual systems design and construction.

TIP Creating a set of customized standard forms that are project-specific and pre-addressed at the beginning of the project can help save time and speed up the approval process. You may need to create these forms entirely, or you may be able to insert the basic project information into the standard contract administration and project management forms already in use in your organization.

Information-Related Documents

Examples of project documents associated with communicating project information include letters of transmittal, requests for information, and progress reports.

Letter of Transmittal

This is a formal cover letter used whenever drawings, samples, submittals, or other formal documents are sent. It clearly states the addressee, the sender, and their contact information, and it also must include a list of what is being sent, along with relevant revision number or date, and any expected action to be taken by the receiving parties. A letter of transmittal should always be used, whether the items are sent by snail mail, courier, e-mail, or file transfer.

Request for Information

A request for information (RFI) is used to formally request clarification or specific answers about a particular aspect of a project. RFIs typically address one of three basic subjects: a design issue, a site issue, or an owner change or request. This process is usually based on a paper or electronic form established for the project that indicates who the RFI is from and to whom it is directed, with spaces to enter the question and the response.

Some RFIs are resolved simply by a written clarification from the RFI's recipient, without the need for contract changes. Others may need resolution through a change in the construction contract. If the resolution of the RFI results in any change of the project's scope, cost, or timeline, a request for change (RFC) must then be generated.

Progress Report

A progress report can be a formal document or an informal report that informs the client of the state of completion of the project as of a specific date. It is important that a progress report be integrated into the project's documentation and project management systems. Such systems may require hardcopy reports. They may be required to be in a specific format used by a business software suite, or they may be required to be uploaded (or directly entered) into a network- or Internet-based construction management system. In some situations, a progress report may have to be provided to your own company's project management system, in addition to the project-wide documentation system.

The progress report should include any issues or concerns that will affect the ability of the AV vendor to complete the project within the agreed-upon schedule and budget. Including images of the current status of an ongoing process may help illustrate any issues that may have arisen.

Punch/Snag/Problem List

A *punch list,* also known as a *snag list* or *problem list,* is a record of project items that are pending completion or correction to ensure that the item conforms to the scope of the project. Often, it is an informal, working checklist maintained by a project manager. Images of the current status of listed items will often simplify the creation of the list. Such a list may also be a formal contract document issued when a project is substantially completed, with final payment for a project depending on the satisfactory resolution of open items. Critical open items on this list (or recently completed ones) are often documented in progress reports.

Timesheets

This form is a simple way to record how many hours you have spent on specific project tasks. It will help you and your project manager track progress on your part of the project. Information from timesheets is applied to the organization chart and can be used to plan future projects.

Change Documentation

Project documents related to changes include requests for change, change orders, and construction change directives. Figure 3-4 shows an example of the change process.

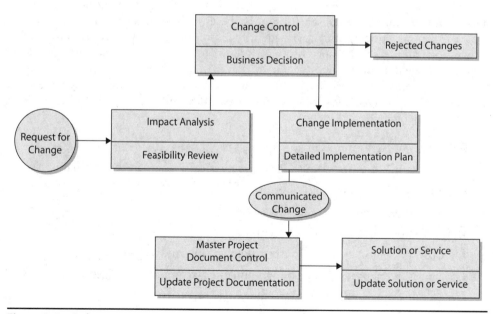

Figure 3-4   A change process

Request for Change

The RFC is a common document used in AV and other projects to change contractual obligations, equipment models, and specifications or to modify a system design element. When an RFC is created by the vendor or when the vendor replies to an RFC created by another party to the contract, all the costs resulting from the requested change and their impact on the project must be included. These impacts may include completion delays, as well as changes in performance or appearance. The following are some issues that can prompt an RFC:

- A change in the intended use of the system
- Manufacturer's discontinuation or significant revision of a specified product
- Architectural, mechanical, or construction changes
- Discovery of system or product incompatibilities or functions

Any member of the project team can submit an RFC, although in an AV project, the integrator or consultant commonly creates it. An approved RFC then becomes a change order.

Change Order

This document is used to document agreed-upon changes to the project specifications or contract as the project progresses. Change orders (COs) are arguably the most important document used during the construction phase because they can significantly change the contract scope and pricing, and clearly identify responsibility for all costs associated with a change. Typically, the issuance of a CO follows an RFC and subsequent agreement between the owner, owner's representative, contractor, and other involved team members (such as consultant, designer, architect, or engineer), but minor COs are sometimes issued based on verbal or informal written requests. Reasons for an AV system CO might include the following:

- Changes or clarifications of anticipated system use
- Architectural, cabinetry, finish, or other physical changes in the installation site
- Design conflicts, omissions, or errors
- Changes in a product's availability or specifications
- Availability of new products or technologies
- Discovery of hidden or previously unknown site conditions
- Changes in the project scope or equipment list required by budget adjustments
- Schedule changes and delays by others

The process for the creation and approval of a CO is established by contract at the beginning of the project. Since changes can dramatically alter the project schedule or budget, any necessary RFCs and COs should be processed as quickly as possible.

COs usually affect the overall value and cost of a project in several ways. They can have an impact on financing, leasing, insurance, and bonding costs. Guarantors should be notified (by those responsible for these aspects of a project) if a CO will cause the project value to exceed the original budget or if the schedule will be affected in a significant way. The projected cost of a CO may be specified in the RFC and confirmed in the CO, or the CO may include an agreement that costs will be charged on a documented time-and-materials basis.

Construction Change Directive

A *construction change directive* (CCD), sometimes called a *field order,* is usually issued as a result of time-critical conditions or events in the field to trigger immediate work related to a necessary change in the contract. This form is typically used by the owner or the general contractor/building contractor to bypass the more formal request for information (RFI)–RFC–CO process. It is important to confirm that the changes discussed in field meetings result in the issuing of a formal CCD.

Duty Check

This chapter relates directly to the following tasks on the CTS-I Exam Content Outline:

- Duty A, Task 1: Review Audiovisual Project Documentation
- Duty E, Task 2: Coordinate with Other Contractors

These tasks comprise 8 percent of the exam (about eight questions).

Types of AV Projects

Although you may be familiar with the different types of AV projects that designers, installers, and project managers work on, it is useful to review the basic functionality that is required in typical room systems and recognize how they differ from unique AV applications. For instance, advanced AV solutions such as command-and-control centers and visualization walls require expertise in networking and control systems. System integration firms specializing in these and digital signage applications usually have programmers in-house. On the other hand, boardrooms and conference rooms involve typical presentation and collaboration systems. Most types of AV spaces, including boardrooms, conference rooms, training facilities, and multipurpose rooms, require end-user device connectivity. Small collaborative meeting spaces, sometimes called *huddle rooms,* may require bring your own device (BYOD) connectivity and collaboration, display, and conferencing capabilities to be packed in a small space designed for a workgroup of as few as four to six people.

Many of the audio, video, and networking principles and skills you study will help you address the needs of a typical fixed installation. But you will have to develop and use critical thinking skills to apply the appropriate installation and project management methods to AV installations in different types of organizations, based on insights gained from experience in the real world. For example, although a training room may have similar AV needs to those of a small classroom, the design and system requirements could be different depending on whether the room is at a corporate facility or at a university's medical school. Similarly, there will be differences between the project documentation required on a government project and that required on a hotel project. Also, bear in mind that the communication patterns or "culture" will differ from one organization to another.

Live events range in complexity from a single presenter standing in a pool of light with a whiteboard, a wired microphone, and a simple voice-reinforcement audio system, to a production with a live orchestra; hundreds of performers with wireless microphones and wireless in-ear monitors; media servers feeding video screens covering set pieces and the entire stage floor; pyrotechnic and atmospheric effects systems; robotic moving scenery; hundreds of robotic lights; complex cueing and communications systems; video cameras feeding an IMAG (image magnification) system and video broadcast streams; sound systems providing high-intensity, high-quality audio to a live audience of tens of thousands; and the control systems to interface, trigger, monitor, and coordinate it all.

The AV-deliverable components of a live-event installation may include only such systems as video servers, screens, and projection, or may extend to include all audio, communications, control, video cameras and their mounting and suspension systems, web streaming, outside broadcast interfaces, vision control, and image generation systems.

There are many types of AV applications, and the systems are selected to provide end users with the functionality that meets the client's purpose or communication objectives. The system design documents will indicate where the systems are installed, but real-world conditions may require modifications.

Key Stages of AV Installation

Let's review the key stages of AV installation. This will serve as a reminder of the activities a project manager or other team members need to track to ensure that the installation is completed with excellence, on schedule, and on budget.

Pre-installation

Many preliminary activities must be undertaken before starting on the actual installation of AV systems. The first of these is project planning, which comprises such tasks as conducting a needs analysis, obtaining scale drawings of the client's space, and gathering other essential information.

Other important pre-installation tasks include conducting a job site survey, evaluating the site environment, de-installing existing equipment, and preparing an AV tool bag. These will be discussed in Chapter 5.

Although project documentation is set up at the start of the project, updating it and tracking changes are continuous and ongoing activities. As noted earlier, AV documentation during a project involves substantially more than the previously discussed project documents. These additional technical documents, such as technical diagrams and drawings, will be discussed in Chapter 4.

System Fabrication

All the tasks and steps involved in building the AV system—from procuring the equipment to turning on the system for the first time—are part of the system fabrication phase of a project. System fabrication involves knowledge of cable handling, hardware mounting, power and grounding, and rack building. These topics are covered in Part III of this book.

AV Installation

The successful installation of an AV system is dependent on essential knowledge of audio, video, control, and networking systems. You will need a thorough understanding of circuit and signal theory in addition to a range of hands-on skills, including installing and mounting AV equipment, soldering, constructing racks, and running and terminating cables and fibers. Several chapters in this book cover the basics and best practices in audio, video, control systems, and networking technologies.

AV Verification

An AV project is not complete until you have verified that the AV installation meets the documented specifications. AV verification requires a solid knowledge of audio and video calibration and verification, standards compliance, control system programming, and system networking. You will need to know how to use a range of signal generators and analyzers to set gain and equalize the audio and video systems. You will also need to know how to troubleshoot problems at this stage. The chapters in Part V of this book cover all these topics. In addition, AVIXA University offers an in-depth course on system verification.

System Closeout

To obtain client sign-off, you must confirm the AV system performance against the design specifications. If troubleshooting or requested changes remain undocumented or incomplete, the project remains open and could require additional staff hours to resolve. This could possibly delay payments and project conclusion. System closeout involves such tasks as cleanup at the job site post-installation and end-user training on the operation of the installed AV equipment. You will find a detailed discussion of these processes in Chapter 16.

Ongoing Project Responsibilities

Once system handover and sign-off have been completed, there are ongoing warranty responsibilities associated with every project, plus the possibility of additional contracts to cover preventive maintenance and equipment service on the completed project. These processes are discussed in Chapter 17.

Chapter Review

This chapter discussed the importance of developing a scope statement and a project plan. It covered the allocation and tracking of resources with the help of project management software. Planning and project management help preempt problems and accomplish the client's objectives for a purpose-built AV system, while staying on schedule and on budget.

Managing an AV project is a team effort led by the AV project manager, but the installer should be proactive in managing installation-related items. Project management activities include managing the budget, monitoring work on a job site, and coordinating AV work with other trades. Information on the various types of AV project communication and management documents were presented.

This chapter also briefly reviewed the key stages of AV installation in preparation for in-depth coverage in subsequent chapters.

Review Questions

The following questions are based on the content covered in this chapter and are intended to help reinforce the knowledge you have assimilated. These questions are similar to the questions presented on the CTS-I exam. See Appendix E for more information on how to access the free online sample questions.

1. Before developing a project plan, it is important to:
   A. Make sure the client has made a deposit payment
   B. Have a clear understanding of the scope of work
   C. Order the AV equipment needed
   D. List all the teams that will work on the project

2. A scope statement should include:
   A. Only the items within the project scope
   B. Only the items within the deliverable scope
   C. All the tasks that have to be performed
   D. Both in-scope and out-of-scope items

3. The key elements of a project should be driven by:

    A. The cost of materials

    B. The overall budget

    C. The client's need

    D. The deadline for completion

4. What is a *dependency* within a project task schedule?

    A. A task that cannot begin before another task has been completed

    B. A task that must be completed by a specific date

    C. A task that will be required if specific conditions occur at the project site

    D. An optional task that the client can determine is needed once the project is underway

5. Project budget refers to which of the following?

    A. The amount of labor and materials the project manager has allocated to complete the project

    B. The total amount that the client has paid the AV company for the project

    C. The amount needed for purchasing the materials and supplies

    D. The actual cost of labor and materials used to complete the project

6. A primary reason for attending construction meetings is to:

    A. Develop a relationship with the other people on the job

    B. Identify changes to the original design that will affect the AV installation

    C. Find out the project's completion date

    D. Submit invoices and other budget documentation

7. In the event of a change in the placement of air-conditioning ducts in a boardroom, which of the following documents should the AV integrator submit first?

    A. Punch/snag/problem list

    B. Change order (CO)

    C. Request for change (RFC)

    D. Request for information (RFI)

8. Which of the following documents should an AV integrator submit in the event that a manufacturer has discontinued a projector model?

    A. Letter of transmittal

    B. Progress report (PR)

    C. Request for change (RFC)

    D. Construction change directive (CCD)

9. The ANSI/AVIXA 2M-2010 *Standard Guide for Audiovisual Systems Design and Coordination Processes* lays out a procedure for determining the _____.

   A. functions and features that are required to support a given set of tasks

   B. project communication that needs to take place on a given project

   C. room types that are suitable for a given set of client needs

   D. verification tests that should be performed during a given project

10. After which of the following activities is an AV project completed?

   A. All hardware is installed

   B. Software integration is completed

   C. Verification that the system is operational

   D. System closeout activities are completed

Answers

1. **B.** Before developing a project plan, it is important to have a clear understanding of the scope of work.

2. **D.** A scope statement should include both in-scope and out-of-scope items.

3. **C.** The key elements of an AV project should be driven by the client's need.

4. **A.** *Dependency* within a project schedule is a task that cannot begin before another task has been completed.

5. **A.** The project budget is the estimated amount of labor and materials the project manager has allocated to complete the project.

6. **B.** A primary reason for attending construction meetings is to identify changes to the original design that will affect the installation.

7. **D.** When a change order is issued by another party and it will have an impact on the AV installation, the AV integrator should first submit a request for information (RFI).

8. **C.** A request for change (RFC) form should be submitted by the AV integrator in the event that a manufacturer has discontinued a projector model.

9. **B.** The ANSI/AVIXA 2M-2010 *Standard Guide for Audiovisual Systems Design and Coordination Processes* lays out a procedure for determining the project communication that needs to take place on a given project.

10. **D.** Only after all the system closeout activities have been accomplished will the AV installation project be considered completed.

CHAPTER 4

Audiovisual Documentation

In this chapter, you will learn about
- Types of drawings
- Audiovisual drawings and diagrams
- Basics of drawings
- Drawing views

As an installer, you will need to read and interpret architectural and audiovisual (AV) drawings and diagrams. The installation of AV projects is integrated with the projects of other subcontractors. The architectural and engineering drawings, created with the use of computer-aided drafting (CAD) software, provide a view of all the areas and various components involving all the trades. The AV drawings package, typically provided by the system designer, will show you where each element goes and how gear should be installed. The various drawings will help you to understand how the AV installation relates to the other disciplines, such as electrical, data networks, telecommunications, fire services, security, lighting, HVAC (heating, ventilation, air conditioning), and so on. These technical documents will help you plan and successfully complete the AV installation.

To read and interpret the drawings and diagrams accurately, you will need to know how the drawings are scaled; how to convert measurements on a scaled drawing to the actual sizes; and how to recognize the symbols, abbreviations, and drawing views. You will need to become familiar with commonly used abbreviations and what they stand for, such as NTS for "not to scale" and EL for "elevation." But first, let's take a look at some of the types of drawings you will be working with on an AV project.

Types of Drawings

It is extremely important for installers to be able to read and interpret architectural and engineering site drawings and AV system plans. Through these drawings, you can understand the vision and scope of the client's project, as well as the AV system installation requirements.

As noted in Chapter 3, the ANSI/AVIXA 2M-2010 *Standard Guide for Audiovisual Systems Design and Coordination Processes* provides a detailed explanation of all the documents typically used on an AV project, including design documents, architectural drawings, and construction drawings.

The two primary groups of drawings that AV professionals work with are

- **Architectural and engineering drawings package**   The drawings in this package illustrate the overall design of the building and rooms, including the room layouts; features and dimensions; locations of the ducts that are used in HVAC and plumbing; and locations of electrical, telecommunications, data networks, security, and lighting systems throughout the building. Small jobs may have only one or two drawings. Larger projects will have entire sets, divided into different groups based on the construction process, such as the following:

  - Mechanical drawings to show ductwork that goes through the building for HVAC and plumbing
  - Electrical drawings to show power and lighting systems
  - Communications/data drawings to show telecommunications, data networks, and security systems
  - Structural drawings to show the wood, concrete, and steel structures of the building support system
  - Furniture and console layout drawings to show the locations for devices that will be mounted in and on furniture

- **Audiovisual drawings package**   This package provides the overall picture for an AV installation. These system design documents will show where and how each hardware component should be installed. The package typically contains functional diagrams, connection details, plate and panel details, patch panel details, equipment diagrams, rack elevation diagrams, and the control panel layout.

Audiovisual Drawings and Diagrams

As an installer, you will receive applicable drawings from the other trades on the project, as well as AV drawings from the system designer. These drawings will help you to see "the big picture" and, more importantly, identify undocumented changes while visually inspecting the site.

Construction Drawings

Also known as *submittal drawings, workshop drawings,* or *shop drawings,* these unalterable documents include sufficient detail to convey the physical configuration of the AV systems. On some projects, other drawings in the construction documentation may be used for reference. For example, the architectural floor plans showing locations of lecterns, racks, screens, loudspeakers, and projectors may be marked "verify in field" or "verify with AV contractor."

The AV team usually checks mechanical drawings (HVAC) that show the ductwork throughout the building to ensure that ducts and piping are not in the area where suspended equipment is planned. Site drawings show compass direction and are useful for placing installation locations. Electrical drawings show power, lighting, lighting control, and electrical screen positions. Communications/data drawings show the locations of telecommunications frames and sockets, data network equipment, patch panels and sockets, and security devices and control panels. Structural drawings showing the building's infrastructure would be useful to refer to when considering mounting equipment. Fire protection drawings show the locations of detection devices, emergency panels, control panels, and the sprinkler or douser system and should be checked against the locations of AV wall panels, ceiling-mounted loudspeakers, cameras, display screens, and projectors. Architectural reflected ceiling plans coordinate all other drawing sets, showing diffusers, fluorescent lights, down lights, sprinklers, wireless access points, security devices, fire detection, loudspeakers, and suspended screens, projectors, and displays.

Facility Drawings

These construction drawings show the location and size of equipment in a given space. Facility drawings include device and equipment shown in plan, elevation, section, and detail views. They are a major source of communication for the project team. For the AV industry, facility drawings would show the locations of objects such as projectors, screens, loudspeakers, and displays.

System Drawings

During the project's design phase, the AV designer will produce system drawings. These detail the audio, video, and control signal switching and routing, as well as other details necessary to convey the complete AV system design. System drawings include

- Block diagrams (also known as functional, schematic, or flow diagrams) indicating signal flow and system interconnection
- Equipment rack elevations
- Plate and panel details
- Wiring diagrams
- Patch panel layouts
- Other items

The system drawings intended for construction/fabrication include all of the information from the design documents. They also have additional information required to complete the fabrication and installation of the AV system. It is extremely important to review these plans before installation. For example, on one project, the design called for a projector to be installed above a conference table, but when the AV team reviewed the drawings, they noticed that pendant lights would hang in front of the projection lens.

Functional Diagrams

The functional diagrams used in the construction phase include all of the information in the AV design documents with the following additions:

- Each device labeled by manufacturer and model number
- Labeling indicating exact connector types and gender

- Each connector and wire clearly labeled indicating cable and connector type and/or manufacturing part number
- Cable numbers corresponding to the schedules
- Any additional information required to instruct the field technicians in the proper installation of the systems equipment
- IDs that are on each device within the system

Labeling schemes vary widely, but device IDs are absolutely necessary and should include information such as room, rack, equipment, and cables. Best practice calls for device IDs even on wire labels and rack-mounted devices. Device IDs should be included on all drawings that reference a given device.

As-Built Drawings

After a project is complete, the installation team hands over extensive documentation to the project manager. Included in the documentation are *as-built drawings,* or drawings of record.

As-built drawings and documents are those drawings that have been modified by the installer to reflect changes made onsite to the original plans. The original plans often cannot take into account product changes that may require changes in the field. By carefully keeping track of any wiring changes, any corrections to levels, equalization, and other system designs, the as-built drawings become valuable references in the future during service calls and routine maintenance.

Duty Check

This chapter relates directly to the following task on the CTS-I Exam Content Outline:

- Duty A, Task 1: Review Audiovisual Project Documentation

This task comprises 4 percent of the exam (about four questions). This chapter may also relate to other tasks.

The Basics of Drawings

To read and interpret drawings, you need to understand drawing scales and be able to convert dimensions on scaled drawings. Construction drawings are drawn in full scale in CAD and output in hard copy at a reduced size. Therefore, it is extremely important that you make accurate conversions when you are using drawing scales.

Drawing Scales

Scale drawings are used to communicate the dimensions of a full-size project on a paper or electronic document. In other words, what is depicted in the drawing is proportionally identical to the full-size room/space/system it represents. The scale is the ratio of the full-size objects to their representation in the drawing.

The scale used is usually found in the title block in the lower-right corner of the drawing but may be located anywhere on the plans. A set of plans may include more than one scale ratio, and you must check each drawing page for its scale. In situations where the plans also include inset detail drawings (described in the "Drawing Views" section), you may find more than one scale employed on a single page.

Some drawings, such as schematic layouts, show relationships and connections between objects but are not drawn to any scale. This is usually indicated by "NTS" (not to scale) in the scale section of the drawing's title block.

Two different types of scales can be found on AV drawings:

- **Metric** These scales are used in most regions of the world. The meter (3.28ft) and the millimeter (0.039in.) are the standard units of length, although some countries may also use the centimeter (0.39in.). There are 10 millimeters in 1 centimeter and 1,000 millimeters in 1 meter.

- **U.S. customary** These scales are generally found on drawings used in the United States, Myanmar, and Liberia. The foot (304.8 mm) and the inch (25.4 mm) are the standard units of length. There are 12 inches in 1 foot.

Scale Measurements

Drawings using metric measurements usually state the scale as a single ratio, such as 1:50, as shown in Figure 4-1, which means that 1 unit on paper is equal to 50 of the same units in the real space. Drawings using U.S. customary measurements will often state a scale using a particular fraction of an inch to represent a foot. For example, you might see that 1/4 inch on the drawing equals 1 foot in the real space, as shown in Figure 4-2. This is a ratio of 1:48, which is similar in size to the international 1:50 scale.

Figure 4-1
Examples of metric scale keys that you may find on a drawing set

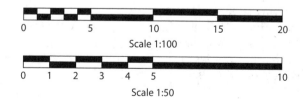

Figure 4-2
Examples of scale keys in U.S. customary units that you may find on a drawing set

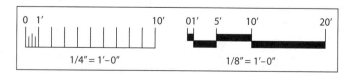

The following are typical U.S. unit scales:

$^3/_{32}$ inch = 1 foot (1:128)	¼ inch = 1 foot (1:48)	¾ inch = 1 foot (1:16)
$^3/_{16}$ inch = 1 foot (1:64)	$^3/_8$ inch = 1 foot (1:32)	1 inch = 1 foot (1:12)
$^1/_8$ inch = 1 foot (1:96)	½ inch = 1 foot (1:24)	1½ inches = 1 foot (1:8)

The following are some common metric (SI) scales:

1:1/1:10	1:2/1:20	1:5/1:50
1:25/1:250	1:100/1:1000	1:500/1:5000

Using a Scale Tool

The word *scale* is used to represent the size ratio in the drawing, as well as the tool commonly used to interpret scale drawings. The scale tool is also known as an *architect's scale* or a *scale ruler*. It provides a quick method for measuring the object on paper and interpreting its true size in the actual space.

Many scale tools are triangular prism-shaped devices having six edges that each bear a different scale; others resemble conventional rulers and carry just four scales. A whole or fractional number to the left or right edge of the measurement tool indicates the scale those numbers represent.

A scale marked 16, 10, or a fraction such as 3/4 or 3/16 is a standard U.S. customary ruler. For metric architectural scales, there are paired ratios, so each side of the scale can be used to measure two different ratios, such as 1:5/1:50.

To verify that you are using the correct scale on your scale tool, select the face of the tool that matches the indicated drawing scale. Lay the 0 point at the extreme end of the drawing's scale key and read the corresponding value at the other end of the line. If the distance read is dramatically different, you may have the wrong scale selected. If the scale is marginally different, it could be that your drawing has not been printed or copied exactly to scale. It is important to remember that if dimensions are written on the document, they take precedence over scale measurements.

Avoid writing on a print or drawing unless you have been authorized to make a change. Keep the drawing clean, neat, and intact.

As shown in Figure 4-3, the title block provides valuable information to help assure you that you are working with the appropriate and most current drawing. Details include the name of the client, project title, sheet title, revision date, and scale. Figure 4-4 shows another example of a title block with similar details.

COMPANY	PROJECT TITLE	SHEET TITLE	01	E.C.O	07/20/05	N	DATE	DRAWN	SCALE
AVIXA	AVIXA University 11242 Waples Mill Rd. Fairfax, VA 22030	SECTION	NO.		DATE REVISION		11/02/02	H.J.K.	1/8" = 1"

Figure 4-3 Example of a title block on an architectural drawing

Figure 4-4
An alternative
title block
configuration

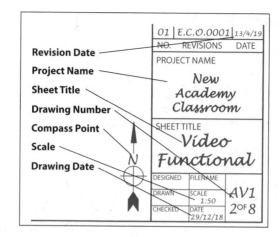

Calculating Dimensions from Scaled Drawings

An essential and significant skill is the ability to correctly apply a drawing scale when interpreting dimensional drawings. For instance, you will need to convert a scale length to the actual length to determine cable run estimates or to determine the actual height, width, and depth of an AV rack.

The sample conversion activities will show you how to make calculations of actual measurements based on scaled dimensions on drawings. CAD systems include tools for measuring off distances on electronic drawings and can be configured to display the results in a variety of formats and units.

If you have a scale rule matching the scale of the drawing, you will be able to read the actual measurements directly from the scale rule.

Sample Conversion Activity 1

On a U.S. customary scaled drawing of 1/8 inch = 1 foot, the length of a measurement is 3.5 inches. What is the actual length?

Step 1 Convert 1/8 inch to a decimal value.

1 / 8 = 0.125 inch

Step 2 The actual length is obtained by dividing the total length of the measurement 3.5 inches by the decimal value 0.125 inch.

3.5 / 0.125 = 28 feet

Answer The actual length is 28 feet.

TIP When solving multistep problems, do not round numbers at each step. Round only your final result.

Sample Conversion Activity 2

On a metric-scaled drawing of 1:50, the total length of a measurement is 100 millimeters. What is the actual length?

Single Step   The actual length for a metric-scaled drawing is obtained by multiplying the total length of the measurement, 100 millimeters, by the scale factor 50.

$$100 \times 50 = 5,000 \text{ millimeters or 5 meters}$$

Answer   The actual length is 5 meters.

Converting Ratios

Because scaled drawings use ratios to represent sizes, you can use the "cross-multiply and divide" method to simplify calculations to determine an unknown value. Let's use this method with a sample ratio conversion activity.

Sample Ratio Conversion Activity

If you measure 8 inches on a scaled drawing where 1/4 inch or 0.25 inch equals 1 foot, what is the actual length?

In the "cross-multiply and divide" method, you will first set up the proportion, as in Figure 4-5, using X to indicate the unknown value (that is, the actual length).

Step 1   Set up the proportion.

The equation in Figure 4-5 represents 8 inches on the drawing divided by the actual length in feet (X), which equals 0.25 divided by 1. The next steps are cross-multiply and then divide to calculate the actual length.

Step 2   Cross-multiply.

$$8 \times 1 = X \times 0.25$$
$$8 = X \times 0.25$$

Step 3   Divide both sides of the equation by 0.25.

$$8 \div 0.25 = (X \times 0.25) \div 0.25$$
$$32 = X$$

Answer   The actual length is 32 feet.

Figure 4-5
Example of
step 1: Setting up
the proportion

$$\frac{\text{inches on drawing}}{\text{actual feet}} = \frac{0.25}{1}$$

$$\frac{8}{X} = \frac{0.25}{1}$$

$$\frac{8}{X} \times \frac{0.25}{1}$$

Convert From	To	Multiply By
meters (m)	feet (ft)	3.28084
centimeters (cm)	inches (in.)	0.3937008
millimeters (mm)	inches (in.)	0.03937008
feet (ft)	meters (m)	0.3048
inches (in.)	centimeters (cm)	2.54
inches (in.)	millimeters (mm)	25.4

Table 4-1   Converting Between U.S. Customary and Metric Measurements

The more you practice converting scaled drawing dimensions into actual measurements, the easier and faster you will be able to accomplish this task. The CTS-I exam includes these and other AV math challenges. As metric scales are simple direct ratios, there is no requirement for ratio conversions.

Converting Measurements

As an AV installer you must be able to convert accurately between the metric and U.S. customary measurement systems. Good-quality unit conversion software is freely available for all desktop and portable device operating systems, and most personal digital devices come with a built-in basic unit conversion application. There are also many websites offering instant online unit conversions. However, if you do need to calculate the conversion from one system to another, use Table 4-1. In the left column of the table, locate the unit of measure from which you are going to convert. Follow straight across the row to the middle column for the desired unit of measure (for example, from feet to meters). Finally, locate the conversion factor for multiplication (such as 0.3048).

Drawing Dates

Drawing dates are critical to ensuring that you are working with the most current information. It is important to recognize that dates may come in one of several formats that can be ambiguous.

- **MM/DD/(YY)YY**   Month/Day/Year is the only format used in the United States. It is also a minor alternative format used in Canada, some parts of Southeast Asia, and some parts of Africa.

- **DD/MM/(YY)YY**   Day/Month/Year is the only format used in South America, most of Europe, North Africa, South Asia, Oceania, and most of Southeast Asia. It is also the major alternative format used in Canada, the remainder of Europe, the remainder of Southeast Asia, and some parts of Africa.

- **(YY)YY/MM/DD** Year/Month/Day is the only format used in North Asia and some parts of Eastern Europe. It is the minor alternative format in some other countries. This format is less likely to be confused with the alternatives.

Dates may easily be confused during some parts of the year, particularly if only two digits are being used for the year number. Care should be taken to verify which format is being used. As the United States is one of the few countries using nonmetric measurements and nonmetric drawing scales, other elements of a drawing's title block give a good indication of the likely date format should the drawing date be ambiguous.

Elements of Drawings

AV installers should be able to read construction drawings. These architectural documents contain several elements, some of which you have already reviewed, such as title blocks, and others that you should become familiar with, such as abbreviations and symbols.

Drawing Abbreviations

Table 4-2 lists some common abbreviations used in architectural drawings and their definitions.

 CAUTION Abbreviations vary substantially between countries.

Drawing Symbols

There is an established AV drawing symbol standard: ANSI-J-STD-710, published jointly by AVIXA, the Consumer Electronics Association (CEA) and the Custom Electronic Design and Installation Association (CEDIA), and accredited by the American National Standards Institute (ANSI). However, each project drawing has different symbols and icons that are used to depict specific elements of the project design or the relationship between multiple drawings, such as where a room depiction in one drawing is continued on another drawing. Symbols do not necessarily follow a single standard, so the meanings of specific symbols are defined in the drawing legend or symbol key. An example of a drawing legend can be seen on the right of the reflected ceiling plan in Figure 4-14. Symbol keys or drawing legends are often located in a separate labeled box towards an edge of the drawing.

The ANSI-J-STD-710 defines the use of standardized and easily interpreted symbols for the complete documentation of an AV system design, including AV and information technology (IT) equipment, services, cabling points, and many electrical devices, including U.S. power distribution equipment. Some of the ANSI-J-STD-710 standard's

Abbreviation	Definition
AFC	Above finished ceiling
AFF	Above finished floor
AS	Above slab
AVC	Audiovisual contractor
CL	Center line
CM	Construction manager
DIA	Diameter
(E) or EXG	Existing
E.	East
E.C.	Empty conduit
EC	Electrical contractor
EL	Elevation
ELEC	Electrical
EQ	Equal
FB	Floor box
FUT	Future
GC	General contractor
ID	Inside diameter
IR	Infrared
LV	Low voltage
LVC	Low voltage control
MISC	Miscellaneous
NIC	Not in contract
NTS	Not to scale
OC	On center
OD	Outer diameter
OFCI	Owner furnished, contractor installed
OFE	Owner-furnished equipment
OFOI	Owner furnished, owner installed
PM	Project manager
RCP	Reflected ceiling plan
SECT	Section
TB	Table box
TYP	Typical
VIF	Verify in field

Table 4-2 Common Architectural Drawing Abbreviations

recommended symbols for AV objects in a drawing, together with some common symbol attribute abbreviations and their meanings, can be seen in Figure 4-6. Figure 4-7 is an example of a floor plan drafted to this symbol standard.

1	Audio - Video Systems				
Category-Symbol #	Name (Abbreviation)	Symbol	Technology Attribute Abbreviation Examples *For more abbreviations see Annex C*	Stretchable	Source
1-1	Loudspeaker (SPKR)		HF – HIGH FREQUENCY M – MONITOR LAR– LINE ARRAY P – POWERED SPEAKER S – SUBWOOFER LCR – LCR BAR ST – STEREO LR – LR BAR	Yes	CEA
1-2	Display Monitor (VMON) (TV)		M – MIRROR TV TV – TELEVISION VM - VIDEO MONITOR WP – WEATHERPROOF TV	Yes	J-STD 710
1-3	Video Projector (PROJ)		LCD – LIQUID CRYSTAL DISPLAY DLP – DIGITAL LIGHT PROCESSING LED – LIGHT EMITTING DIODE LCOS – LIQUID CRYSTAL ON SILICON	No	CEA
1-4	Projection Screen (SCRN)		F –FIXED M – MOTORIZED R – REAR PD – PULL-DOWN PU– PULL-UP MLB – MOBILE	Yes	CEA
1-5	Video Camera (CAM)		D – DOCUMENT IP – IP CAM PT – PAN/TILT PTZ – PAN/TILT/ZOOM	No	CEA
1-6	Remote AV Source (R-AV)		A – AUDIO SOURCE AV – AUDIO VIDEO SOURCE V – VIDEO SOURCE	No	J-STD 710
1-7	White Board (WB)		A – ACTIVE I – INTERACTIVE OVLY – OVERLAY PAS – PASSIVE	Yes	J-STD 710
1-8	Microphone (MIC)		B – BOUNDARY CLP –CLIP GNK – GOOSENECK HH – HANDHELD SGN – SHOTGUN ST – STEREO WLS – WIRELESS	No	J-STD 710
1-9	Junction Box (JBOX)		AV – AUDIO VIDEO D – DATA J – JUNCTION BOX WP – WEATHER PROOF	No	NCS

Figure 4-6 Some AV object symbols from the ANSI-J-STD-710 standard

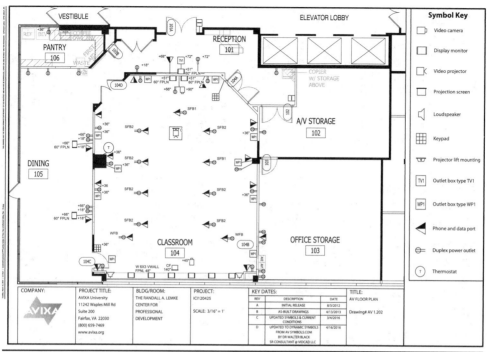

Figure 4-7   A floor plan drafted to the ANSI-J-STD-710 symbol standard

Architectural Symbols

Architectural symbols are standardized by country or region, and normally a drawing set will include a legend. As long as they are defined in the legend, the actual object does not matter. Typical symbols you will see include column lines, match lines, and three types of reference flags (elevation flags, section cut flags, and detail flags).

Grid System Symbols

A grid system is used to indicate the locations of columns, load-bearing walls, and other structural elements within the building layout, prior to the room locations being defined. Grid lines are used to reference schedules and for dimensioning.

Vertical grid lines should have designators at the top and be numbered from left to right. Horizontal grid lines should have designators at the right and be alphabetized from bottom to top. Figure 4-8 shows an example of this system.

Figure 4-8

Example of grid system symbols

The grid system can be used to find your way around a work site in the early construction phase. Contractors will refer to a point in the space with respect to where it exists on the drawing, as in "4 meters west of B6." Because the space may not yet be divided into rooms, identifying a point by the grid system ensures that everyone at the site understands the location.

Match Lines

Match lines are used to line up, or rebuild, a drawing that has been cut into separate drawings because it does not fit onto a single page. For example, it may take multiple pages to depict one area of a building. To assemble the separate drawings so you can see the whole picture, the individual pieces are aligned using the match lines as a guide.

There may also be a single drawing, at a much larger scale, that indicates the relative positions of the pages in such a drawing set. In Figure 4-9 the drawing number reference is the side that is considered.

Figure 4-9
Example of a match line

MATCH LINE
SEE TA106

Reference Flags

In some drawings you will find symbols that refer you to another drawing, such as an elevation, a section, or a detail drawing.

Elevation Flags   As shown in Figure 4-10, elevation flags are used on plan drawings to indicate related elevation drawings. The center text indicates the identification number of the elevation drawing and the sheet where that elevation drawing appears. The text at the apex of the triangles identifies the elevation drawing

Figure 4-10
Example of an elevation flag

on that page. The direction of the apex of the triangle gives an approximate orientation of the elevation from the viewpoint of the symbol on the plan.

Each elevation view for a room is drawn in sequential order and presented in a clockwise manner. This way, you can "look around the room" as if you were standing on the elevation flag and viewing the walls.

Section Cut Flags   In Figure 4-11, a section cut flag indicates which section drawing depicts a section of a master drawing in more detail.

The bottom number (A3.0) indicates the page number where the section drawing can be found. The top number (D1) is the section drawing identification because multiple section drawings may be on the same page. A line extends from the center of the symbol indicating the path of the section cut. The right angle of the triangle is an arrow indicating the view direction of the section drawing.

Figure 4-11
Example of a section cut flag

Section views are similar to elevation views. They depict specific wall sections, based on the section cut line indicators on the master floor plan drawing.

Detail Flags   As shown in Figure 4-12, detail flags indicate which small item detail needs to be enlarged. These items are too small to draw or see at the project's typical drawing scale. In a large system, there may be hundreds of areas or items requiring more detail that need to be read accurately, so each item is numbered.

Figure 4-12
Example of a detail flag

Box	Description	Type	Mount	Location	Cover		Note
					Type	**Source**	
AV1	Presenter input	4-gang	Flush	@ 18" AFF	Custom	AV contractor	
AV2	Presenter input	4-gang	Flush	@ 18" AFF	Custom	AV contractor	
CT1	Conduit termination point	12×12	Surface	@ 48" AFF	Blank	Electrical contractor	Verify location with AV contractor
V1	Projector	3-gang	Flush	In ceiling	Custom	AV contractor	Verify location with AV contractor
V2	Document camera	2-gang	Flush	@ 18" AFF	Custom	AV contractor	
V3	Auxiliary input	1-gang	Flush	@ 18" AFF	Custom	AV contractor	

Table 4-3   Example of a Drawing Schedule

The bottom number indicates the page number where the detail drawing can be found. The top number is the detail drawing identification because multiple detail drawings may be on the same page. The symbol in Figure 4-12 refers to detail number 1, where you can find more information about drawing AV1.06.

Most drawing sets also have a schedule that lists materials needed during construction. A schedule will call out the details and specifications of the installation. In the example shown in Table 4-3, the schedule identifies the location and type of boxes containing AV terminations.

Drawing Views

When reading a drawing, the first step is to notice the point of view from which it is drawn. All of the drawing sets mentioned (architectural, engineering, mechanical, AV systems, and so on) typically include drawings that depict the structure from many different points of view. Are you looking at a drawing depicted from a ceiling view or a side view? The perspective of a drawing is referred to as a *view*.

To convey the design vision and layout concepts, architectural drawings provide a number of different views of the site. Drawing views depict the site from the top, front, side, and back. Take a look at a few of the different types and subsets of drawings that the AV team typically works with.

Plan View

Plan views provide an orientation to the space. This is a "top view" taken from directly above, such as a floor plan or site plan. A floor plan is typically represented as a slice through the building at 1m above floor level (4ft in the United States).

As shown in Figure 4-13, the plan view identifies the room locations and layout dimensions, including locations of walls, doors, and windows. A floor plan view typically contains indicators to other detailed views of the site. For example, the arrows in the plan drawing indicate the direction of view to other detailed views of the site.

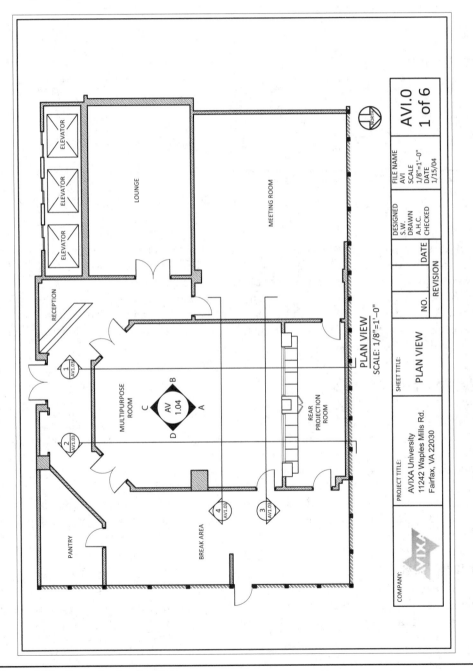

Figure 4-13 Example of a plan view

Reflected Ceiling Plans

A reflected ceiling plan is used to illustrate elements in the ceiling with respect to the floor. It should be interpreted as though the floor is a mirrored surface, reflecting the features within the ceiling.

As shown in Figure 4-14, the reflected ceiling plan view shows the locations of non-AV elements such as ventilation diffusers and light fixtures, as well as the position of AV components such as ceiling-mounted loudspeakers, display screens, and projectors. This view is of particular interest to the AV system designer and installer because it helps ensure that the AV systems will meet performance standards. It also indicates which other trades the AV installer may need to coordinate with on specific rooms or areas during the pre-installation phase of the project.

Elevation Drawings

An elevation drawing looks at built surfaces such as walls, doorways, and windows from a front, side, or back view. As shown in Figure 4-15, elevation drawings provide a true picture of anything that might be on the wall, including display panels, cameras, electrical outlets, AV faceplates, or chair rails.

Section Drawings

A view of the interior of a building in the vertical plane is called a *section*. A section drawing shows the space as if it were cut apart, and the direction you are looking is indicated by an arrow in the plan drawing, as shown in Figure 4-13. It shows the space with bisected walls, which allow you to view what is behind the wall or the internal height of the infrastructure. A section drawing can be rendered at an angle, so study it carefully. Figure 4-16 is an example of a section drawing.

 NOTE   It may help to understand the difference between elevation drawings and section drawings by thinking of elevations as pictures and sections as cutouts.

Detail Drawings

A detail drawing view shows items that are too small to see at the project's typical drawing scale. For example, Figure 4-17 indicates where the projection screen housing should be installed and fastened to the grid. Details may show how very small items should be put together or illustrate mounting requirements for a specific hardware item.

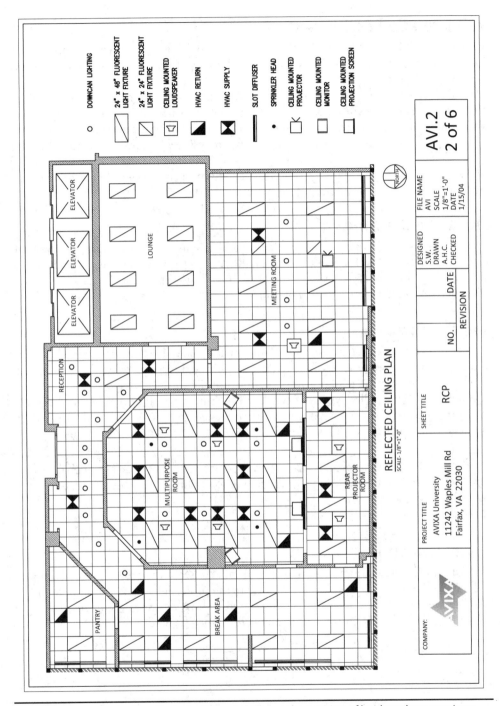

Figure 4-14 Example of a reflected ceiling plan (shows the position of loudspeakers, monitors, and rear-projection system screens). Note the symbol key on the right.

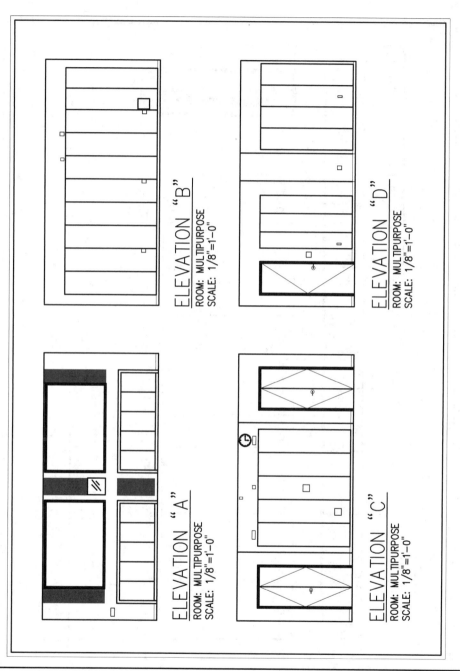

Figure 4-15 Examples of elevation views of the plan in Figure 4-13

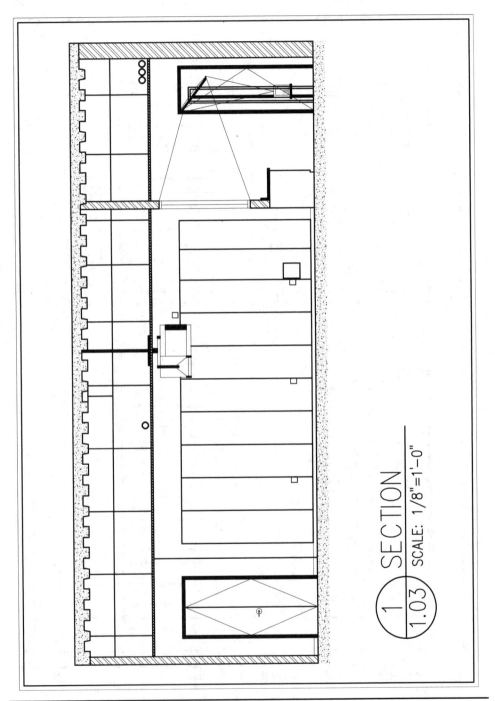

Figure 4-16   Example of a section drawing of the plan in Figure 4-13

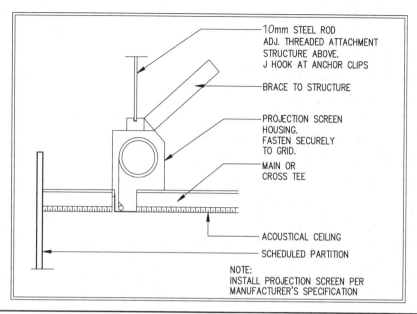

Figure 4-17 Example of a detail drawing

Chapter Review

This chapter provided an overview of the types of drawings you might have to read and interpret while working on an AV installation project. It also provided a list of common abbreviations with definitions used in AV documentation. Some of the symbols used in construction drawings and their characteristics were discussed, and you learned how to convert scaled dimensions on a drawing into actual measurements.

With this basic knowledge of how to interpret architectural drawings and site plans, you should be able to convert a scaled measurement or ratio into its actual equivalent, as well as identify the type of drawing and the scale view.

Review Questions

The following questions are based on the content covered in this chapter and are intended to help reinforce the knowledge you have assimilated. These questions are similar to the questions presented on the CTS-I exam. See Appendix E for more information on how to access the free online sample questions.

1. Which set of drawings should you reference to view the ductwork that goes through the building?

   A. Architectural drawings

   B. Mechanical drawings

   C. System drawings

   D. Electrical drawings

2. Which set of drawings would you reference to locate the power outlet for a video display?

    A. Structural drawings

    B. Mechanical drawings

    C. Electrical drawings

    D. System drawings

3. Which of the following must the as-built drawings show?

    A. The client's vision for the finished system

    B. Details of the system as finally installed

    C. Original system design specification

    D. Mechanical drawings

4. Write the correct name for each symbol in the column on the right.

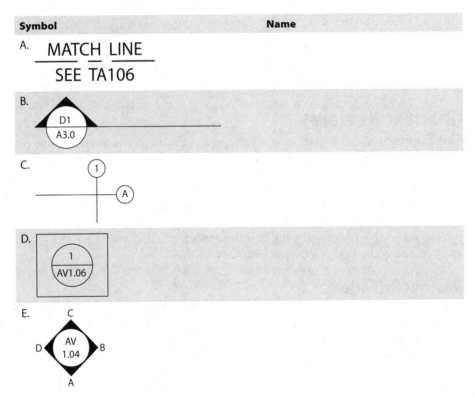

5. An architectural drawing was created using the U.S. customary scale of 1/8 inch = 1 foot. You measure a distance of 4.25 inches on the drawing. Which of the following is the actual distance in feet?

   A. 34ft

   B. 16ft

   C. 28ft

   D. 68ft

6. An architectural drawing was created using the SI scale of 1:200. You measure a distance of 13 millimeters on the drawing. What is the actual distance in millimeters?

   A. 5,200mm (5.2m)

   B. 260mm

   C. 1,300mm (1.3m)

   D. 2,600mm (2.6m)

7. Which abbreviation on a drawing tells you that you will not need to install a specific display screen?

   A. OFOI

   B. OC

   C. OD

   D. NTS

8. What does an elevation flag on a plan drawing indicate?

   A. An interior view of a wall structure

   B. Related elevation drawings

   C. Detailed view of the elevator or lift

   D. Loudspeaker mounting details

9. What type of drawing view shows the layout of items on the ceiling?

   A. Reflected ceiling plan

   B. Reflected floor plan

   C. Detail ceiling drawing

   D. Elevation drawing

10. What type of drawing would you use to find out exactly how the projector should be mounted to the ceiling?

   A. Reflected ceiling plan

   B. Mechanical drawing

   C. Detail drawing

   D. Section drawing

Answers

1. **B.** Mechanical drawings depict the ductwork that goes through the building.

2. **C.** Electrical drawings show the power outlets that could be used for a video display.

3. **B.** The as-built drawing should show details of the system as finally installed.

4. A = Match line, B = Section cut flag, C = Grid system or column lines, D = Detail flag, E = Elevation flag.

5. **A.** The actual distance is 34ft; when on a U.S. customary scaled drawing where 1/8 inch = 1 foot, you measure a distance of 4.25 inches.

6. **D.** The actual distance is 2,600mm (2.6m); when on an SI scaled drawing using a ratio of 1:200, you measure a distance of 13 millimeters on the drawing.

7. **A.** The abbreviation OFOI refers to "owner furnished, owner installed" and indicates that the installer will not have to install the specific display screen.

8. **B.** Elevation flags on a plan drawing indicate related elevation drawings.

9. **A.** Reflected ceiling plans depict the layout of items on the ceiling.

10. **C.** A detail drawing would depict exactly how the designer wants the projector mounted to the ceiling.

Pre-installation Activities

In this chapter, you will learn about
- Conducting a site survey
- De-installing existing equipment
- Safety requirements
- Working with other trades
- Steps to escalate a problem

Conducting a site survey is one of the key activities in the pre-installation phases of an audiovisual (AV) systems integration project. Your objective during the site visit with the client is to discuss their general AV project expectations and review the characteristics of the various rooms and areas in which the AV components and systems will be installed. Make notes of all items that will help in the technical evaluation. The information you gather will also help in the evaluation of facility conditions and site preparation before the installation. Attend construction meetings and note changes to the original design that will affect the AV installation.

On projects that involve an upgrade of systems, it may be necessary to de-install existing equipment, and in some cases, reuse some of the components. During these stages and the actual installation phases, you will need to comply with safety regulations and take personal precautions against potential calamities.

There are many things to organize and bear in mind while preparing for an installation job. You will need to review architectural and AV system drawings before your site visit. It is also necessary to be mindful of your communication with key stakeholders and project workers in allied trades.

Duty Check

This chapter relates directly to the following tasks on the CTS-I Exam Content Outline:

- Duty A, Task 1: Review Audiovisual Project Documentation
- Duty A, Task 2: Conduct Technical Site Survey

- Duty A, Task 3: Prepare for Audiovisual Installation
- Duty A, Task 4: Evaluate Overall Facility Readiness
- Duty A, Task 5: Conduct On-Site Preparations for Installation
- Duty B, Task 1: Deinstallation of Existing Equipment and Cabling
- Duty E, Task 2: Coordinate with Other Contractors

These tasks comprise 28 percent of the exam (about 28 questions).

Types of Audiovisual Job Sites

Two types of job sites involve AV installations: construction sites and spaces that are already occupied by the client. Whether in new or existing construction, most large AV jobs have an assigned systems designer and/or project manager, whereas smaller installations may require the lead installer to make decisions based on available information or documentation. On small-scale jobs, it is critical for the installer to gather as much information as possible on the job site and the project. You may need the help of the project salesperson to facilitate the process. The salesperson may have already gathered initial information about the job site, client needs and expectations, and so on. At the least, the salesperson may have relationships with client stakeholders that you can leverage to gather the information you need.

Construction Site

From coordinating your installation tasks with other construction activities to job site safety issues, it is essential for AV installers to understand the specific issues associated with a new construction job site. A construction job site may not have a roof or walls, may be open to weather, will likely be dirty and noisy, may have other trades working on the site at the same time, and may present dangerous conditions. The types of tasks that AV installers perform at this stage can include running cables and installing mounts for AV equipment.

Client-Occupied Space

A client-occupied space is a building that is currently occupied. It can be an office, hotel, conference center, church, sports venue, or other facilities. The typical AV installation tasks in a client-occupied site include installing and testing systems and equipment. When working in such locations, remember that the site is being used for ongoing business, so you must behave in a professional manner.

Many large AV installation projects are both a construction site and a client-occupied space. The installation may begin during the new construction phase of a project and continue until after the project has turned into a client-occupied space. For example, you could run cables while the site is under construction and return to install video display screens after the client has moved into the site. In these cases, installers must make the mental transition from new construction to client-occupied space and conduct themselves in the appropriate manner.

In live-event situations, installation (bump-in), and after the event, during the bump-out, the event site is very much like any other construction project, with many other trades present, although the time scales are often more compressed, requiring minute-by-minute detailed planning and coordination.

Conducting a Site Survey

A site survey is your opportunity to make sure you fully understand the client's needs for installing the AV systems. The client is depending on your company's expertise to determine what hardware will meet their communication needs in each AV-equipped room or space and the best way to deliver specific AV solutions.

Examine the site and note features relevant to the AV installation in your documentation, such as room dimensions, layout, electrical capacity, HVAC (heating, ventilation, and air conditioning) component locations, existing data network and telecommunication facilities, and any issues that may have an impact on the AV system design and installation tasks. The more comprehensive your site survey, the better your plans and preparations will be for the actual installation. For example, if the project involves ceiling-mounted projectors, you will want to measure the distance between the ceiling tiles and the joists to make sure you bring the right ladder and mounting gear during the installation phase of the project.

Technical Site Survey

Review drawings with the AV designer and/or project manager before you set out to conduct a technical site survey. Review the equipment list and the locations where system components will be installed. Make a list of the items you need to collect information about. Here are some examples of technical items you should check on:

- Identify loudspeaker configuration.
- Note the ceiling height and construction method for installing display screens.
- Determine whether there are custom plates or floor boxes fabricated for the job.
- Review wiring methods.
- Review network infrastructure.
- Review the arrangements for the site works that will be undertaken by other trades, including wall blocking and structural reinforcement, installation of raceways, carpentry work, and mains power distribution.
- Verify connector, equipment, and tool checklists with the warehouse.
- Record model and serial numbers on your project documentation for all equipment used in the job.

The information you gather will help you and your team with planning for the installation. Be sure to convey specific information to the appropriate person on the AV team. For example, all technical information recorded and notes on installation challenges that you anticipate should be discussed with the AV project engineer or system designer. When reality does not match the original design, the AV project manager will want to use formal documentation such as a request for information (RFI) or request for change (RFC) to get approval on a change order (CO).

Evaluating Facility Conditions

You will also want to gather general information, such as building characteristics and construction styles. These general characteristics can clue you in to the types of codes and regulations with which you might need to comply. Observe and make notes on the current state of site conditions as they relate to client needs because they may lead to changes and additions to the infrastructure. Take pictures and document all existing conditions, including damages in a finished space or room that is ready for installation. Make sure that the site is clean and dust-free before AV equipment is brought to the site.

While collecting site information, start a list of the extra gear that you may need during the installation, such as additional safety equipment, extension ladders, elevated work platforms, coring tools, and hole saws. You will also want to note if it is necessary to acquire the services of specialty craftspeople. For example, if devices have to be mounted into timber/wooden lecterns, it may be necessary to acquire the services of carpenters for finishing details.

It is also important to survey areas that may affect logistics, such as limits on parking or loading vehicles. Record the sizes and capacities of access doors, loading dock heights, corridor widths, stairways, ramps, elevators, and doorways, to confirm that it is possible for all large or heavy items of equipment to be transported into the project area.

In a client-occupied site, installation work may be restricted to certain hours of the day or days of the week. Buildings and facilities with high security may require you and your vehicles to obtain special identification to gain access to the site. Attention to these and other constraints right at the start will enable you to plan effectively and complete work successfully at the client site without delays and additional costs.

Knowing what to look for comes from experience. Some installers have learned the value of making detailed checklists the hard way—by missing important details on a site survey. One installer reported showing up at the client's building to unload the gear at the loading dock and found that the loading dock was open only in the morning, making the trip a waste of time.

Prior to going to the job site, you should make certain preparations. Print out directions to the job site or use a global positioning system (GPS) device. Confirm that loading docks and room keys will be available upon arrival. Ask whether you can reserve freight lift/elevator hours if you plan to use them to transport heavy equipment. Investing some time before leaving can save time, money, and resources. It can also prevent delays and ensure that you meet your project deadlines. The best way to make sure you are collecting all the necessary information is to develop and use a checklist.

TIP It is helpful to take extensive pictures of the site for reference in case there are questions later in the project. Get permission before taking pictures.

Drawing Review Process

For any AV integration project, documents and samples almost always need to be prepared and submitted to the owner or the design team for review and approval. These are often called *submittals* in construction contracts. The parties that receive submittals vary according to the project. You may submit your AV system drawings to a technology manager, project manager, design consultant, general contractor/building contractor, or other authorized professional. For example, on a traditional design/bid/build project, the integrator would need to submit AV system drawings to the design team, which would in turn submit infrastructure drawings to other disciplines. On a design/build project, the integrator submits infrastructure drawings directly to other disciplines. On a technology manager–led project, the integrator may submit AV system drawings to the technology manager, who would in turn coordinate with other disciplines, potentially including internal clients such as department heads.

It is a best practice for the AV integrator's team to review and/or approve all the AV system drawings internally prior to submitting to other trades, so that all contributors have input with respect to their specific tasks. These internal contributors include those responsible for rack fabrication, test/alignment, site supervision, interface design, software programming, and quality assurance. You should compare your equipment list, specifications, and drawings to make sure that they all match. This will help you find any discrepancies in your work before you actually start spending money on the tools and equipment. It will also help you get the AV system drawing approval you need from the client before beginning work. Written approval should be obtained before proceeding with acquiring equipment or engaging contractors.

After this review process is complete, you should know exactly what you need for the installation. You must communicate your equipment needs, including any changes approved by the client, to the project manager or purchasing agent so they know what you need and can place the order for you. This communication should always be documented for inventory purposes and should include both the needed equipment and specialty and custom items, such as furniture, rear projection systems, special lenses, custom wall plates, or custom screens. Specialty and custom items can take some time to be delivered. Make sure you are aware of the lead time necessary for ordering each piece of equipment.

De-installing Existing Equipment

Sometimes you may be required to de-install existing equipment in an AV space, especially if you are tasked with upgrading an existing system. Do not be tempted to use quick and easy methods that may damage the equipment and void an existing warranty. Instead, remove the equipment carefully and sort it by what needs to go back to the customer and what you can dispose of properly. Your customer should have given you a document that lists what equipment they want to keep and what they no longer need. If you are required to reinstall some of the existing components, make sure the equipment that will be reused is fully operational and confirm that it will be compatible with the new system. Similarly, when removing existing cable, identify and mark what is to be reused. In some jurisdictions it is mandatory to remove all cables that are not functional.

 TIP When de-installing equipment, remove the connecting cables. Do not leave wires hanging inside the building infrastructure. Removing the wires prevents potential fire hazards, keeps the area tidy, and allows the materials to be recycled.

Reusing Existing Equipment

If the job is an upgrade or replacement, test and document the existing system in the room before making changes. You may need to record some benchmark tests, including applicable signal results.

The tests you need to perform depend upon the system functions involved. The ANSI/AVIXA 10:2013 Audiovisual Systems Performance Verification standard includes a list of performance verification tests by system function. This list is divided into tests that should be performed before, during, and after system installation. The standard is intended to help you evaluate a system's performance against its documented goals. However, you could use the standard's list of tests performed after installation to help generate a checklist of tests to evaluate the performance of an existing system.

Document every problem revealed while testing equipment that is to be reused. Report it to the AV engineer and other responsible parties to determine what to do next. It is possible that the client may want the previous installer to correct any problems before new components are added. Verifying and documenting that an existing system performs as expected is validation of your professionalism, which will help prevent misunderstandings, will improve customer relations, and could even lead to new business opportunities.

Job Site Safety Guidelines

Safety is a primary concern when working on a job site. A construction site is typically riskier than a client-occupied space, but hazards could be present in either type of location. You will almost certainly encounter hazardous environments in every installation job, and that's why it is important with each job to know who to contact on safety issues and how to reach them in an emergency.

You will need to wear protective gear, even while conducting a site survey at a construction site, because you may have to climb ladders to examine ceilings, there may be overhead risks, or there may be sharp protruding objects elsewhere that you might not notice.

Safety Regulations

Ask the client about safety requirements before your first visit to a job site. In most cases, compliance with safety rules is also a requirement of the client's insurance provider.

The AV team must comply with all applicable safety regulations set forth in the administrative rules in the project documentation, as well as local, regional, and national safety codes and regulations. It is also your responsibility to assess and respond to any additional safety risks that your specialized AV knowledge and experience may make you aware of.

Applicable rules are defined by what is known as the *authority having jurisdiction* (AHJ), or regional regulatory authority. On construction and event-related jobs, the AHJ is frequently, but not always, a government agency. For example, the local fire marshal or fire authority in a town where you are working may be the AHJ over fire codes, while a national, state, regional, or even town power utility may be the AHJ for electrical safety codes. It is important to establish precisely which body is the AHJ for each aspect of a project. In many jurisdictions, the AHJ requires a licensed engineer or architect to submit appropriate documentation if there are going to be structural changes to a facility. Resulting rules or cautions that the AV team needs to be aware of should be noted in the project documentation. Note that rules and regulations vary significantly from one town, city, region, prefecture, or state to another.

Project Safety Induction

The general contractor/main contractor/building contractor, and/or the authority having jurisdiction over a building site may have specific site-safety training that all contractors and subcontractors on the project are required to undertake before commencing work on the site. This may take the form of an in-person induction session or an online tutorial followed by an assessment of your understanding.

Best Practice for Code Compliance

The AV team should follow the most restrictive code for the region in which they are working. If you find that the interpretations of applicable codes and standards are in conflict, follow the requirements of the local AHJ.

Safety Equipment

There are several ways to prevent accidents and provide immediate care for minor injuries. If it is necessary for you to carry heavy pieces of equipment or furniture on the job site, get help and never lift alone. In many jurisdictions there is a maximum weight that can be manually lifted by an individual worker—usually in the region of 23kg (50lb) for males and 16kg (35lb) for females. Follow common guidelines for lifting safely, such as making sure your footing is steady and the weight is balanced between your feet. Remember to keep your back straight, knees slightly bent, and lift slowly without making sudden moves or twists as you move.

Always carry a first-aid kit with items such as antiseptic cleansing wipes, antiseptic gel/cream/spray, gauze dressing pads, hypoallergenic first-aid tape, adhesive plastic and fabric bandages, sterile saline sachets/tubes, tweezers, and latex-free vinyl gloves. A basic first-aid kit should help you take care of most minor scrapes and wounds.

Protect yourself from various job site–related injuries by using personal protection items and standard safety equipment.

Personal Protection Equipment

By wearing personal protection equipment (PPE), you will be able to prevent personal injuries.

- Hard hats help protect your head from unexpected injuries. For example, when inspecting a ceiling, a hard hat can protect you from sharp support elements that may be hard to see. Hard hats also protect your head if you fall from the access equipment.
- High-visibility garments that increase your visibility to vehicle, crane, and site machinery operators.
- Work boots or protected-toe footwear that meets applicable standards can help protect your feet from being crushed or cut.
- Work gloves protect your hands from cuts, scrapes, burns, or contact with harmful materials.
- Safety glasses help protect your eyes from dust and other hard-to-spot debris.

Approved Ladders and Work Platforms

Make sure you use the correct type of ladder or work platform at a work site. Never use electrically conductive ladders, even if you are not working with electrical components. Someone else at the site may make a mistake that results in an electrically charged surface, which could cause electrocution of anyone standing on a conductive ladder. Fiberglass ladders are preferred, and wooden ladders are generally acceptable.

In some cases, conditions in the insurance policy for the job site, or in policies held by individual contractors, will restrict the use of ladders and work platforms to those who have provided them.

Fall Protection

Fall protection gear is a requirement for many site inspection and installation tasks. Since falls are the leading cause of worker fatalities at work sites, your AHJ will usually require that workers use approved fall-protection harnesses and static lines when working above a specified height.

Audiovisual Tools

System integrators must have the necessary tools to complete a project correctly and on time. The quality of the selected tools will affect their reliability. You may prefer to use tools from certain manufacturers because you trust the quality and performance of their products. Ultimately, you will need to select the tools that you are the most comfortable with and can also get the job done correctly. Table 5-1 lists items such as a verification checklist and some important tools that integrators should have in their tool bag when setting up and verifying a system. Make a list of your own that includes these and other items you will find useful on the installation job you are working on. Because each job will require a different set of tools, you will need to make a checklist for each project.

Some items of test equipment may be replaced with a portable computer or a mobile communications device, equipped with the appropriate test or analysis applications and special-purpose interfaces or adapters.

General		
Verification checklist	Screwdrivers	Wrenches
Cable crimpers	Solder and de-solder tools	Pliers and cable cutters
Saws	Drills	Laser level/measuring tool
Cable ties	Level	Electrical tape
Cable testers	Multimeter (DMM or analog)	Data storage devices
Laptop/notebook computer and signal adapters	Notebook	Marker pens
Self-adhesive labels	Scissors	Knives/cutting tools
Measuring tapes	Flashlights/torches	PPE
Nut drivers	Wire strippers	Camera
Audio		
Impedance meter	Signal generator	Signal analyzer
SPL meter	Real-time analyzer (RTA) hardware or software	Dual-channel FFT analyzer
Measurement microphone	Cable adapters	Headphones
Video		
Oscilloscope	Signal generator	Signal analyzer
Adapter cables	Test monitor	Replay device

Table 5-1 Essential Items in an Installer's Audiovisual Tool Bag

Working with Stakeholders and Other Trades

Even though you have an extremely important job in executing the installation, remember that you and your team are only some of the people who will be on the job site. Showing courtesy to the other trades and specialists occupying the space is both polite and professional.

Working with allied onsite trades is particularly important when working in a client-occupied space, where your customer and their staff may be present throughout the process. No matter the type of job site you are working in, you still must be courteous to the people who share your space. Communicating clearly, and with integrity, is key to working successfully with stakeholders and teams in allied trades.

Typical Participants

On a job site, AV installers will interact and coordinate with these typical participants:

- **Owner**   Your company's client is the "owner." Your responsibility is to help ensure the owner will be pleased with both the quality of your work and the manner with which you undertake it.

- **Architect**   The architect is responsible for the overall vision of the structure and the overall integration of elements within the structure. The architect has a large role in determining how the AV system elements are installed.

- **General contractor/prime contractor/building contractor**   The prime contractor coordinates the activities of all workers, including internal staff and outside contractors. The general contractor may be your direct client, having hired your company to complete the AV elements of the project.

- **AV designer/project engineer**   The AV designer or project engineer keeps track of the progress of the installers to verify that the system design is working out in the field. The designer/engineer will also visit to verify the quality of the installation, and usually participates in at least some aspects of system verification.

- **Acoustic consultant**   This consultant suggests the best acoustical arrangement for the room, but may also recommend acoustical wall treatments or seating layouts. Their acoustic solutions may include active sound masking, which is usually installed by the AV contractor.

- **Electrical contractors**   AV installers must work with the electrical contractor to ensure conduit and raceway requirements are met and electrical systems are compatible with AV installation needs. The electrical contractor is also responsible for the provision of power to AV equipment and the installation of the lighting and lighting control systems, which may interface with AV control systems.

- **Telecommunications/data network contractors**   You will need to work closely with the data network and telecommunications contractors to coordinate the timely provision of the communications infrastructure and data network access required for the AV installation.

- **Mechanical contractors**   Mechanical contractors are responsible for HVAC, plumbing, sprinkler, and other infrastructure items. AV installers may need to interact with mechanical contractors when planning runs of conduit and raceways for AV cabling and mounting infrastructure for loudspeakers, screens, projectors, videowalls, and display panels.

- **Ceiling contractors**   You will frequently interact with the ceiling crew, since you often need to discuss and coordinate the installation of screens, projectors, and loudspeakers with the ceiling contractors.

- **Plaster/drywall contractors**   You will often work with the plaster/drywall contractors to coordinate the installation of control panels, wall boxes, videowalls, monitors, display panels, and screens.

- **Floor contractors**   You may be required to work at the same time in the same area as the floor installers because of time frame issues that require the coordination of activities. You may need to coordinate with them on the installation of underfloor cabling, floor outlet boxes, and cable traps.

- **Painters**   You may also work simultaneously in the same rooms as the painters because of time frame issues. For example, painters may be completing touch-up work as you install mounts and mount substructures.

- **Other specialties**   Since you will be performing many installation tasks as the project is nearing completion, you will also need to coordinate your activities with contractors who may be installing security, intercom, building management information, and other systems.

- **Building receptionist or security team**   The building receptionist or building security team often has control of scheduling who can use different areas of a building. Let them know early when you plan to be onsite and when you're going to need access to service elevators, freight docks, alarmed escape doors, or other areas for loading and unloading equipment.

- **Fire suppression contractors**   This contractor makes sure that the space is built in a way that minimizes fire damage and adheres to the applicable fire codes. Often, this means installing fire-detection networks and overhead sprinklers and maintaining the integrity of fire doors.

- **Building cleaning crew**   This team will come in to clean and tidy the area, especially in a client-occupied space.

Professional Behavior

Project stakeholders need special attention. These are the people who are most invested in the successful outcome of the AV project, and they often have the most say in decisions about the system. Because they could control many factors of your project, you need to take steps to make sure they are on board with the AV solution you will be installing.

To better understand your client and other stakeholders on the project, you can create a stakeholder analysis chart by answering the questions in Figure 5-1.

PART II

Stakeholder Analysis

Stakeholder Name and Role	Stake and Motivators	Likes *Enablers*	Dislikes/Fears *Barriers*	Actions to Take
Who are they, and what do we know about them?	What is important to them? What do they want?	What would they like about this solution/ product/ service approach?	What would they dislike about this solution/ product/ service approach?	What actions will let us keep our stakeholder happy while still getting the job done?

This could be:
Position, status, authority and influence, access to resources, decision-making autonomy, comfort, future prospects, opportunity to succeed, working relationships, risk tolerance.

Figure 5-1   Constructing a stakeholder analysis chart

By understanding stakeholder interests, preferences, and concerns, you will be able to identify enablers and barriers they may present. This will help you take considered actions and better communicate with them.

Outward Appearance

As you work with customers and other project stakeholders, both during the planning stages and during the installation itself, you need to act professionally. A customer will evaluate your behavior directly from face-to-face conversations, or indirectly through your nonverbal communication and job site manner. As a representative of your company, you need to know how to conduct yourself professionally.

It is easy to feel invisible to customers when you are hidden behind large stacks of equipment, but customers are constantly paying attention to you, before, during, and after the installation. Your appearance and manners determine the way customers perceive you, even if you are not the project manager. Let's review some ways to convey your competency, reliability, honesty, and sincerity to the customer.

Signs of Active Listening   When speaking to any project participant, practice active listening. Stop what you are doing and look at the person speaking to you. If you are wearing mirrored or dark sunglasses or face masks, remove them so the person can see your eyes. If you understand what they are saying, convey that to your customer by nodding your head. Don't interrupt; listen until the customer has stopped talking.

Body Language Sends a Message   A smile conveys a positive attitude. Smile when you greet the customer and when you close your business. Introduce yourself clearly, indicating your job and responsibility if necessary, and shake the customer's hand.

Do not wait for the customer to make the first move. Standing up straight is an indication of your confidence.

Observe the customer's body language. If the customer seems nervous or confused, give them time to respond. Simplify your explanation and then ask whether they have questions.

Personal Appearance   Bathe regularly and avoid heavy colognes or perfumes. Hair should be neat and clean. Dress professionally. Your company may provide a uniform or have an "event" dress policy. Ask your supervisor about the dress code before you start job site tasks. Tuck in your shirt and wear a belt. Wear modest clothes that cover you up when you are working. With the exception of your organization's branded apparel, avoid clothing bearing images, slogans, logos, or advertising.

Your client may have higher expectations for professionalism and behavior in certain environments than your own company does. You need to provide your clients with the professional service they expect.

Adherence to Policies

Installers must observe the client's general policies for dining and smoking. You should not eat on the job unless the customer has agreed and then only in designated areas.

When working in a live-event setup, do not consume any food or beverages unless you are invited.

Also be aware of your company's policies on dining and alcohol consumption. In the interest of both workplace safety and insurance liabilities, it is advisable to avoid alcohol while at work.

Follow smoking policies. This means either smoking in designated areas only or waiting until you are off the job site.

Customer Service

Several indicators of professionalism are expressed by following common rules of etiquette, demonstrating an understanding of ethical service, and utilizing effective communication skills, especially with difficult clients.

In the context of AV installation jobs, demonstrating professional behavior involves all the topics discussed earlier, including commonsense ones such as obeying parking rules when you arrive at the job site and not leaving equipment in places that block emergency egress.

In a client-occupied space, it is necessary to minimize any disruption to the client's business activities, including noise that can disrupt ongoing work activities. Check with adjoining spaces before beginning loud work activities, such as drilling or operating motorized work platforms. Make sure talking loudly, testing audio equipment, or playing music will not disturb anyone who works at the client site.

Protect the furniture at the client site by moving it to a safe area or protecting it with padded material. Remember that placing tools and equipment on furniture or conducting work activities such as soldering or drilling can damage office furniture. For your own safety, and for your client's peace of mind, avoid standing on furniture to reach ceiling areas. Protect the entire work area from damage by using a drop cloth, plastic sheeting, or padded shipping blankets to protect furniture and carpets.

Clean the site when you are finished working. Pick up all scraps, remove empty boxes and rubbish, vacuum and dust the area, and replace any furniture or other items to their original locations.

Sustainability and Environmental Stewardship

Following best practices in environmental stewardship and sustainability is expected from a professional AV installation team. This should include employing low-impact chemicals and processes in preparing, installing, and cleaning up from a project and the careful handling and disposal of any hazardous waste substances that may arise from demolition and/or installation. Waste materials, packaging, and surplus equipment should be recycled where possible or disposed of thoughtfully, using approved or mandated practices.

Steps to Escalate a Problem

Despite how much planning you do or how careful you are, something will probably go wrong during your installation project. Maybe you and your stakeholders agreed on installing a specific microphone, only to find that the manufacturer discontinued that model a year ago. Maybe some equipment was not packed correctly on the truck and was crushed in transit to the job site. Or maybe you arrived on the job site only to find that the sprinklers were installed right where you planned to install a projector.

No matter what goes wrong, you need to know how to resolve the problem in the most efficient and cost-effective way.

 CAUTION If you spot a potential design flaw, *do not* quietly make a system change without talking to the AV designer first. Any change could have far-reaching consequences. Following established change management procedures will ensure that the right people weigh in on all decisions.

If you encounter a potential problem on the job site, the first thing you should do is talk to your project manager. Often, the most effective way to communicate your need is by filling out the proper paperwork. If you see something that needs to be changed, mark it on your project documentation with a red pencil or pen. These drawings are often called *redlines* or *markups*. Once you have marked your requested changes, your project manager will translate this into a formal RFI or RFC document. This document will go out to the designers and architects, who can approve the change with a CO document. Then, and only then, should the design be changed.

Following the established change management procedure may seem like a lot of tedious back-and-forth, but it could save you money and pain. Any design change you make could affect not only the AV system but other trades or systems in the space as well. The RFI process helps ensure that dependencies are identified before any decisions are made. Some design changes result in a significant increase in effort or require additional materials or equipment changes. If your client does not approve these changes, they are not necessarily obligated to pay for them. The RFC and CO process ensures that all parties understand any additional costs and responsibilities incurred as the result of a change.

Procuring Equipment and Supplies

Depending on the size of the installation firm, one or more people will be responsible for ordering and allocating equipment, supplies, and tools for an installation project.

Most installation companies have a person who is responsible for ordering equipment from suppliers, such as manufacturers or distributors. This person is usually in the purchasing or ordering department. The purchaser uses the design documents, equipment lists, and installation timeline or plan to obtain the equipment necessary. Sometimes the equipment will be pulled from the installer's stock, but more than likely, the equipment will be ordered and scheduled for delivery to the installer's workshop location.

The equipment should be staged in the warehouse or workshop. Staging is where the equipment is gathered in an organized manner and inventoried against the equipment list. Either the purchasing agent or the project manager will monitor the arrival of equipment and keep track of the pieces yet to arrive. Any items that have not arrived as scheduled will be noted and escalated with the supplier for estimated arrival dates. When the installation schedule dictates it, the equipment will be moved to the workshop for rack mounting, workshop assembly, initial setup, and testing.

Supplies may or may not be ordered for a specific project. Many installation companies will hold inventory of cable, connectors, solder, cable ties, electrical tapes, and other items that are used in most projects. The workshop manager or lead installer will keep track of general supply levels and will request that more supplies be ordered through the purchasing agent. The lead installer or project manager must take note of any unusual connectors, cables, mounts, or other items that are not normally kept in stock so they can be ordered in a timely manner. When you receive any ordered equipment, make sure the equipment is properly received and unpacked and confirm that it meets project specifications.

It is important to test all equipment before it is delivered to the project site. Much of the equipment will be tested during rack assembly and preliminary commissioning in the workshop. Readily-damaged items such as displays and furniture are best inspected and tested when initially received to check for shipping damage and "out of the box" warranty issues. This may save time by eliminating a round-trip delivery if an item has to be returned, replaced, or repaired. After testing, repackage equipment in its original shipping configuration. Collect all of the accompanying documentation, such as warranties, instructions, owner manuals, and serial numbers. Many AV firms gather all manuals and warranty documents for security and reference during the job and then create a customer binder with the information when the project is complete.

Larger items or special situations will require a "drop ship," where the equipment is sent directly to the installation location. An example might be projection screens or rear-screen material. Because of the size of the equipment, it does not make sense to ship it to the workshop only to put it on another truck to deliver it to the customer. Drop shipping's disadvantage is that someone has to be responsible to receive the shipment, and the customer may not have the personnel or expertise to handle the incoming freight. This means the installer or project manager will have to greet the shipment and handle the logistics. This can be very time-consuming.

Chapter Review

In this chapter, you studied the characteristics of construction sites and client-occupied spaces that an AV installer would typically work in. You learned how to conduct a site survey and reviewed items that should be covered in a technical site survey and in an evaluation of facility conditions. Key factors in the de-installation of existing equipment and reuse of components were discussed. Since safety is of critical importance, the requirements that must be followed to ensure safety at your work site were reviewed. The short list of essential items an installer's AV tool bag should include will help you assemble your own. Effective communication skills that will help facilitate relations with stakeholders and workers from allied trades at the work site were discussed. You also reviewed methods to escalate problems and the guidelines for the procurement of the AV systems.

Review Questions

The following questions are based on the content covered in this chapter and are intended to help reinforce the knowledge you have assimilated. These questions are similar to the questions presented on the CTS-I exam. See Appendix E for more information on how to access the free online sample questions.

1. Which of the following is the AV team likely to perform during new construction?

   A. Install the sprinkler system

   B. Run cables and install mounts

   C. Mount the drop ceiling

   D. Build the firewall

2. What is the primary reason an AV installer should attend weekly construction meetings?

   A. Learn about safety requirements

   B. Establish relationships with allied trade teams

   C. Identify changes that will affect the AV installation

   D. Find out when the cafeteria is open

3. Which of the following is the most important consideration during the installation of AV systems in a client-occupied space?

   A. Securing access to any office supplies or tools belonging to the client that may be useful in completing the installation

   B. Acoustically treating any spaces adjacent to the area where work will take place

   C. Ensuring your attire is as formal as the client's

   D. Avoiding disruptions to your client's business

4. What is the first step in conducting a technical site survey?

    A. Reviewing drawings with the AV designer

    B. Recording model and serial numbers of all equipment

    C. Setting up your GPS with the site location

    D. Ordering the equipment from the warehouse

5. What are best practices when evaluating facility conditions of a space that is ready for installation? (Select all that apply.)

    A. Note areas that may affect logistics

    B. Make sure the site is clean and dust-free

    C. Take pictures and document all existing conditions

    D. Note if you will need an extension ladder

6. Which statement accurately describes what you should do when de-installing existing equipment?

    A. Dispose of equipment right away

    B. Remove the connecting cables

    C. Smash the hardware so that data cannot be stolen

    D. Call an electronic waste recycling contractor

7. If existing components have to be reused in an AV system, what is a critical task that the installer must perform?

    A. Draw a schematic showing where the old equipment fits

    B. Call the manufacturer to extend the warranty

    C. Test the equipment to make sure it is operational

    D. Discard items that do not meet specifications

8. Which organization's general safety codes and regulations must you follow when visiting a job site?

    A. The client

    B. The AHJ

    C. The client's insurance company

    D. Your company

9. What will a stakeholder analysis help you identify?

    A. The original timeline for the project

    B. The changes that were made to the architectural drawings

    C. The client's concerns and fears about the AV system

    D. The GC's budget for the AV system

10. If the audio digital signal processor you have to install is discontinued by the manufacturer, what should you do? (Select all that apply.)

A. Install another similar model and record the model number on the project documentation

B. Discuss the need for change with the AV designer and your project manager

C. Redline the change on the appropriate project document, send it to your project manager for submission of a formal RFI or RFC, and wait for a CO

D. Try to find the specific model through an online vendor

Answers

1. **B.** During new construction, the AV team could run cables and install mounts.

2. **C.** The primary reason an installer should attend weekly construction meetings is to identify changes that will affect the AV installation.

3. **D.** Of the considerations listed, avoiding disruptions to your client's business is the most important consideration when installing systems in a client-occupied space.

4. **A.** The first step in conducting a technical site survey is to review the drawings with the AV designer.

5. **A, B, C, D.** All items listed are best practices when evaluating facility conditions.

6. **B.** Remove the connecting cables when de-installing existing equipment.

7. **C.** If old equipment has to be reused in an AV system, the installer must test it to make sure it is operational.

8. **B.** The authority having jurisdiction (AHJ) establishes general safety codes and regulations. However, an installer must follow the safety requirements of all the organizations listed. In case of conflict, best practice is to follow the most restrictive requirement.

9. **C.** The stakeholder analysis will help you identify the likes and dislikes (concerns and fears) of stakeholders, including the client.

10. **B, C.** These two statements recap the steps involved in escalating a problem and making a change, such as substituting a different digital signal processor model for one that has been discontinued by the manufacturer.

PART III

System Fabrication

Cable Essentials

In this chapter, you will learn about
- Identifying cables and connectors
- Plotting a cable route and pulling cable
- Conduit capacity
- Terminating cables

System fabrication involves preparing, assembling, testing, and configuring all of the audio, video, and control systems in advance of an audiovisual (AV) installation. It includes mounting equipment, running the cables, building and wiring the racks in preparation for the installation. Before wiring an installation, you will need to know how to install cable, as every piece of wired AV equipment depends on its cables. Without the cables that feed it power and AV signals, the most cutting-edge piece of AV equipment is nothing more than an expensive paperweight. Having the skill to judge which cable you need for a particular application and then being able to plan for the installation will allow you to do your prep work quickly and give you more time to install the system.

Installers should be able to identify the different types of cables and connectors, know where each type is best applicable, and determine cable lengths based on project documentation. You will need a basic understanding of wire and cable, including cable support, routing, pulling, and termination. Improper cable termination is the root of a majority of AV problems. Therefore, an important task is to make sure that the AV system you are working on has the right cables and that they are terminated correctly.

You will also occasionally need to use AV math to calculate conduit capacity and jam ratios for one, two, or three or more cables. Your knowledge and skills in cable handling will be put to good use during the system fabrication stages of any AV installation project you will work on.

Duty Check

This chapter relates directly to the following tasks on the CTS-I Exam Content Outline:

- Duty B, Task 3: Pull Cable
- Duty C, Task 5: Terminate Cables

These tasks comprise about 9 percent of the exam (about nine questions). This chapter may also relate to other tasks.

Wire and Cable

The basic function of wire and cable is to transmit a signal from point A to point B without altering it in any way. Good wiring is akin to using good fuel for an expensive motor vehicle. In an AV system with high-end professional equipment, it does not make sense to carry that signal on a cable highly susceptible to interference and noise.

The terms *wire* and *cable* are often used interchangeably, but each has a distinct meaning. A wire contains only one conductor, and that conductor may be either solid or composed of strands. Cable contains any number of insulated wires in a protected bundle and is the result of combining two or more wires in a manner that conforms to certain requirements or standards.

Conductors

A *conductor* is a material that allows a current to pass through it continuously. Some metals make excellent conductors because they are inexpensive and easy to work with. The conductor in a wire carries the signal by conducting current between a source, such as a microphone, and a load, such as a mixing console.

Conductors may be classified by the following:

- Size (gauge or cross-sectional area)
- Construction
- Conductive material

A *solid conductor* is a single-strand conductor and costs less than a multistranded conductor. Solid conductor wire is less flexible than stranded wire and has more restrictive bend-radius limitations. The bend radius dictates how much you can bend the cable before it deforms and alters its electrical properties. Although this type of cable is relatively inflexible, it can generally withstand more strain than a stranded conductor of similar cross-sectional area. Single-stranded cables are not suitable for applications where regular flexing is required, but they are completely suited to infrastructure wiring.

A *stranded conductor* uses multiple smaller, solid conductors that are twisted or braided together. The finer the strands, the more flexible the conductor, which enables easier handling and installation. Stranded conductors are more flexible and easier to handle than solid conductors, but they are more expensive. They are used where flexibility is required, such as for patch leads or in a live-event environment, where the cable needs to be flexed and moved around frequently. In situations like these, the flex life of a cable becomes an important consideration. *Flex life* is a general term that describes how long a wire may last before it physically fails to conduct signals properly.

The cross-sectional area or gauge of the conductors in a wire must be sufficient to handle the intended signal's current without offering too much electrical resistance or producing too much heat.

Some insulation displacement connectors (IDCs) such as the RJ45 (8P8C) have different versions for solid or stranded wire. The version for solid wire wraps around the wire of the conductor to make the electrical connection, while the version for stranded conductors pierces into the strands making up the conductor to create the electrical connection. While the version for solid wires will also work reasonably reliably with stranded wire, the version for stranded wires does not necessarily make a reliable connection with solid-core wire. An AV installer should always work with the appropriate type of connector for the wire being terminated; however, using a connector for solid wires on a stranded conductor is reasonably reliable for short-term connections.

Insulation

The function of insulation is to prevent physical contact between multiple conductors and to avoid signal interactions between different conductors. If bare wires touch, they can create short circuits that prevent signal transmission and damage equipment. To prevent such problems, insulation, which is a highly resistive material, surrounds the conductors.

In general, insulation is made from materials such as PVC, silicone, rubber, or PTFE (Teflon). Different types of insulation can affect the performance of wire. The ability of an insulating material to block electric fields is known as its *dielectric strength*.

Heat has an opposite effect on an insulator than it has on a conductor. While most conductors increase in conductivity with temperature, higher temperatures result in lower dielectric strength in an insulator.

The selection of the type of insulation on cables used in AV applications also requires consideration of how the material breaks down when overheated or in a fire. Many insulation materials produce toxic gases and smoke at high temperatures and are therefore unsafe to use in situations where those combustion products may reach spaces occupied by humans. Only certain types of insulation materials are considered suitable to be used on cables located in ceiling spaces that may form part of the HVAC system for a building. Such cables are identified as being plenum-rated.

Wire and cable installation is often the most labor-intensive portion of any AV system installation.

Types of Cable Used in AV Systems

Audiovisual cables vary in size, color, number and type of conductors, and other specifications. Whether they are shielded or not, the type of shielding and other factors, such as robustness, flexibility, strength, resistance, and capacitance, will determine which ones you will select for your project. In addition to checking the specifications, check the label and ratings marked on the cable jacket. Ease of termination is also a significant factor to consider, but maintaining signal integrity throughout its transport is of primary importance. Let's take a look at cables most commonly used in AV installations.

Best Practice for Cable Selection

When selecting cable, it is critical to use cable that is rated for its application (firesafe and UV/weather stable) by the relevant authority.

Coaxial Cable

Commonly used for both analog and serial digital (SDI) video and radio frequency (RF) signals, coaxial cable contains a single center conductor that is surrounded by a dielectric (insulating material with special electrical characteristics). Around the dielectric can be one or two types of conductive shields and a jacket. As shown in Figure 6-1, the first shield is a solid-foil material; the second shield is made of a metal stranded or braided material. Both the shielding and the center conductor are coaxial (they share the same axis), which is why it is often referred to as *coax*.

Coaxial cable carries an unbalanced signal. The single wire running down the center is the signal conductor, and the shield acts as the return. The impedance of the signal conductor is different from that of the shield, which is at ground potential, so they are not balanced.

Unshielded Twisted-Pair Cable

Twisted-pair cable is composed of one or more pairs of insulated wires twisted around each other. As shown in Figure 6-2, each wire is individually insulated, and each pair of wires is twisted together. The twisting of the pair facilitates common-mode noise rejection when balanced signals are carried by the paired wires. Unshielded twisted-pair (UTP) cable is commonly used for high-speed data transmission as well as many proprietary signal transport schemes. Some variants of UTP cable used in very high-speed data transport include insulated spacers to separate the pairs and thereby reduce the crosstalk between the signals being carried.

Figure 6-1
Coaxial cable

Figure 6-2
A Category 5
cable with four
unshielded
twisted-pairs

Twisted-Pair Network Cables (Cat x)

Since the adoption of four-pair, 100Ω, UTP Category 3, 16MHz-rated cable for tele-phony, and later for 10Mbps Token Ring and Ethernet networking, the noise immu-nity and data speed ratings of four-pair 100Ω twisted-pair data cables has progressively improved, with each generation tested and assigned a data category rating by the Tele-communications Industry Association (TIA) and the Electronics Industries Association (EIA). The most widely used of these are

- **Category 5 (Cat 5)**   Bandwidth 100MHz for data speeds up to 100Mbps over distances up to 100m (328ft).

- **Category 5e (Cat 5e)**   An enhanced version of the 100MHz Cat 5 cable standard that adds specifications to reduce far-end crosstalk (FEXT) for data transmission up to 1Gbps over distances up to 100m (328ft).

- **Category 6 (Cat 6)**   Cat 6 has even more stringent specifications for crosstalk and system noise than Cat 5e, with a bandwidth of 250MHz and data speeds up to 10Gbps over distances up to 55m (180ft). Sometimes used with an overall shield around the four pairs (S/FTP).

- **Category 6A (Cat 6A)**   Constructed using a mechanism to physically separate the twisted pairs, with more stringent specifications for alien and near-end crosstalk (NEXT), with a bandwidth of 500MHz and data speeds up to 10Gbps at distances up to 100m (328ft). Sometimes used with an overall shield around the four pairs (S/FTP).

- **Category 7 (Cat 7)**   To reduce crosstalk each twisted-pair is shielded, and the entire cable is shielded, for a bandwidth of 600Mhz and data speeds up to 10Gbps at distances up to 100m (328ft). Cat 7 is not recognized as a standard by the TIA or EIA and is not terminated with an RJ45 (8P8C) connector.

- **Category 7A (Cat 7A)**   To reduce crosstalk, each twisted-pair is wrapped in a foil shield, and the entire cable is covered with a braided shield, for a bandwidth of 1GHz and data speeds up to 10Gbps at distances up to 100m (328ft). Cat 7A is not recognized as a standard by the TIA or EIA and is not terminated with an RJ45 (8P8C) connector.

- **Category 8 (Cat 8)** Constructed with a shield around each twisted-pair and a further shield enclosing all twisted pairs, with a bandwidth of 2GHz and data speeds up to 40Gbps for runs up to 30m (100ft). Cat 8.1 is terminated with an RJ45 (8P8C)–compatible connecter, Cat 8.2 uses a lower-noise connecter that is not RJ45 compatible.

Shielded Twisted-Pair Cable

Shielding isolates and protects a signal from sources of electromagnetic (EM) and RF interference. Shields can be implemented in a variety of ways. They can be used to provide overall coverage around a single insulated conductor, or around individual insulated conductors or twisted pairs of conductors in a multiconductor cable, as shown in Figure 6-3. Shielded twisted-pair (STP) cable is commonly used for analog audio transmission. Shielding is also used to provide all types of data with protection from electromagnetic interference (EMI) and radio frequency interference (RFI) on longer cable runs or within severe RF fields.

Shielded Twisted-Pair Network Cables

Overall shielding is often added to twisted-pair network data cables to both exclude external EM and RF noise from affecting the data on the twisted pairs inside and to prevent the emission of RF from the signals on the twisted pairs into the outside world.

Shielding may also be placed around the individual twisted-pairs to further isolate them from exterior EM and RF interference, but more importantly, to isolate each pair from the emissions of the other pairs in the cable (crosstalk). Cat 5, Cat 5e, Cat 6, and Cat 6A are available as either unshielded or shielded cables, but Cat 7, Cat 7A, and Cat 8 must be shielded to meet to their specifications.

The shielding may be constructed of either foil tape (type F) or braided wire (type S), or sometimes both (type SF). Unshielded cable is identified as type U. The variations of shielding on twisted-pair data cables are identified by the ISO using the following format:

X/YTP
Where X is the overall cable shield type
Y is the shield type on the individual twisted pairs

TP simply indicates that the cable contains twisted pairs

Figure 6-3
Shielded twisted-pair audio cable

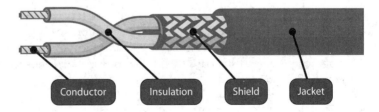

Conductor Insulation Shield Jacket

ISO Format	Common Industry Names	Overall Shielding Type	Twisted-Pair Shielding Type
U/UTP	UTP	None	None
F/UTP	FTP, STP, ScTP (screened twisted-pair)	Foil	None
S/UTP	STP, ScTP	Braiding	None
SF/UTP	SFTP, S-FTP, STP	Braiding and foil	None
U/FTP	STP, ScTP, PiMF (pairs in metal foil)	None	Foil
F/FTP	FFTP	Foil	Foil
S/FTP	SSTP, SFTP, STP, PiMF	Braiding	Foil
SF/FTP	SSTP, SFTP	Braiding and foil	Foil

Table 6-1   Some Twisted-Pair Cable Shielding Types and Their Common Industry Names

As you can see from Table 6-1, there are often several different industry names for the same cable, and sometimes the same industry name for different cables. It is important that before installation the AV installer verifies that the type of cable supplied is actually the one specified in the system design and that appropriate connectors, stripping, and termination tools are available onsite to terminate the cables.

Fiber-Optic Cable

As shown in Figure 6-4, fiber-optic cable uses conductors made from transparent glass or plastic fibers. These cables are commonly used for both long-distance and high-bandwidth data transmission. They are also used to provide galvanic isolation between devices where high voltages such as lightning strikes are possible, for RFI immunity where high levels of RF are present in locations near high-power transmitters, and where EMI is present near high-voltage electrical transmission lines.

Combination Cables

Combinations of the four types of cable you have just reviewed are used for various AV applications. For example, coaxial and UTP are sometimes combined in a single cable to simplify installation.

It is also important to note that just because a cable has a "typical" use, it is not the cable's only purpose. For instance, you can run audio and video signals on unshielded twisted-pair cable and audio signals on coaxial cables.

Figure 6-4

A multicore fiber-optic cable with an outer sheath

Figure 6-5
A cross-section of
optical fiber

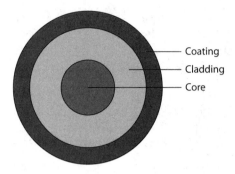

— Coating
— Cladding
— Core

More About Fiber-Optic Cable

Fiber-optic cable offers many benefits for the transport of audiovisual content and other high-bandwidth data over long distances.

Fiber has a transparent glass or plastic core used for carrying modulated light, as shown in Figure 6-5. The transparent cladding is the next layer of the fiber, which is used to reflect the light back onto the core, which keeps the signal moving down the fiber. The next layer is the coating, which protects the fiber from damage. A sheath is usually added to provide mechanical strength and additional protection to the cable in harsh environments.

There are two modes or paths of light transmission used in fiber-optic cable, known as single-mode and multimode. As shown in Figure 6-6, single-mode fiber is narrow in diameter, which makes the signal path mostly straight, with little of the light reflecting off the walls of the glass.

In multimode transmission some of the signal goes straight down the fiber, while the rest repeatedly reflects off the interface with the cladding. As a result of these multiple reflection paths, the signals take slightly different times to reach the end of the fiber, making it more difficult to detect the edge of each data pulse being received. As signal dispersion increases with the length of a multimode fiber, signals can typically travel more

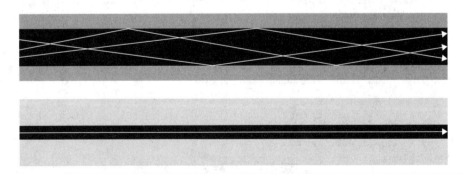

Figure 6-6 Longitudinal sections of multimode (top) and single-mode fiber-optic cable showing light paths

reliably over extended distances using single-mode fiber than with multimode. However, the wider multimode fiber is more easily able to carry multiple signals at different wavelengths (colors), a technique known as wavelength division multiplexing (WDM), which enables multiple channels of data to be carried.

Distance Limitations of Fiber

There are distance limitations with fiber-optic cables based on the type or mode of the cable, as shown in Table 6-2. Single-mode fiber is loss limited, meaning you will run out of light before the quality is compromised. Multimode fiber is bandwidth limited because although there is a lot of light, there is high dispersion.

Fiber-optic cabling can be used in many different environments: indoor, outdoor, hybrid, or composite. Indoor fiber cabling is similar to copper cabling, with plenum, riser-rated, and *low smoke zero halogen* (LSZH) cable options. Riser-rated cable can support its own weight as it rises up the building floor by floor.

Outdoor fiber cabling is mainly used for aerial or in-ground installations, but is also used for rental and staging applications.

Hybrid cabling combines both fiber and copper. For example, you could use hybrid cabling for a cable run to a high-definition camera, where fiber will carry the video signal and copper will power the camera electronics.

Composite combines both single-mode and multimode fiber. For instance, you might need to send 12 fibers down a cable. Six would be for single-mode, and six would be for multimode. Typically, you would use only one of these modes, but if two companies are sharing a space, the composite option would be more appropriate. Your selection should be based on the application.

 NOTE It is important to check that the fire-safety rating of the cables you are installing matches or exceeds the requirements of the AHJ over an installation.

Type	Typical Jacket Color	Max Range at 100Mbps	Max Range at 1Gbps	Max Range at 10Gbps	Max Range at 40Gbps	Max Range at 100Gbps
OM1 multimode	Orange	2km (1.24mi)	275m (900ft)	22m (72ft)	Not specified	Not specified
OM2 multimode	Orange	2km (1.24mi)	550m (1,800ft)	82m (270ft)	Not specified	Not specified
OM3 multimode	Aqua	2km (1.24mi)	550m (1,800ft)	300m (980ft)	100m (330ft)	100m (330ft)
OM4 multimode	Aqua or violet	2km (1.24mi)	1km (3280ft)	400m (1,300ft)	150m (490ft)	150m (490ft)
OM5 multimode	Lime green	2km (1.24mi)	1km (3280ft)	400m (1,300ft)	150m (490ft)	150m (490ft)
OS1 single mode	Yellow	10km (6.2mi)	10km (6.2mi)	10km (6.2mi)	10km (6.2mi)	10km (6.2mi)
OS2 single mode	Yellow	200km (124mi)	200km (124mi)	200km (124mi)	200km (124mi)	200km (124mi)

Table 6-2   Optical Fiber Types and Data Ranges

Fiber Applications

Optical fiber cables are constructed for specific applications and to meet specific installation requirements.

Risers

To allow fiber-optic cables to be safely suspended in such places as vertical shafts in a building core or in runs between different floors of a building (risers), some cables have strengthening elements (often flexible metal-wire or heavy non-optical glass fiber) included to carry the weight load of the cable without stressing the optical communication fibers. If the cable is also fire-resistant, it considered to be riser-rated.

Whether or not there are current-carrying conductors in a riser cable, those using conductive metal wire as the load-carrying element may be rated as "conductive" cables by fire-safety authorities.

Plenums

Every cable, electrical or optical, that is to be used in ceiling spaces, air ducts, and spaces that carry air for HVAC systems (plenums) must be constructed of materials that are fire-resistant and that produce no toxic gases on combustion. These are often identified as low smoke zero halogen (LSZH) cables. This construction reduces the risk of toxic fumes entering a building's ventilation systems during a fire outbreak. These are classified as plenum-rated cables.

Fiber cable types may be identified in a number of ways, as listed in Table 6-3, with the ratings in the U.S. National Electric Code being one of the more common naming schemes.

Table 6-3 Some Common Abbreviations for Fiber Cable Types		
	OFC	Optical fiber, conductive
	OFN	Optical fiber, nonconductive (can replace OFC)
	OFCG	Optical fiber, conductive, general use
	OFNG	Optical fiber, nonconductive, general use (can replace OFCG)
	OFCP	Optical fiber, conductive, plenum
	OFNP	Optical fiber, nonconductive, plenum (can replace OFCP)
	OFCR	Optical fiber, conductive, riser
	OFNR	Optical fiber, nonconductive, riser (can replace OFCR)
	OPGW	Optical fiber composite overhead ground wire
	ADSS	All-dielectric self-supporting
	OSP	Optical fiber, outside plant
	MDU	Optical fiber, multiple dwelling units

CAUTION   The improper handling of fiber-optic cable can lead to serious injury. Take precaution to avoid splinters of the tiny fiber needles piercing your skin. Use a black surface (workbench or cloth) to work on. To prevent eye damage always use a power meter to verify that the cable is working, not a microscope or magnifying glass. The camera in a personal digital device may also detect a working fiber.

Identifying Connectors

For a reliable connection, the type of connector attached to or terminated onto a cable must be compatible with the connector type used on the AV device.

Choosing the right connectors for your application is of critical importance, especially when terminating cable. You need to be able to identify the different types of connectors and understand where and how they should be used. Selecting the wrong connector could result in system failure. Table 6-4 shows images (see the copyright page for image credits related to this table) of commonly used connectors and their AV applications.

PART III

Connector	Type and Applications	Termination Method
	BNC Typically used for antenna, power, video, timecode, sync on wireless equipment, and SDI digital video/audio	Crimp or linear Compression
	Captive Screw (Phoenix or Euroblock) Control/audio/power Typical on permanently installed control devices	Screw down Compression Clamp
	DB-9 Control Typical on legacy and some current serial data control systems	Solder Crimp Euroblock
	DisplayPort Digital HD video/audio	Typically provided by manufacturer
	Mini DisplayPort Digital HD video/audio on small form-factor devices, also used for Thunderbolt 1 and 2	Typically provided by manufacturer
	DVI-D (top = male, bottom = female) Digital video signals	Typically provided by manufacturer but may be soldered

Table 6-4   Connectors with Typical Applications and Terminations *(continued)*

Connector	Type and Applications	Termination Method
	DVI-I (top = male, bottom = female) Digital and analog video Has analog pins for use with VGA adapter Similar to DVI-D but note additional pins	Typically provided by manufacturer but may be soldered
	F-type (male) Control/power/video/audio RF modulated audio and video Typically on antennas, TV, VCRs	Linear compression or crimp
	HDMI Digital HD video/audio	Typically provided by manufacturer
	RCA (phono) Typically used for video, audio, and RF on consumer devices	Solder or linear compression
	RJ-45 (8P8C) Modular connector Typically used for network, control, power over Ethernet (PoE), VoIP, AV over IP, HDBaseT digital video, digital audio, system and network control devices	Insulation displacement
	Spade Lug Audio/power	Crimp
	Speakon Typically used for loudspeaker-level audio	Screw Clamp
	TRS (tip, ring, sleeve) or phone jack 6.5mm (1/4in) variant typically used for balanced mono or unbalanced stereo on mixers and headphones 3.5mm (1/8in) (mini) variant typically used for unbalanced stereo headphone output for computers and multimedia devices Tip Ring Sleeve	Solder or crimp
	TS (tip, sleeve) or phone jack 6.5mm (1/4in) variant typically used for unbalanced mono audio on musical instruments and instrument amplifiers Tip Sleeve	Solder or crimp
	HD15 (VGA) (top = male, bottom = female) Sometimes incorrectly identified as a DB15 Typically used for legacy RGBHV video signals	Solder or crimp

Table 6-4 Connectors with Typical Applications and Terminations *(continued)*

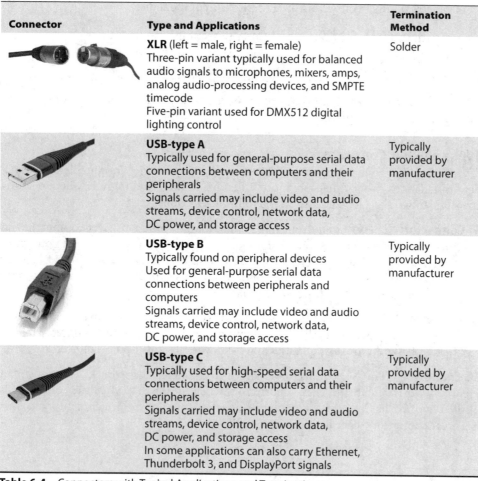

Connector	Type and Applications	Termination Method
	XLR (left = male, right = female) Three-pin variant typically used for balanced audio signals to microphones, mixers, amps, analog audio-processing devices, and SMPTE timecode Five-pin variant used for DMX512 digital lighting control	Solder
	USB-type A Typically used for general-purpose serial data connections between computers and their peripherals Signals carried may include video and audio streams, device control, network data, DC power, and storage access	Typically provided by manufacturer
	USB-type B Typically found on peripheral devices Used for general-purpose serial data connections between peripherals and computers Signals carried may include video and audio streams, device control, network data, DC power, and storage access	Typically provided by manufacturer
	USB-type C Typically used for high-speed serial data connections between computers and their peripherals Signals carried may include video and audio streams, device control, network data, DC power, and storage access In some applications can also carry Ethernet, Thunderbolt 3, and DisplayPort signals	Typically provided by manufacturer

Table 6-4   Connectors with Typical Applications and Terminations

Fiber Connectors

Many of the termination methods and cable types can use connectors interchangeably; however, when selecting a fiber connector, look for these additional characteristics:

- Low insertion loss (attenuation) depending on frequency and wavelength
- Good repeatability
- High reliability
- Ease of installation

Although these qualities are equally important in all connectors, the performance of each type of connector varies.

In general, all fiber connectors have these basic components:

- **Ferrule**   Holds the fiber in place after the termination procedure
- **Connector body**   Holds the ferrule and fiber in place
- **Cable**   Attaches to the connector body to be the point of entry for the fiber
- **Coupling device**   Mates connectors of the same gender, which is a common practice in fiber optics

Types of Fiber Connectors

A variety of fiber connectors are used in AV applications. Some are suitable for permanent installations, while others are more suited for flexible, and more hostile, operating environments. Take a look at each in the following sections.

ST Connector   The ST connector can be found on transmitter-receiver equipment and is similar to the BNC connector. Take a look at Figure 6-7. It is a bayonet connector, meaning that all you have to do is "stab and twist" to lock it into place, which keeps the fiber and ferrule from rotating during connection. This connector can be used on both multimode and single-mode fiber-optic cable.

LC Connector   As shown in Figure 6-8, the LC connector is much smaller in diameter than the ST and is used for basic wiring applications and AV systems from many manufacturers. It has great low-loss qualities and is known as a "push-pull" connector.

SC Connector   The SC connector is larger in diameter than the LC, as shown in Figure 6-9. It is a "stab and click" connector, which means that when the connector is pushed in or pulled out, there is an audible click because of the attachment lock. This is a great connector to get in and out of tight spaces.

Figure 6-7
ST connector

Figure 6-8
Duplex LC
connector

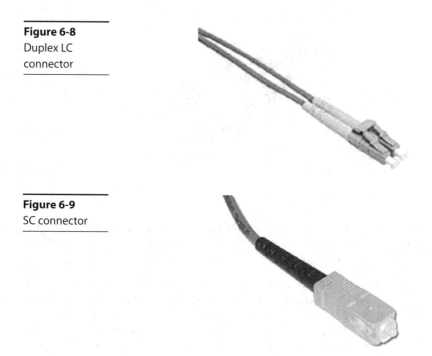

Figure 6-9
SC connector

MT-RJ Connector   The MT-RJ (Mechanical Transfer Registered Jack) connector shown in Figure 6-10 is a compact, small form-factor, dual-fiber (duplex) connector, similar in style to the RJ45 (8P8C). It was designed as a lower-cost, more compact replacement for the SC connector in data network applications.

Hermaphroditic Connector   Although connectors are usually either male or female, a hermaphroditic connector could be considered genderless. A hermaphroditic connector contains identically paired male and female parts to allow the connector to be mated with another connector sharing the same identically paired male and female parts. The hermaphroditic connectors shown in Figure 6-11 are limited to fiber-optic applications.

Figure 6-10
MT-RJ duplex
connector

Figure 6-11
Hermaphroditic
connector

Figure 6-12
Expanded beam
connector

Expanded Beam Connector   Shown in Figure 6-12, the expanded beam connector is easy to use in an outside environment because it is much more tolerant to dust and dirt and easier to clean than the other connectors.

opticalCON Connector   Shown in Figure 6-13, the opticalCON is a rugged, easy-to-use connecter designed for repeated connection and disconnection in hostile environments such as live production, roadshows, outside broadcasts, and exhibitions. The panel connector is actually a pass-through connector that accepts one (duo variant) or two (quad variant) duplex LC connectors.

 VIDEO   You can view step-by-step demonstrations of various termination techniques in the AVIXA Termination Video Library using the link in Appendix D.

Figure 6-13
opticalCON
connectors

AVIXA courses on system fabrication provide hands-on practice with the different termination techniques for various connectors.

Plotting a Cable Route

The importance of plotting a cable route should not be minimized. Where does the cable start and where does it end? That's the basic premise behind routing cables. Before you start planning where your AV systems are going to be installed, you have to plan where your cables are going to go. As obvious as this seems, many installers who have "done this a million times" skip this step and run into obstacles that could have been prevented if they had just taken a look at the drawings and formed a plan. Plotting a cable route is not a task you can undertake in isolation—you should verify the locations of the work being undertaken by other trades to ensure that your route choices will be effective and practical.

The process of cable route plotting falls under the Electrical category of verifiable tasks in the Pre-integration section the AVIXA System Verification Process specified in ANSI/AVIXA standard 2M:2010 and should be verified and signed off by the relevant stakeholders when the plotting is complete.

Take a look at Figure 6-14, which shows three drawings from the AV system designer. Let's say you want to run cable from the rear projection room (shown at the bottom center in all the drawings) to the pantry (shown in the top left). The most direct path to route a cable from the rear projection room to the pantry is to go through the break area on the left, inside the hallway, along the wall, and then into the pantry.

This is the best route because of its direct nature; also, if you need to do maintenance, the break room is always going to be open with minimal traffic, so you will always be able to get to it whenever you need to, although it may be full of people taking a break.

Here is a brief overview of what is involved in cable routing. First, you have to know where the cable has to come from and where the cable has to go to. Next, you decide how to get it from point A to point B. To do this, use your drawings showing the available cable support pathways and the locations of intermediate and terminating boxes. When determining the cable's path, avoid difficult routes to get the most direct pathway possible. Avoid EMI sources, such as mains power feeds, motors, switch mode power devices, dimmer racks, and fluorescent lighting, which can be the sources of undesirable interference. Scrutinize the cable route; verify the location, type, and cable capacity of the cable supports and boxes; and look for potential problems. The lead technician should mark the cable route on the project plans.

In some projects another contractor will be required to supply and install all cabling as specified in the AV documentation. This not only requires accurate cable routing information but also necessitates that the routing plan includes details of the lengths of cable tail to be left at each end of the run to allow for good cable management in boxes and racks and for correct cable service loops and terminations.

PART III

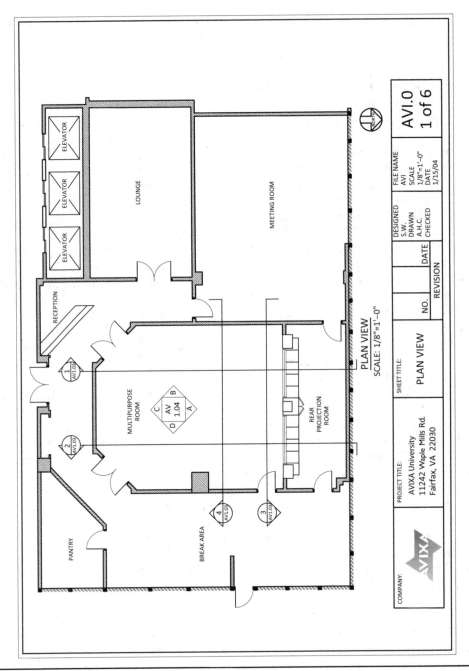

Figure 6-14 Floor plans used in cable routing *(continued)*

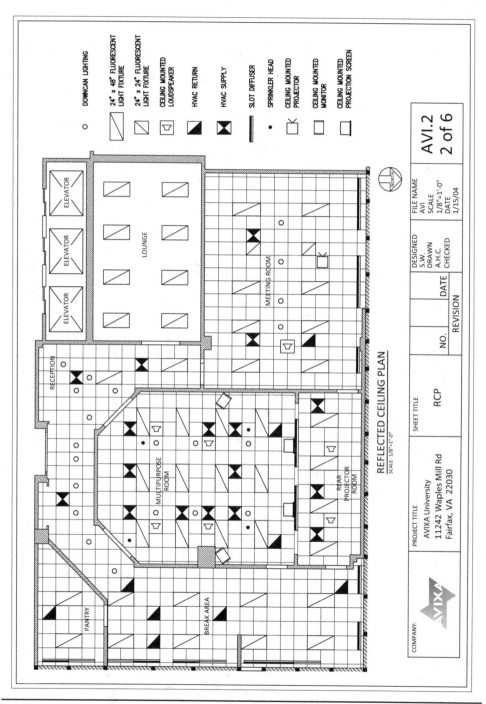

Figure 6-14 Floor plans used in cable routing *(continued)*

PART III

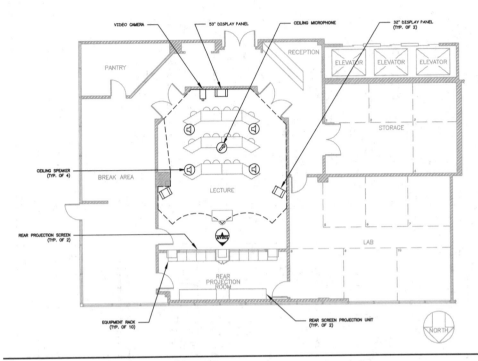

Figure 6-14 Floor plans used in cable routing

Cable Support Systems

A *raceway* or cable support is a device or mechanism that supports cables by equalizing the tension throughout the length of the cable. Raceways vary in construction from full cable enclosures to partial enclosures and cable suspension systems. Some common cable raceway systems are listed in the following sections.

Cable Tray

A cable tray, as illustrated in Figure 6-15, is an assembly or grouping of sections of noncombustible materials (usually metal) to provide rigid support for cables. Cable trays are usually open-topped structures and frequently include slots or holes that allow

Figure 6-15
Slotted cable tray
(left) and cable
ladder (right)
(Image: vav63 /
Getty Images)

PART III

Figure 6-16
Various types
of conduit

cables to be anchored to the structure after being run out. Cable trays for vertical cable runs often take the form of a cable ladder. They provide no EMI or RFI shielding for the supported cables.

Conduit

Conduit is a tube, usually circular in cross-section, that completely encloses the cables that are run through it, supporting them and protecting them from damage. Conduit, as shown in Figure 6-16, may be constructed of metal or a range of polymer materials that may be weatherproof, suitable for burial in the ground or in concrete, or even ultraviolet-resistant for use in direct sunlight. Steel conduits may act as EMI and RFI shields for cables carrying low-level signals. Conduits are used for permanent installations or are sometimes left unpopulated for temporary use in production or exhibition spaces.

Cable Duct

Cable ducts are structures that enclose and support cables on at least three sides. They may be rigid or partially flexible and may be constructed of metal or a range of plastic materials. Many ducts have a capping mechanism that allows them to be closed after completing cable installation and reopened for service access. Some duct systems have internal dividers to separate the different services being carried. In enclosed metallic ducting systems, the cables are isolated from external EMI and RFI sources. In internally divided metal ducting systems, the separate sections offer EMI and RFI isolation between sections.

Hook Suspension Systems

A wide range of hook-type suspension systems have been developed for use in situations where other cable support systems may be unnecessary or difficult to install. They consist of a series of hooks (as shown in Figure 6-17) or rings suspended from the building structure. The bundles of cables are then looped between hooks. While generally lower in cost and easier to run cables through, hook suspensions offer little mechanical support and no physical or signal protection to the cables they carry. This is not usually an issue for cables run through spaces that are infrequently accessed.

Figure 6-17
A hook cable
suspension

Cable Handling

As an installer, you may occasionally need to route cable through pull boxes and conduit. Typically, electricians install cable support systems, including electrical conduit and pull boxes, and AV installers use these pathways to route cables where they need to go in a facility. A maze of conduit connects to pull boxes at certain locations. When the cable reaches its destination, you will terminate it to the appropriate wall box or gear. Therefore, you will need to be familiar with the different types of electrical boxes, cable labeling, and cable pulling techniques.

Electrical Boxes

Electrical boxes vary in size and function. They include custom boxes, such as specialized enclosures for floor access, or in-wall boxes that may house power outlets, data outlets, and small AV devices. The ones you may be working with include

- **Ganged boxes**   As the name suggests, these boxes can be connected with one another to create a double or even triple box capacity.

- **Pull boxes**   These electrical boxes enable the installer to pull cable through the conduit. They are intermediate points within a long or complex conduit run and should be installed every 30m (100ft) or 180 degrees of bends.

- **Electrical enclosures**   The AHJ will define standards for various grades of electrical enclosures. These vary from enclosures that are suitable for general-purpose applications indoors, where atmospheric conditions are normal, to watertight enclosures suitable for locations where they may be subjected to drenching water from any angle.

- **Equipment racks**   These are centralized housing units that are used to organize and protect electronic equipment.

Sometimes your AV components, such as button panels, small touch panels, or input plates, will not fit the standard electrical boxes used in your region. In such instances, you will need to use an adapter to allow proper mounting. Some AV equipment will come with appropriately sized adapters for a range of regions. You need to supply the electrical contractor with the appropriate adapter and organize to have them fitted while laying out the room's electrical system.

Cable Labels

Although the "no cable left behind" thinking process is not an official standard, it is what every good installer should remember when working with cable. This involves labeling cable on both ends and numbering each and every cable before you enter the site for installation. This step will help you keep track of your cables, making sure each gets to where it needs to go.

A label ensures that each end of the cable is appropriately terminated with the right connector and plugged into the right input or output. The process for labeling is as follows:

1. Label one end of your cable and label the spool on the other end.

2. Pull your cable.

3. Take the label off the spool and put it on the cable.

4. Cut the cable from the spool.

There are two different kinds of cable labeling: temporary and permanent. Temporary labels are made on write-on tape, as shown in Figure 6-18, or sometimes marked

Figure 6-18
Temporary
cable label

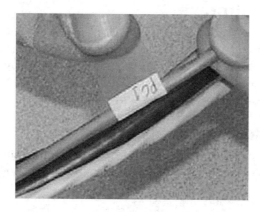

Figure 6-19
Permanent cable
label to the
F501.01 standard

directly on the cable jacket. Once the cables are pulled and terminated, permanent labels should be applied as resources for future technicians. You should save the more expensive, professional-looking labels, such as the one shown in Figure 6-19, to permanently mark each cable after termination.

Some information that is helpful on the temporary cable is not needed on the permanent one. Some technicians include the destination row and column coordinates from the architectural drawing to visualize where the cable is headed.

Permanent labels should conform to AVIXA Standard F501.01:2015 *Cable Labeling for Audiovisual Systems,* which defines requirements for audiovisual system cable labeling for a variety of venues. The standard provides requirements for easy identification of all power and signal paths in a completed audiovisual system to aid in operation, support, maintenance, and troubleshooting. Label information should match the labeling schema defined for the project. It should include a unique number for each cable and may also identify the number of conductors in the cable, the cable type, the equipment connected, the input/output name, and the signal that is carried.

Securely fasten the temporary label about 600 millimeters (2 feet) from the end of the cable so it will not be lost or damaged when the cables are bundled, pulled, or terminated. Attach permanent labels between 25 millimeters (1 inch) and 300 millimeters (1 foot) from the connector on each end of the cable.

Keep a master document of all label information. Record as you go; document everything accurately. This document should identify the path the cable will take and the physical location of each cable in the path.

Record all temporary label data on the cable pull and termination plans used during the install. If you make changes during the install, be sure to note them. Record all permanent label data on the drawings of record. In the end, the drawings of record should reflect the actual location of each cable and connection.

Cable Pulling

Before you start pulling cable, you need to make sure you have the appropriate tools. On a difficult pull, you might have to use creative pull arrangements or a specific tool. A clear understanding of pulling tension and experience to know when tension has reached critical limits will help you on every pull. You will also want to be aware of safety issues, particularly the risks associated with the fine glass fibers contained in optical fiber cables.

Cable Pulling Tools

There are several ways to make it easier to guide the cable from point A to point B. Make sure you have the essential tools for the job. The following are some potentially useful tools.

Pull String

One method is to insert pull string into the conduit. A pull string is nylon cord that goes around corners well. Sometimes the nylon cord will already be in the conduit because an electrician may have provided it when the conduit was installed. Note that this is against code in some areas. Where uncabled conduits are provided for temporary cabling, it is considered good practice to leave a pull string behind when you remove your temporary cables during the bump-out/load-out/strike at the conclusion of a project.

Fish Tape

Fish tape is a long, flexible metal or fiberglass tape, spooled like measuring tape, used to pull cable through conduit. Fish tape comes in a variety of lengths with a hook on the end. The tape is "fished" into the conduit at the far end. You should roll up the fish tape immediately when you are finished.

Push/Pull Rod

A push/pull rod is a thin, extendible, semi-rigid rod system made from sections of insu-lated material (usually fiberglass). It is used to push or pull cables or pull-strings through straight sections of duct, cable tray, or wide conduit. The rod is usually made from mul-tiple short sections that can be fitted together to extend the rod for the task. A range of hooks and pulling loops can be fitted to the ends of the rod, as befit the job.

Mouse or Rat Kit

A mouse or rat kit is made of foam or plastic and is shaped like a cylinder or ball. It is sized to fit in a pipe. A fine plastic string is attached to one end of the ball, and it has a steel eye fitted at the other end. A vacuum or blower sucks or blows the mouse (and attached pull string) to the other end of the conduit. A mouse is used to pull wire through long feeder conduit, such as underground ducts between manholes. This method requires pressure fittings or a completely sealed conduit, and it works best on metallic conduit. A drawback is that it may not work through pull boxes. It is used on pulls of more than 60 meters (200 feet).

Fish Pole or Gopher Pole

A fish pole or gopher pole is a telescopic pole of light-weight but rigid material, similar in construction to the fish pole use by sound recordists on location. The hook or loop on the far end of the pole is used to lift cables into place or hook them over cable trays or cable hooks without having to climb a ladder or bring in an access platform. Fish poles must be made from nonconductive materials to eliminate the potential risk of electric shock from unseen electrical devices overhead.

Cable Snake or Duct Rodder

A cable snake or duct rodder is a long (at least 20m/65ft), flexible, semi-rigid rod, usu-ally made from fiberglass. It is used to push or pull cables or pull-strings through long cable ducts or large-diameter conduits. The snake is usually stored rolled up on a drum or feeder system to keep it tidy and avoid tangles.

Conduit Measuring Tape

You can use a conduit measuring tape if you need to know the length of a conduit run. You can also use it as a pull string that doubles as a measuring tape. This can be helpful in calculating the amount of cable used on a job.

Cable Lubricant

Cable lubricant can be applied to cables to make them slide more easily through ducts and conduits. It is designed not to damage the sheath materials of most cables, but can act to accumulate dust and dirt on the cable, making it difficult to handle during termination and maintenance. After the cable has been pulled, lubricant should be wiped off the cable sheath on service loops and other exposed cables.

 TIP When pulling a group of cables through conduit, it is advisable to tape them together.

Pulling Tension

Pulling tension or *tensile strength* is the actual stress you can apply to cables when pulling. It is important to handle cable gently when pulling cable through a run because cable can be damaged by stretching. Each type of cable has a different pulling tension, but bear in mind that too much tension will stretch the copper or fiber and damage the effectiveness of the wire.

All cable specifications state pulling tensions, which is the weight the cable wire can sustain before damage occurs. But it is not the number that is significant. The important part is learning, through practice and experience, how to gauge pulling tension for the cable you will be pulling.

Table 6-5 shows the maximum amount of tension you can exert on some typical electrical cables (single conductors) without causing stress damage. Notice how the larger-gauge cables can handle more force. Pulling tension limits for optical fiber cables are specific to cable applications and the manufacturer's published data.

American Wire Gauge Wire Size	Cross-sectional Area (approx. metric equivalent)	Diameter	Maximum Weight Sustained
24 AWG	0.21 (0.2) mm²	0.020in/0.51mm	4lbs/1.8kg
22 AWG	0.33 (0.35) mm²	0.025in/0.64mm	7lbs/3kg
20 AWG	0.5mm²	0.032in/0.81mm	12lbs/5.4kg
18 AWG	0.82 (0.75) mm²	0.040in/1.02mm	19lbs/8.6kg
17 AWG	1.04 (1.0) mm²	0.453in/1.15mm	24lbs/10.7kg
16 AWG	1.31 (1.5) mm²	0.051in/1.29mm	30lbs/13.6kg
14 AWG	2.08 (2.5) mm²	0.064in/1.63mm	48lbs/21.7kg
12 AWG	3.3 (4.0) mm²	0.0818in/2.05mm	77lbs/35kg
Category 5e UTP	8 × 0.21 mm²	8 × 24 AWG	25lbs/11.3kg

Table 6-5 Weight Loadings for Pulling Electrical Cables

When estimating tension, you will need to factor in not just the gauge of the wire but also that of the shield, the number of conductors in the cable, and any additional mechanical strengthening elements that are part of the cable's construction. These are the factors to consider:

- The diameter of the copper
- The dielectric and jacket of the cable
- The weight of the cable being pulled
- The length of the pull
- The friction of the cable against the conduit or other support system
- The bends along the pull

The number of cables you want to pull should also be factored in your calculations. As an example, a 22 AWG ($0.35mm^2$) microphone cable with a 24 AWG ($0.2mm^2$) cable shield drain wire would be able to sustain about 8 kilograms (18 pounds) of pulling tension before damage occurs. To arrive at that conclusion, assume that there are two conductors on the microphone cable; each is 22 AWG (in other words, 3kg + 3kg or 7lb + 7lb pull strength) in addition to the shield wire of 24 AWG (in other words, 1.8kg or 4lb pull strength).

It is critical to *not* exceed the allowable tension stated by the manufacturer of the pulling device. Do not yank, jerk, or force a cable; if you have to do that, you are pulling too hard or trying to put too much cable into the conduit. The force used to pull cable must be equally distributed among all conductors.

Cable Pulling Techniques

Whether you are a novice or an expert in the AV industry, you have probably picked up a few tricks for pulling cable. The best way to know whether you are doing it right is to follow established best practices.

Best Practice for Pulling Cables

The following are some best practices for pulling cable recommended by AV installers:

- Be aware of your safety and code requirements.
- Wear safety goggles, gloves, and hard hats; governments or insurance companies usually require it.
- Use a fall protection system such as a harness whenever employing a ladder or other raised platform at heights above 2 meters (6 feet). Consult your local AHJ for fall protection and other safety requirements.
- Set up barriers around work areas.

- Feed the cable as straight as possible into the conduit. You want to avoid having the cables cross over each other when they enter the conduit.
- When cable gets hung up in conduit or bunches up, pull it back.
- Avoid twisting cable during installation.
- It is often easier to deliver the cable from the bottom of the cable spool.
- Leave some slack if the cable is to be pulled through an intermediate pull box. Be sure to identify the cable at the pull boxes to make sure it is your cable.
- Watch for sharp edges on all of the pull boxes and the ends of conduits and raceways.
- Do not pull the cable too quickly.
- Pull smoothly; pulling cables creates friction, friction generates heat, and heat burns cables.
- When cables cross paths, they should do so at 90 degrees.
- Try to run low-level signals by the most direct route because they are easily susceptible to interference and loss.
- In ceiling runs, keep the cable off the tiles.
- Cabling is not to be laid directly over the top of a suspended ceiling. It must be carried by a support system above the suspended ceiling.
- When installing cable under a floor, clean out under the floor before you start, using a vacuum cleaner if necessary.
- Use a cable spool/drum holder or cable caddy. This will keep your cables neat and orderly and allow for proper dispensing of the product. It may be helpful to have someone manually assisting to feed the cables of the drums.
- Do not lay the spool/drum on its side to remove the cable, as this will cause kinking, making it hard to pull, and may damage the cable.
- When installing cable without the support of conduit, cable trays, or ducts, use a hook support system

Cable Snout

The tedious task of having to install each individual cable along a pull route has been eliminated with the ingenious method of using a cable snout. A cable snout will enable you to pull cable easily by bundling together all of the cables and attaching a pull string at the end. Follow these steps to build a cable snout:

1. Taper the ends of the cables to be pulled.

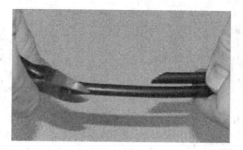

2. Stagger the cables of the bundle. Start by selecting a lead cable. It should be the longest and strongest (the largest gauge) cable in the bundle. Put the end of the second cable about 75mm to 100mm (3in to 4in) after the lead cable, and stagger the rest of the cables in even steps of about 75mm to 100mm (3in to 4in) until all of the cables are bundled together. Wrap some electrical tape around the bundle to hold it together.

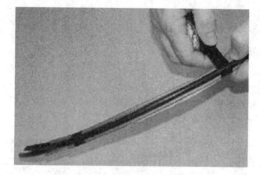

3. About 600 millimeters (2 feet) from the end, loop and tie a pull string around the entire bundle with a self-tightening knot (slip knot). Secure it with electrical tape.

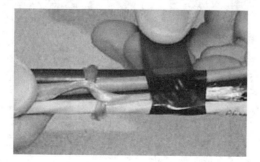

4. Make another loop and slide it down the bundle, and repeat three or four times. Try to position the knots in the groove between the cables.

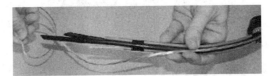

5. Stretching electrical tape, tightly wind it over the bundle and pull the string. You want the wrap to be uniform and smooth at the points where the knots and staggered cables appear. Use extra wraps at the transitions for a smoother snout and be consistent and even all the way up.

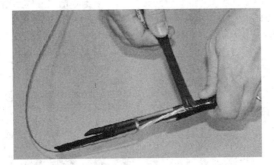

6. Continue the tape to the end of the bundle to complete the snout. The final bundle should be flexible and not stiff.

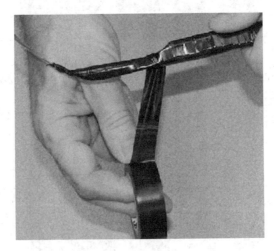

Although it is quite simple to build a cable snout, practice is the only way to achieve dexterity and speed. With the use of a cable snout, pulling cable will be easier and faster.

Conduit Capacity

To maintain cable and signal integrity, conduits should not be overfilled. As shown in Figure 6-20, when cables are inserted into conduit, the remaining space, known as *permissible area*, ensures that the cables will not be jammed together and damaged. Permissible area rules are usually identified in electrical codes.

Typical examples of permissible area percentages for cables are listed in the U.S. National Electrical Code:

- One cable may occupy 53 percent of the conduit's inside area.
- Two cables may occupy 31 percent of the conduit's inside area.
- Three or more cables may occupy 40 percent of the conduit's inside area.

For example, if you want to run three cables in a piece of conduit, you will first have to calculate the capacity using the conduit capacity formula to know whether it will meet that 40 percent permissible area. Permissible area in any given region is determined by the AHJ.

Calculating conduit capacity requires these values:

- Inner diameter of the conduit, D
- Outer diameter of each conductor, d
- Permissible area of the conduit

Use the appropriate formula for calculating conduit capacity based on the number of cables.

Figure 6-20
Conduit showing space for cables and the permissible area

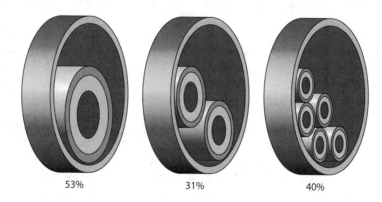

53%   31%   40%

Conduit Capacity Formula for One Cable

$$D > \sqrt{\frac{d^2}{0.53}}$$

Conduit Capacity Formula for Two Cables

$$D > \sqrt{\frac{d^2 + d^2}{0.31}}$$

Conduit Capacity Formula for Three or More Cables

$$D > \sqrt{\frac{d^2 + d^2 + d^2 \ldots}{0.40}}$$

Jam Ratio

In addition to conduit fill percentage calculations, you need to calculate a jam ratio. You will need to use a jam ratio to ensure that when three cables are installed in a conduit, they do not slip or jam. When three cables are installed in a conduit, one of the cables may slip in between the other two, especially at a bend where the conduit may be slightly oval because of the bending process. If the third cable slips between the other two, the three cables may line up and jam at the exit of a conduit bend.

To find the jam ratio, calculate the ratio of average cable outer diameter to inner conduit diameter. If the result is between 2.8 and 3.2, the cables are at risk of jamming. If the calculation falls within that range, the next larger size conduit should be specified to avoid any potential jam.

A jam ratio of 3.0, for example, would mean that the value of the combined diameters of the three lined-up cables equals the inner diameter of the conduit.

$$Jam = ID \Big/ \left(\frac{OD_1 + OD_2 + OD_3}{3} \right)$$

Cable Termination

Cable termination is an essential skill for an AV installer. Poorly executed terminations are the sources of many signal quality problems in an AV system.

To terminate a cable properly, you need to first prepare it through *stripping*, a process of preparing the cable for the connector. Different types of cables have different stripping methods.

After stripping, you will select the appropriate termination method for your cable and connector. Once you have terminated the cable, you should then test it to make sure you terminated the cable correctly.

Correctly preparing the cable is key to a good termination. If you nick any of the wires and you do not realize it until you have already installed the cable, you may find yourself troubleshooting signal loss problems after your installation is complete. Consider how much it would cost to purchase new materials and pay for the labor for the repair. You may also lose the customer's business because of a lack of confidence in your installation techniques and the possible damage to their equipment, not to mention the time it will take to redo your work.

Termination problems are hard to troubleshoot because they can be intermittent and difficult to locate. By choosing the appropriate tools and connectors, as well as taking a little bit of time to make sure the cables were terminated correctly, you can easily steer clear of these termination problems.

Stripping Cable

Before you can start the termination, you must prepare the cable. Stripping is the process of preparing the cable for the connector by removing the jacket, and oftentimes insulation, from the end of a cable to reveal the conductor.

To keep errors to a minimum, it is important to use the stripping tool designed for the specific cable and connector you are terminating. Cable strippers are the most common tools for stripping cable. Make sure your stripping tool has a sharp cutting edge. You should replace the blades when they become dull or replace the entire tool if necessary.

Steps to Strip Cable

Step 1   Get your wire strippers ready.

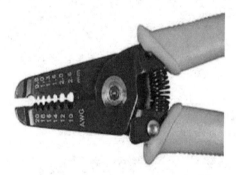

Step 2   The outer jacket of the cable is the first thing to be removed. This needs to be accomplished without disturbing the insulated wire underneath the cable jacket.

Using your stripping tool, cut around the jacket in a location appropriate to the connection type you will be completing. Slightly bend the cable to inspect the depth of your cut. Continue to cut just until the jacket is cut through. If you cut too deeply, you risk damaging the inner conductors. On shielded cables avoid cutting or nicking braid wires when stripping the jacket. Damaging braid wires reduces the mechanical strength of the connection.

If the cable contains multiple separately sheathed and possibly even shielded pairs (such as in multicore audio cables), the process of stripping the outer jacket will need to be repeated for each sheathed inner cable.

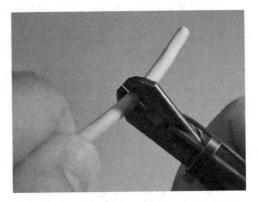

Step 3　For some connector types (IDCs such as RJ-45/8P8C), only the outer jacket is removed. However, for many other connector types, you will need to continue by removing the individual insulation from each conductor.

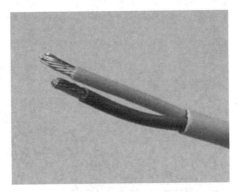

Step 4　When evaluating the quality of your stripped cable, check the edges of the stripped cable to see whether they are clean or frayed. If your stripper blade is not sharp, it could result in frayed edges and stray strands of wire.

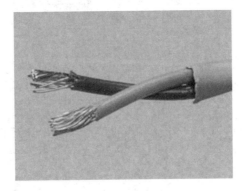

Step 5   If you nicked or damaged the solid center conductor or cut through any of the strands on a stranded conductor when stripping the insulation, trim the cable and begin again. The trim should be neat and even.

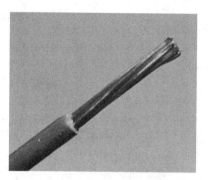

Stripping Coaxial Cable

Different types of cable require different methods of termination. Therefore, you will need to be able to strip the different types of cable to prepare for the termination. For example, coaxial cable often requires a crimp termination. Coaxial cable is constructed by surrounding a center conductor (solid or stranded wire) with dielectric material to maintain a constant distance between the shield and the center conductor. The outer jacket protects the shield from shorts and provides additional mechanical strength.

Stripping coaxial cable is a little bit different from the method you just learned. A typical coaxial cable stripping tool will have spring-tensioned jaws that wrap around the cable. Depending on the connector specification, there can be either two (for Type F) or three (for BNC) cutting blades.

Some tools are designed with the blades in cassettes, which can be easily changed to keep the tool sharp, as dull blades will cause inconsistent or uneven cuts and waste time and money. These razor-sharp blades usually require some cutting-depth and spacing adjustment before a proper cut can be completed. The tool manufacturer will provide specific adjustment instructions.

An improper termination or severe bend or pinch in the cable can disrupt the distance between the center conductor and the shield, changing the electrical properties of the cable and causing part of the signal to be reflected back, resulting in poor signal transfer.

When stripping a coaxial cable, remove portions of both insulation materials as well as some of the shield. All this must be done without damage to the conductors.

Precise stripping ensures a good mechanical and impedance match with the connector. Stripping the coaxial cable improperly can distort the shape of the insulation, thereby altering the impedance properties of a cable.

   VIDEO   Check Appendix D for a link to *Stripping Coaxial Cable*.

Steps for Stripping Coaxial Cable

To ensure accuracy, follow these steps when stripping coaxial cable.

Step 1   Have your coaxial stripping tool ready. The tool must be constructed to work with the size/type of cable and connector. Often, the connector package will provide exact dimensions for each cut. Use a scrap piece of the cable to test the adjustments you have made. Do not proceed until the tool is cutting the cable properly.

Step 2   Cut the cable to the proper length.

Step 3   Insert the cut end into the coaxial stripper by opening the cutting jaws. Make sure the cable goes all the way through the cable stripper. Some strippers may have a stop point intended for the cable to rest against so it is cut to the correct length. Otherwise, you can adjust the length of the center conductor after the stripping is complete.

Step 4 With the cable in the stripper, rotate the stripper several times so the blades can cut cleanly through the jacket, shield, and dielectric at the proper points. You generally will hear and feel the wires being trimmed.

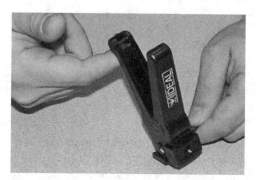

Step 5 Remove the cable from the tool by opening the jaws. Carefully remove the cut layers of jacket, shield, and dielectric. Off-cuts of braided shield wire can be messy and should be disposed of carefully. Do not allow it to get into carpeting, fabric, or electronics.

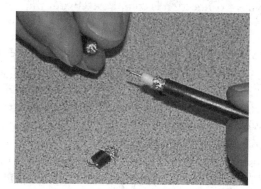

Step 6 Inspect for clean cuts to the insulation and shield wire. Make sure that the center conductor was not scored by the blades of the cutter, which would weaken it. Check that no braid wires have become caught around the center conductor.

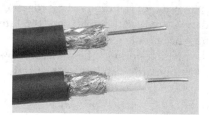

Cable Termination Methods

It is important to understand the correct connector and termination technique required for cables and signal types you will be installing. Most connections involve one of the following termination methods:

- **Screw terminal** Also known as *screw-down termination*, screw terminal termination directly connects a wire to a metal screw. This connection type is used for permanent connections and is typically less expensive than separable connectors, although it usually requires more labor. Screw terminal termination allows a large number of wires to be terminated and interconnected in a small space at a low cost.

- **Crimping** By applying pressure with a hand tool, this method enables a connector to be properly attached to a cable. It provides good reliability and good mechanical and electrical connections when done properly. Crimping is a permanent action. Use a little more cable than you need in case there is an error in termination and you have to cut the cable and start over.

- **Compression** This method bonds multiple wires for circuits such as speaker wiring. The point of connection is compressed by turning a screw that drives a moveable metal plate to tightly squeeze the copper conductors against a fixed metal plate. The fixing screw may be reversed to relieve the compression, allowing the connected wires to be moved or replaced before being recompressed.

- **Linear compression** In this method, the cable is stripped, prepared, and inserted into the special linear compression connector and tool. The tool completes the termination with a squeeze of the hand. The entire process takes about one minute.

- **Insulation displacement** This method uses a slot into which an insulated wire is placed and then pressed down using a special tool. Using this method, you do not strip the wire insulation for termination. As the wire is pushed through the slot, the insulation is cut, and a metal strip makes contact with the conductor. The pressure displaces the insulation and tightly grabs the conductor. Insulation displacement can be used on solid and stranded conductors. Typical examples are the punch-down process on 8P8C patch panels and line connectors.

- **Soldering** This is a method of bonding metals together by using heat and a metallic alloy (solder). To reduce the toxicity of the fumes released during soldering, lead has now been removed from the mix of metals in the solder. This has increased the melting point of the solder by approximately 30°C (60°F). Higher soldering iron temperatures and more heat-resistant components are currently in use.

VIDEO Check Appendix D for the link to several videos illustrating cable termination using crimping, compression, and soldering methods.

Continuity Testing

Now that you have stripped your cable and terminated the cable using the appropriate method and connector, you should test each cable to make sure it will pass signals as intended. The first step is continuity testing.

There are different kinds of continuity testing devices such as cable testers, toners, and UTP/FTP/STP data cable testers. You can perform a continuity test using a multimeter, sometimes called a *volt-ohm-milliamp meter* (VOM). Although multimeters have a wide variety of uses, the two main functions used for verifying continuity are

- **Identifying and verifying pin-outs (that is, order of connections)**   When terminating multiconductor cables, this test tells you which should be connected to which connector pin.

- **Measuring resistance**   This test verifies that a cable and its connections fall within the acceptable resistance range for the application. If your terminations are not properly completed or the cable is damaged, the meter will show a higher-than-acceptable resistance.

As shown in Figure 6-21, the multimeter has two color-coded wires terminating in insulated-handle *test probes*, which you connect to the electrical points to be measured. Hold the probes by their handles only, and do not touch the circuit while measuring.

Most general-purpose multimeters include a continuity or "beep test" mode, which in addition to displaying resistance values in the low-ohms range, sounds an audible tone (beep) if there is a low-resistance path between the probes. This avoids the need to read the meter display while juggling the probes to confirm that you have identified the correct connections.

When using a multimeter to measure resistance, first test the device's ability to give an accurate measurement. Choose the appropriate scale setting on the multimeter. (Typically, you will use the "Auto Scale" or the "R × 1" scale to check the resistance of a wire.)

Figure 6-21
Digital multimeter
for testing
continuity

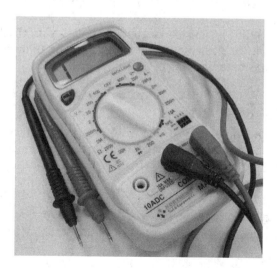

Next, touch the probes together. When touching the probes together, the meter should measure zero ohms. If the result is non-zero, your meter may have an adjustment facility to enable measurements that take into account the resistances of the probes, the connecting wires, and the state of charge of your meter's batteries.

When you measure resistance with the meter, you are measuring the ability for electrons to flow. Higher resistance causes less electron flow (less current). When you terminate a cable, there should be very low resistance from one end to the other. If the termination has been correctly performed, the only resistance measured should be the resistance of the cable itself.

The key to measuring the resistance is anticipating what the reading should be before taking the measurement. The resistance of the cable is directly related to its length and the gauge or cross-sectional area of the cable. A common 22 AWG microphone wire would have about 17 ohms of resistance per 300 meters (1,000 feet), or approximately 2 ohms per 30 meters (100 feet).

To test resistance, touch the ends of the probes firmly to each end of a cable. For example, if you are measuring a 30-meter (100-foot), 22-gauge microphone wire, what should you anticipate to be the amount of resistance from complementary points on each end of the cable? It should be approximately 2 ohms. If you measure 10 or 20 ohms, this indicates a problem. The problem could be a cold solder joint or broken strands in the cable, or even a bad connection with the meter probes. It is also important to check the other conductors in the cable to see whether any of them are shorted together.

Many special-purpose cable testers are available, including devices designed to test multiple conductors simultaneously, which allows for the checking of pin-to-pin alignment (polarity) in addition to continuity. Many AV technicians carry a tester that checks continuity and polarity for a variety of the common cables used in AV systems, such as XLR 3, XLR 5, USB, RCA, 6.5mm (1/4in) jack, 8P8C, etc.

Specialized data cable and fiber testers are now widely available with functions that include characterization and certification functions for testing the data rate capabilities of a cable in addition to basic continuity and pin alignment. Some models include a time domain reflectometer (echo tester) to assist in locating cable breaks.

 CAUTION   When conducting continuity tests, check resistance and continuity with no power to the cable you are testing. Power to the cable can damage the meter (and the operator).

Chapter Review

You learned the importance of correctly identifying cables and connectors, as well as gained skills that are necessary in system fabrication. Upon completion of this chapter, you should be able to do the following:

- Identify the common types of AV cables
- Identify the common types of AV connectors
- Plot a cable route that avoids difficult pathways on a given architectural drawing

- Identify common cable support systems and their applications
- Explain the process for pulling cable while following best practices regarding labels, snouts, tools, and techniques
- Calculate conduit capacity and jam ratios for one, two, or three or more cables
- Select the appropriate connector and termination method for a given AV application

With this knowledge you will be ready to learn about mounting AV hardware and building AV racks.

Review Questions

The following questions are based on the content covered in this chapter and are intended to help reinforce the knowledge you have assimilated. These questions are similar to the questions presented on the CTS-I exam. See Appendix E for more information on how to access the free online sample questions.

1. What type of cable is shown in the following image?

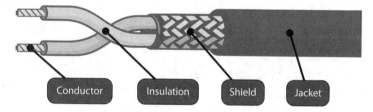

A. Unshielded twisted-pair cable

B. Fiber-optic cable

C. Shielded twisted-pair cable

D. Coaxial cable

2. _____ can transmit AV signals and other data up to a distance of 1km (3,280ft) or more before signal quality is compromised.

A. Shielded twisted-pair (STP) cabling

B. Multimode fiber-optic cabling

C. Unshielded twisted-pair (UTP) cabling

D. Coaxial cabling

3. Single-mode fiber-optic cables can transmit AV signals up to a distance of _____ before quality is compromised.

A. 300km (186.4mi)

B. 300m (984ft)

C. 10km (6.2mi)

D. 200km (124mi)

4. When working with fiber-optic cable, what is a major concern?

   A. It takes too much time.

   B. It bends easily.

   C. Improper handling can lead to injury.

   D. It is highly susceptible to loss of signal quality over distance.

5. What type of connector is shown in the following image?

   A. XLR

   B. RCA

   C. BNC

   D. VGA

6. What type of termination technique is used for a BNC connector?

   A. Solder

   B. Crimp

   C. Insulation displacement

   D. Stab and twist

7. What is a best practice when installing cables where there is no conduit or other raceway for ceiling-mounted projectors?

   A. Interlacing the cables

   B. Resting the cables gently on the ceiling tile

   C. Stacking the cables along one side of the ceiling grid

   D. Securing the cables with bridle rings mounted to bar joists

8. How much pulling tension would a 24 AWG microphone cable with a 24 AWG signal cable shield drain wire be able to sustain?

   A. 8.2 kilograms (18 pounds)

   B. 5.4 kilograms (12 pounds)

   C. 10.9 kilograms (24 pounds)

   D. 7.2 kilograms (16 pounds)

9. An RG59 coaxial SDI cable with an outer diameter of 6 millimeters (0.24 inches) is being installed in a conduit, along with four Cat 6A UTP data cables with an outer diameter of 8 millimeters (0.32 inches). If the allowable fill percentage for conduit with three or more cables in the region is 40 percent, what is the minimum inner diameter of the conduit?

    A. 44 millimeters (1.75 inches)

    B. 33 millimeters (1.28 inches)

    C. 27 millimeters (1.06 inches)

    D. 60 millimeters (2.38 inches)

10. Which test verifies that a cable can transmit a signal from one end to another?

    A. Polarity test

    B. Impedance test

    C. Continuity test

    D. Signal sweep test

Answers

1. **C.** The image shown is of shielded twisted-pair cable.

2. **B.** Multimode optical fiber is suitable for transmitting AV signals over distances up to 2km (6,560ft or 1.24mi).

3. **D.** Single-mode fiber-optic cables can transmit AV signals up to a distance of 200km (124mi) before quality is compromised.

4. **C.** When working with fiber-optic cable, a major concern is that improper handling can lead to personal injury.

5. **A.** An XLR connector is shown in the image.

6. **B.** The bonding method for terminating a BNC connector can be either crimp or linear compression.

7. **D.** A best practice when installing cables where there is no conduit or other raceway in the ceiling is to secure the cables with bridle rings mounted to bar joists.

8. **B.** A 24 AWG microphone cable with a 24 AWG signal cable shield drain wire should be able to sustain 5.4 kilograms (12 pounds). Assume two conductors at 24 AWG in the microphone cable plus a 24 AWG shield drain (that is, 1.8 +1.8 + 1.8 = 5.4 kilograms or 4 + 4 + 4 = 12 pounds).

9. **C.** 27 millimeters (1.06 inches) is the minimum inner diameter for a conduit carrying those cables with an allowable fill percentage of 40 percent.

10. **C.** The continuity test verifies that a cable can transmit a signal from one end to another.

Mounting AV Equipment

In this chapter, you will learn about
- Preparing for mounting
- Mounting types
- Mounting equipment
- Mounting malfunctions

Mounting audiovisual (AV) and networking devices is a critical aspect of an installation job. Typically, projectors and projection screens are mounted to the ceiling or wall, and flat-panel displays are mounted to the ceiling, wall, or floor. Some components are mounted on AV racks.

Secure and safe mounting is a major concern in mounting AV gear. Components such as projectors can weigh more than 100 kilograms (220 pounds) and are often mounted above an audience, creating a potential safety issue. Flat-panel displays that are improperly mounted to the floor have resulted in fatal accidents. Therefore, it is crucial that you know the weight limit of the building structure you will be mounting a device to and how to safely and securely mount components to various building structures.

The mounting of components takes place at both the system fabrication and job site installation stages of a project. During system fabrication, some equipment is mounted into a rack. You will learn about rack mounting in the next chapter. Key factors affecting the mounting of specific components, such as loudspeakers, projectors, and display screens, will be addressed in Chapters 9 and 10.

This chapter will guide you through the process of mounting AV gear in general; it focuses on key factors to consider while preparing to mount the gear, selecting mounting hardware, installing mounts on different types of surfaces, and mounting AV components. You will also pick up some useful tips on how to drill a pilot hole and the stress and safety factors you will need to consider when mounting heavy or lightweight equipment.

Duty Check

This chapter relates directly to the following tasks on the CTS-I Exam Content Outline:

- Duty A, Task 5: Conduct On-Site Preparations for Installation
- Duty B, Task 2: Mount Substructure
- Duty C, Task 4: Mount Audiovisual Equipment

These tasks comprise about 12 percent of the exam (about 12 questions). This chapter may also relate to other tasks.

Mounting Process

The process for mounting AV components consists of three steps. As shown in Figure 7-1, the mount is first fixed to the building structure. Next, the AV component is fixed in place on the mount; Figure 7-2 shows the second step of mounting a flat-panel display. Finally, as shown in Figure 7-3, the component is leveled, cabled, and dressed for aesthetic appearance.

Figure 7-1
Step 1: Fix the mount to the structure (Image: zayatssv / Getty Images)

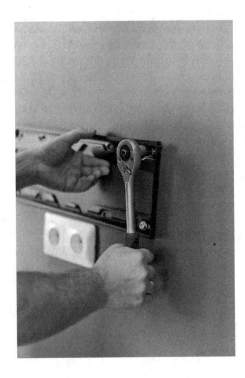

Figure 7-2
Step 2: Fix the AV
component to the
mount (Image:
Naypong / Getty
Images)

Figure 7-3
Step 3: Level the
component and
connect and
dress the cables
(Image: zayatssv /
Getty Images)

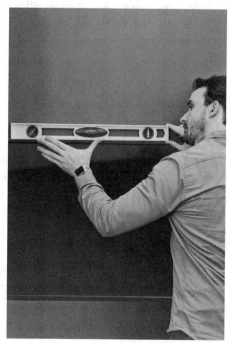

Mounting Preparations

To accomplish the three-step process, you will need the AV device that has to be mounted, the mount with appropriate fasteners, and your tool bag. You will need to know what types of surfaces and structures you will be mounting the AV product onto and the total weight of each component that will be mounted.

Those topics are discussed a little later in this chapter. Proper planning takes time but will help you successfully complete the install. It is important that you review the system drawings to determine what you need to bring to the job site.

NOTE   All mounting decisions for heavy equipment must be taken in consultation with either the owner's representative, an approved structural engineer, or both. It is important to get clear, signed-off, written approval.

Review Tools Checklist

Review your tools checklist. Make sure you have all the connectors and any special tools that you will need for the current project that you do not normally keep in your AV tool bag. Before you leave for the job site, open the product boxes and check for manuals, hardware, and special adapters. Ensure that you have all of the necessary equipment on hand and organized for efficient use.

A checklist is useful before and after the install has been completed. If you get to the site to mount and you realize that you do not have the proper screws for this particular mounting job, you will have to send someone off the job site to get the screws, which will result in loss of time and money. After the install is completed and it is time to pack up, the checklist helps to make certain you have accounted for all of your tools. Keeping your tools organized also helps you keep track of them. If a pouch or slot in a drawer is empty, you know something is missing.

Covering all of your bases at the beginning will help prevent major issues from happening, but be aware that even if you do everything perfectly, there will be a few situations that will hinder you.

Inspect the Building Structure

You probably would have noted any significant observations about areas where equipment had to be mounted in your site survey report. Review them again. The two important questions to consider before mounting equipment are

- Will the building structure hold it?
- Will the mounted support system hold it?

To answer these questions, you should check to see what building material is behind a finished wall or ceiling. Looks can be deceiving. What looks like a solid wall may in fact be old plaster with no structural integrity. Verify by inspection how the building components such as structural beams and wall frames are fastened in areas where you will be mounting AV components.

When mounting lighter-weight components, such as a 5-kilogram (11-pound) loudspeaker, be sure to anchor the equipment into a stud or preinstalled blocking.

When mounting heavy equipment, such as a 20-kilogram (45-pound) projector, always anchor the mounting system into the building's structural support. The building structure is the part of a building that is capable of supporting all the other building materials. It may be made of structural steel, concrete, or wood trusses. Sometimes, the portions of the building you can see are not part of the true structural components.

Always follow the instructions and drawings from a structural engineer. If these are not provided to you, create a mounting procedure and have it approved by a structural engineer. Define the specific installation parts that you will be using, and use only those approved parts. Changing mounting parts or procedures on the job site without a change order can create a potentially deadly situation.

Include the following information:

- Review the size and weight of what you will be mounting, along with all the parts and their weights.
- Verify the location of the mounting.
- Look at the structure to which you will be mounting and assess any potential problems.
- Look at the structural designs of the room or building to ensure safety.
- Review all manufacturer guidelines and drawings.

A structural engineer should review all mounting plans and advise on any difficult situations that you may encounter. When in doubt, always ask for help. If you modify another contractor's work, you become responsible.

Load and Weight Considerations

The load limit is an important part of the equipment's specification when you are considering mounting it to a wall or overhead surface because the information will tell you the weight at which the item will structurally fail.

Part of the load limit is the Safe Working Load (SWL) or the Working Load Limit (WLL) rating, which is the weight that must not be exceeded.

The SWL is determined by dividing the load limit by 5. For example, if the rating for structural failure is 500 kilograms, then the safe mounting load is no more than 100 kilograms (in other words, 500 ÷ 5 = 100).

Likewise, when assessing the weight of equipment to be mounted, it is best practice to multiply the published design weight by 5. For example, if a loudspeaker with mounting hardware weighs 50 kilograms (110 pounds), you should use mounting criteria that will handle 250 kilograms (550 pounds). Five is the safety factor. The term *safety factor* is sometimes referred to as *load factor, design factor,* or *safety ratio*.

CAUTION The published load factor for a mount is only the mount, not the bolts or any other parts used in the mounting. Each individual part has its own load factor. The weakest point in an assembly is the maximum strength for the entire assembly. If you use one underrated bolt, the mount may come crashing down.

Mounting Hardware

When setting up a typical hanging mount, the object to be mounted is attached to backing plates. For suspension, these backing plates are connected to threaded rod or black iron pipe, which is connected to brackets. The brackets are attached to the surface being used for the mounting.

To select the correct mounting hardware, look at the holes or slots on the mount (holes are stronger than slots) and their location. The number of bolts used in a mount determines the structural strength per attach point. Pay attention to the quantity, size, type, and length of the bolts, as well as the length of the threads.

A typical mounting job will consist of the following:

- The object or AV component to be mounted
- Mounting hardware, including backing plates, the connection between the two such as a metal pipe or threaded rod, and brackets
- The surface to which the mount and object are to be attached
- Cable dressing, as needed

Usually, a pipe or threaded rod is used to suspend a device. A threaded rod should be used only for vertical mounts. For long runs, use only a single pipe wherever possible, rather than two pipes joined with couplings, because each coupling gives the mounting another two weak spots or points of failure. Factors to consider when selecting mounting hardware include size, couplings, length and thickness of pipe or threaded rod, and gauge of steel. When the term *schedule* is used with pipe, it refers to the thickness of the material (similar to wire gauge). If you have a choice in pipe material, choose black iron for its strength.

Best Practice for Mounting Hardware

When using pipe, any threaded junction should have some means of being locked. For example, a set screw or an alignment hole with a through-bolt will prevent the assembly from becoming unscrewed over time and use.

Figure 7-4
ISO property classification markings on bolt heads

Selecting Fasteners for Mounts

Fasteners are chosen by the surface you are mounting to. What works in concrete will not necessarily work in hollow-block bricks. Hardware specifications will be categorized by type and manufacturer and how the fastener needs to be sunk into the material.

Standards organizations such as the International Organization for Standardization (ISO), SAE International (formerly the Society of Automotive Engineers), and ASTM International (formerly the American Society for Testing and Materials), have developed ways to categorize mounting hardware by composition, quality, and durability.

A common fastener used for mounting AV components is a bolt. The ISO has a numerical marking system called a *property classification,* as shown in Figure 7-4. The property class numbers relate to the strength of the material that the hardware is made from. The higher the number, the stronger the bolt.

The ASTM International and SAE International use a different marking technique for bolt heads. As shown in Figure 7-5, the raised bumps on the head of the bolt indicate the bolt's grade. If there are none, then it is Grade 2 (see the first image on the left of Figure 7-5), and the metal is usually softer than in higher grades. One raised bump is Grade 3 (not shown). The more bumps there are, the stronger the bolt. Selecting the proper hardware for mounting is critical for a sturdy and safe installation.

It is best to get your fasteners from a reputable organization. A local hardware store may not have ISO 8.8, or ASTM and SAE Grade 5, or a higher-grade fastener in stock. As noted earlier, the weakest point in the assembly is the strongest the entire assembly can be. If you use one underrated bolt, the mount may come crashing down.

Figure 7-5
ASTM and SAE grade markings on bolt heads

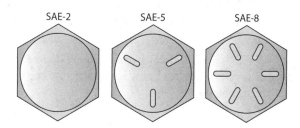

PART III

> ### Best Practice for Fasteners
> When mounting AV equipment, use bolts that are no less than ISO-rated 8.8 or ASTM and SAE Grade 5.

Drilling a Pilot Hole

Before you start mounting, you will need to drill a pilot hole to establish a reference point for mounting screws with precision. It will help prevent too many holes on the surface caused by drilling mistakes.

Here are some tips and best practices to create a pilot hole:

1. Determine the locations for the mounting screw.
   - Mark the locations with a pencil.
   - When drilling through finished material, it is an industry best practice to drill from the finished side. Often, when a drill bit penetrates a wood surface, the wood will splinter. If both sides are finished surfaces, start with a much smaller starter hole and then enlarge the hole from both sides.

2. In some hard materials (metal), it is a best practice to mark the surface with a sharp awl struck by a hammer. This provides a spot for the drill to grab. If you do not do this, the drill may skate off the mark, causing the hole to be drilled in the wrong location. Select the proper screw type and size for the task.
   - First, consider whether the job requires a sheet-metal screw, wood screw, lag bolt/lag screw, concrete anchor, or other type.
   - Second, determine the proper screw length for maximum effectiveness. A screw that is too long can cause great damage.
   - Third, determine the proper diameter by reviewing the instructions of the devices you are mounting.

3. Select an appropriate drill bit. The correct type and size will depend on the material to be drilled.
   - For concrete, you will usually use an insert of an expansion anchor. Use a hammer drill for a better hold, and make sure the hole is cleaned out prior to inserting an anchor.
   - For wood, use a high-speed steel bit for the pilot hole. It should be slightly smaller than the concrete criteria. The wood will spread, allowing the screw to follow the path of the pilot hole.
   - For plastic, follow the suggestions for wood, but drill slowly. Slowly drilling into plastic reduces the possibility of heating and melting the plastic material.

4. Tighten the drill bit into the drill chuck. For wood and plastic, a common rotary drill works best. For concrete, using a hammer drill speeds up the process. The hammer drill will quickly alternate between drilling and pounding. Only use drill bits designed for a hammer drill in such devices. Note: If using a mains-powered electric drill, the industry best practice is to unplug the cord from the electrical supply prior to changing drill bits. It is common to retain the drill chuck tightening tool close to the power-cable connector, promoting its removal from the electrical power source before any adjustment is made.

5. Test the drill action by holding it in a safe place and pressing the trigger. Verify that the motion is in the proper direction for cutting. If using a hammer drill on concrete, check the position of any hammer activation switches. Always follow the tool manufacturer's directions.

6. Verify the item you are drilling will not slip as you attempt to drill into it. This can cause an accident.

7. Line up the point of the drill bit over the mark. Press the drill bit against that point. Start drilling at a slow speed to ensure accuracy, and then slowly increase the speed. Keep the drill bit perfectly straight and begin to drill. Moving the drill from side to side will cause a larger-than-intended hole to be created, weakening the holding capacity of the hardware.

8. Drill only to the predetermined depth. Some drills will provide a depth guide that you can set. This will let you know when you have gone far enough. If your drill does not have a depth guide, use a piece of tape on the bit for a measurement guide.

9. Protect yourself with appropriate personal protective equipment (PPE).

   - Keep drilling dust and debris from eyes by wearing eye protection.
   - Protect your lungs by using an appropriate mask.
   - Drilling can produce high sound pressure levels. Hearing protection is essential.

10. Control drilling debris. Protect the surrounding area and equipment from dust damage.

11. Remove any residual markings from visible areas.

 CAUTION Always investigate the possible path of the drill, making sure there are no pipes, ducts, cables, electrical supply lines, or other objects that can be damaged or cause an electric shock.

Measuring Bolt Depth

After you have drilled the pilot holes, you must determine the bolt depth on the electronic device you are going to mount. You use the bolt depth to select the right size bolt for the job.

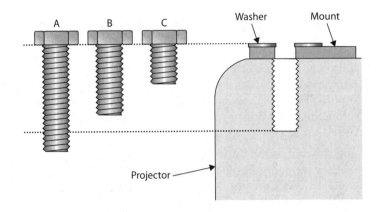

Figure 7-6
Bolt options
for mounting
a projector
showing B as the
appropriate bolt

Here is the process for measuring bolt depth for an electronic device:

1. Locate the threaded mounting holes on the item to be mounted. Refer to the manufacturer's manual as needed.

2. Determine the proper thread standard for mounting. The mounting sockets could be threaded with metric (ISO) or SAE threads.

3. Put the device safely on a flat, padded surface; place the mount and any other hardware items (washers and so on) on top.

4. Use a small screwdriver, sliding-caliper depth gauge, or other thin device to measure the depth of the mounting holes, including the mounting hardware. Compare that depth with the mounting screws.

Figure 7-6 shows three bolts marked A, B, and C for mounting a projector. Bolt B is the proper choice. Bolt A is too long and will cause damage to the projector. Bolt C would actually fit, but there are too few threads to connect the device securely, and the screw may fail.

 TIP When mounting any AV component, it is essential to select screws and bolts of the correct length to ensure that the mount and device assembly is held up.

Mount Placement

Although the placement of AV equipment to be mounted will be determined in the design phase of the project, you will have checked the location, surface materials, and building structure during a site survey. Additionally, you will have asked these two questions: Will the building structure hold the mount and AV device? Will the mount support system hold the device? The answers will help confirm the right placement for an AV device. If not, you will review the mounting location options, determine the placement, and get approval for the change before you start the mounting process.

You will typically have three mounting locations you can choose from to ensure good results: floor, wall, or overhead (such as ceiling). Take a look at each type and learn what is involved with each.

Floor Mounting

Floor mounting is the simplest form of mounting because gravity is working in your favor. Floor mounting is typically used for equipment racks, kiosks, projection pedestals, rear projection pedestals, and other AV components.

This gear is mounted to the floor, ensuring it stays where placed. This adds security from theft and mistreatment. For example, if you spend several hours aligning a multiprojector array, you do not want someone to accidentally move it and inadvertently destroy your work. It is proper to attach the assembly to the floor permanently once the alignment is complete. This, however, should not be done without the permission of the building owner.

Wall Mounting

Audio and video components are routinely mounted to walls. The typical components include projection screens, flat-panel displays, video walls, touch screens, monitors, video cameras, loudspeakers, and small equipment racks.

Let's say you want to mount a remote-controlled video camera to the wall of a training room. The camera assembly weighs 15 kilograms (33 pounds), and the mounting bracket gives a footprint of 150 millimeters by 150 millimeters (6 inches by 6 inches) for support. It may seem that attaching the mounting bracket to the wall is a simple solution, but there are other considerations. The weight of the camera assembly and the footprint of the mounting bracket will help you determine whether wall mounting is a viable solution. Address these questions: Can the wall withstand the load of this camera assembly? Will the weight of the camera on the end of the mounting bracket pull the assembly out of the wall? As you can see in Figure 7-7, the camera suspended out from the wall on a bracket acts as a lever.

Figure 7-7
Camera
suspended
out from the wall
on a bracket

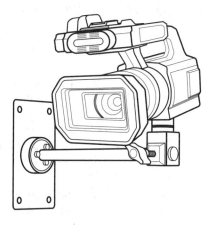

Figure 7-8
Blocking
provides a secure
mounting point

Avoid wall-mounting heavy objects wherever possible. Heavy components should preferably be mounted overhead or to the floor. Also, never hang anything on plasterboard, wallboard, drywall, or gypsum board without connecting the mount to a stud.

Blocking to Secure Wall Mounts

Proper blocking is critical to wall mounting. Blocking is the support system or construction material that is added to the wall, typically before the wall finish is applied. As shown in Figure 7-8, blocking consists of pieces of wood or other load-bearing materials that have been inserted between structural building elements to provide a secure mounting point for finish materials or products.

The building contractor, prime contractor, or general contractor may include reinforcement material into the wall assembly across three or more studs prior to the application of the plaster wall-sheeting. Once the finish is applied to the wall, the blocking is invisible to the eye. This can be done only with previous planning and coordination between the AV designer, the project manager, the architect, and the builder.

It can be very helpful to the installation crew if you take some photos of the blocking before it is covered with the final wall finish, so that they can be aware of the underlying structures.

Ceiling Mounting

Audiovisual components such as projection screens, projectors, flat-panel displays, monitors, lighting fixtures and loudspeakers are often mounted overhead, typically from the ceiling.

Anytime you suspend or hang something overhead, think about what needs to be done to support the weight. You must attach mounting support to the structure of the building and not the ceiling itself. Therefore, your concern is not the ceiling type, but the structure of the building.

You will want to know the type of materials used in the construction of the building. If the building is made of concrete, find out whether it is poured two-way joist slab (waffle-like concrete), reinforced construction (smooth concrete with steel reinforcing mesh placed before pouring), or post-tension construction, which is also smooth concrete, but with high-tension cables running through it for support. In both reinforced and post-tension construction, the enclosed reinforcements act as a mesh wrapped in concrete.

Construction factors are important because if you drill through a cable in a post-tension ceiling/flooring, it will snap. This can be dangerous. The cable can snap with such great force that it could even tear off the outside of the building. Anything in its way is in danger of damage or serious injury.

Always verify the depth and location of these cables with the building owner or building contractor before attempting to mount anything to the ceiling or floor. If this information is not available for the building, then an x-ray of the proposed drilling or trenching site must be conducted and interpreted by specialists.

A building's beams have a weight load they can handle. It is possible that the beams are just strong enough to hold the roof and that any additional weight could cause structural damage. While mounting overhead, never drill into a universal beam/rolled steel joist (RSJ)/I-beam. It is better to weld or clamp a piece of steel-channel strut (for example, Unistrut, Kindorf, or EzyStrut) to it.

Support Installation for Ceiling Mounts

When many items need to be installed in the ceiling, the position of each has to be coordinated with the other items. Using an adjustable mounting structure can allow some flexibility in the location of the mount. The following general instructions describe the procedures to suspend a 70-kilogram (155-pound) projector support from a concrete ceiling, but can be applied to any other ceiling-mounted device, such as loudspeakers, projection screens, or flat-panel displays.

1. Determine the desired location of the audiovisual device that is to be mounted.

2. Verify the space that is required in the ceiling area. This assembly requires an area of approximately a 1.2-meter (4-foot) square. In Figure 7-9, the middle of the square is intended for the center point of the manufactured mount.

Figure 7-9
The center point of the manufactured mount is in the middle of the square

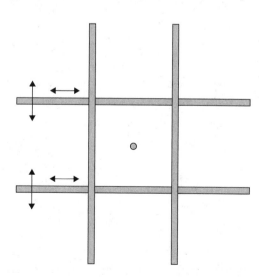

3. Make an erasable reference mark on the concrete ceiling for the center point of the manufactured mount you will be supporting.

4. As shown in Figure 7-10, draw a straight line, perpendicular to the projection screen, through this reference point. The line should extend about 450 millimeters (18 inches) on either side of the reference point.

5. Make two marks on this line, one each about 450 millimeters (18 inches) on either side of the reference mark.

6. Make four marks by measuring from these new marks a distance of approximately 200 millimeters (8 inches) on either side of the line drawn in step 5. These four outer points will form the positions of the upper mounting pieces of steel-channel strut.

7. Cut four pieces of steel-channel strut to a length of about 1.2 meters (48 inches) or of sufficient span to connect to the building structure.

8. Hold one of the steel-channel strut pieces up to the ceiling aligned with the marks you made. Locate at least three mounting points for the 1.2-meter (48-inch) piece of steel-channel strut.

9. Repeat for the second steel-channel strut rail.

10. Prepare your drill and make appropriate sized mounting holes for the hardware you are using.

11. As shown in Figure 7-11, fasten the top two steel-channel strut rails to the ceiling using only rated new hardware of at least ISO Grade 8.8 (SAE Grade 5).

12. Calculate the amount of drop required for the manufactured mount assembly.

13. Cut four pieces of threaded rod (all thread) of an appropriate length to suit the need calculated in step 12.

14. As shown in Figure 7-12, attach the lower steel-channel strut pieces to the upper pieces using the threaded rod and associated hardware (lock nuts, star or split washers, and so on).

Figure 7-10

The line should extend about 450 millimeters (18 inches) on either side of the center point

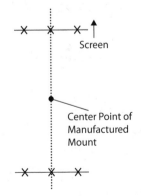

Screen

Center Point of Manufactured Mount

Figure 7-11
Fasten the
top two steel-
channel strut
rails to the ceiling

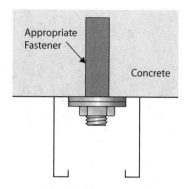

Appropriate
Fastener

Concrete

Figure 7-12
Attach the lower
steel-channel
strut pieces
to the upper
pieces using the
threaded rod
and associated
hardware

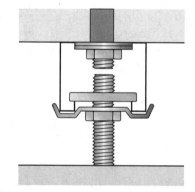

15. Adjust the mount assembly to suit the manufactured mount and fasten them together. Minor shifts of forward-backward as well as lateral positions are easily accommodated after the projector is suspended to the mount.

16. After final adjustments, tighten all hardware.

Mounting Malfunctions

As noted earlier in the section "Load and Weight Considerations," load limit is an important specification. If you exceed the load limit, the device you are mounting will not just fall to the floor but will result in damage to the device and the site, as well as seriously injure people in the area. The stress on the surface the device is mounted to or on the mounting gear can result in mounting failure.

Mounting malfunctions can be because of the following:

- **Improper calculation of loads**   This occurs when only the device load is taken into consideration. Load calculation must also include the weight of all mounting hardware.

Figure 7-13
Shearing occurs
when the
weight of the
component
pushes on a bolt

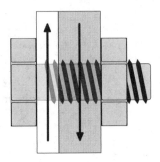

- **Shearing**   The load or weight of the mount and component can result in gravity pulling the mount down and shearing off the bolt (see Figure 7-13).
- **Tensile strength**   This refers to the maximum amount of tension the mount hardware, such as fasteners, can withstand before it fails (see Figure 7-14). Inferior-quality bolts could stretch, twist, or break because of the load.
- **Pull-out**   This occurs when force caused by the load pulls a bolt from the wall or ceiling to which the mount is attached (see Figure 7-15).
- **Placement of bolts**   Bolts placed in the top of the mount carry more stress than bolts placed lower. Follow the manufacturer's instructions on bolt placement.

Figure 7-14
Tensile load
pushes a bolt

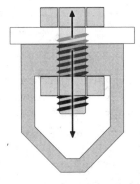

Figure 7-15
A bolt pulling
out of a wall

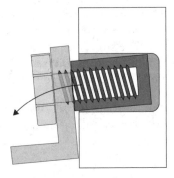

Best Practice for Minimum Safety Factor

Multiply the published design weight, plus mounting weight, by no less than 5. Five is the safety factor. The term *safety factor* is also referred to as *design factor, load factor,* or *safety ratio.*

Safety Considerations

When mounting, always wear safety glasses. Review the list of items discussed earlier in this chapter in "Inspect the Building Structure." Before mounting overhead, take a peek above the ceiling tiles to see whether there are obstructions above the tiles. Do not tip the tiles toward your face when removing them because there may be loose materials or something up there from a previous job that could fall on you.

Fire Safety

AV professionals were once notorious for carelessly drilling holes through firewalls to pull cable. Although drilling through a firewall is sometimes unavoidable, there are ways to do so safely. For instance, always use firestop materials to refill the hole you have made when penetrating a firewall. Firestop comes in many forms, such as foam, blocks, or plugs, but it all has the same basic function. It acts as an impenetrable, nonflammable barrier.

When mounting AV equipment in the ceiling, you may find firestop materials in the way of your installation. Scrape off as little firestop as necessary, and make sure you replace it immediately.

Chapter Review

In this chapter, you studied preparations that need to occur before you install a mount for an AV component, the basic process for mounting the mount hardware and AV component, and specific mounting locations (floor, wall, and ceiling). Key considerations in mounting AV devices include load limit of both the device to be mounted and the mounting hardware. When assessing the weight of equipment to be mounted, it is a best practice to multiply the published design weight by 5. Proper placement and the selection of appropriate mounting hardware, such as fasteners, are crucial to success. A detailed study included steps to drill a pilot hole and the process of installing support hardware for ceiling mounts. Adherence to safety procedures and best practices for mounting gear will help ensure that there are no installation malfunctions.

Upon completion of this section, you should be able to do the following:

- Given an installation project, create a tools checklist to mount AV components
- Drill a pilot hole to install the mounting hardware
- Given a typical mounting job, select the hardware needed to mount AV components
- Given an installation project, decide where you can safely mount an AV device
- When mounting equipment, identify the stress and safety factors you must consider

You have completed a study of general mounting considerations and procedures. You will learn about mounting specific audio, video, and control system components in the following chapters.

Review Questions

The following questions are based on the content covered in this chapter and are intended to help reinforce the knowledge you have assimilated. These questions are similar to the questions presented on the CTS-I exam. See Appendix E for more information on how to access the free online sample questions.

1. What is a best practice when selecting fasteners for mounting AV equipment?
   A. Use bolts with smooth heads.
   B. Use standards-rated bolts.
   C. Use bolts that are available near the job site.
   D. Use bolts that are extra-long.

2. What does *load limit* refer to?
   A. The maximum wattage capacity of the power supply
   B. The weight rating of a component that should not be exceeded to avoid mounting failure
   C. The permissible weight that can be carried onto a job site
   D. The maximum impedance in a speaker circuit

3. What can cause a mounting malfunction? (Choose all that apply.)
   A. Exceeding the load limit
   B. Improperly placing bolts
   C. Installing a mount directly to a ceiling tile
   D. Connecting a mount to blocking

4. Who should evaluate all mounting plans and be consulted on difficult mounting situations?

    A. Design consultant

    B. Mechanical contractor

    C. AV project manager

    D. Structural engineer

5. The published design weight of a speaker array with mounting hardware totals 96 kilograms (212 pounds). What minimum weight should the substructure and mounting assembly be able to support?

    A. 227 kilograms (500 pounds)

    B. 962 kilograms (2,120 pounds)

    C. 480 kilograms (1,060 pounds)

    D. 192 kilograms (424 pounds)

6. Which of the following factors is not related to the correct drilling of a pilot hole for mounting AV equipment?

    A. Location of wires

    B. Ambient temperature

    C. Length of screw

    D. Diameter of screw

7. Raised bumps on the head of a bolt indicate which of the following?

    A. Defect in manufacturing

    B. Length of bolt

    C. Grade of bolt

    D. Size of head

8. The ISO numerical marking scheme on materials is used to _____.

    A. indicate the strength of the material the hardware is made from

    B. calculate the length of the hardware

    C. test the amount of impurities present in the material the hardware is made from

    D. calculate the weight of the material the hardware is made from

9. A projector weighing 195 kilograms (429 pounds) should be mounted only to what type of surface?

    A. Ceiling tile of an auditorium

    B. Front wall in a classroom

    C. Drywall in conference room

    D. Building structural support or blocking

PART III

10. When using black iron pipe, how can you prevent the mounting assembly from coming apart?

   A. Tape parts of the pipe together.

   B. Lock threaded junction points.

   C. Weld parts of the pipe together.

   D. Solder parts of the pipe together.

Answers

1. **B.** Use bolts that are rated by standards organizations.

2. **B.** The load limit rating refers to the weight of a component that should not be exceeded to avoid a mounting failure.

3. **A, B, C.** Exceeding the load limit, placing bolts improperly, and installing a mount directly to a ceiling tile can all cause a mounting malfunction. Connecting a mount to building blocking is the correct way to install a mount.

4. **D.** The structural engineer should evaluate all mounting plans and be consulted on difficult mounting situations.

5. **C.** The correct mounting criteria should be 480 kilograms (1,060 pounds), which is five times the published design weight of 96 kilograms (212 pounds).

6. **B.** Ambient temperature is not related to correctly drilling a pilot hole for mounting AV equipment.

7. **C.** Raised bumps on the head of a bolt indicate the SAE/ASTM grade of the bolt based on composition, quality, and durability.

8. **A.** The ISO numerical marking system is used for indicating the strength of the material the hardware is made from.

9. **D.** Heavy equipment should be mounted only to the building structure or blocking.

10. **B.** When using black iron pipe, any threaded junction should have some means of being locked.

Rack Build

In this chapter, you will learn about
- Types of racks
- Ergonomics and weight distribution
- Building a rack
- Wiring a rack
- Finishing touches

In addition to the material that is covered in this chapter, it is wise to take advantage of the vast body of knowledge assembled by AXIVA's panels of industry experts in the standards developed as guidelines for good practice in working with equipment racks:

- AVIXA F502.02:2020 Rack Design for Audiovisual Systems
- AVIXA F502.01:2018 Rack Building for Audiovisual Systems

The key objectives for mounting audiovisual (AV) equipment into a rack are to protect the integrity of the system and to ensure security. Based on rack elevation and signal routing block diagrams, you will need to plan for the placement and interconnection of AV devices. If the AV designer has not provided these diagrams, your first task is to create them. Deciding where to place equipment in a rack requires knowledge of ergonomics, weight distribution, and heat dissipation. You will also need to know the basics of signal separation, RF and IR reception, power distribution, and thermal management within a rack. When preparing for rack building, you will need to create an equipment list and ensure that you have all the tools needed. You and your team will be involved in assembling an equipment rack and will perform tasks such as mounting devices, terminating connectors, wiring the system, and dressing cables.

Many of the rack-building tasks will be performed at your company's workshops before the rack is moved to the job site; hence, the term *workshop build* or *shop build* is often used. A workshop build is a great way to cut down the time it takes to build a rack, test the equipment, and troubleshoot system failure due to device malfunctions or design errors. It is always more efficient to do these tasks at the workshop than on the site.

Duty Check

This chapter relates directly to the following tasks on the CTS-I Exam Content Outline:

- Duty C, Task 1: Assemble Audiovisual Rack
- Duty C, Task 2: Wire the Audiovisual Rack
- Duty C, Task 3: Distribute Audiovisual Equipment

These tasks comprise about 13 percent of the exam (about 13 questions). This chapter may also relate to other tasks.

Types of AV Racks

The selection of an equipment rack is based on the space requirements of the components to be mounted and wired. Special consideration is given to how much heat a device emits, the types of signals carried to the equipment, whether the equipment has an interface component for a typical user, and the security of the equipment. You will need to make sure there is enough space on the rack for all the components. You should also verify that racks specified meet all safety and access requirements for the intended location and application. Some jurisdictions require racks that are certified safe for specified levels of seismic activity.

Rack Styles

Many configurations and options are available for racks, including doors, additional mounting rails, lacing strips and bars for cable management, fans and filters, electrical receptacles and sockets, and blank and vented panels.

Rack options vary by weight, height, and application purpose; these factors work together to protect the rack's principal functions. Basic rack formats include the following:

- **Wall-mounted** These racks attach to any type of wall with the proper hardware. As shown in Figure 8-1, the swing-frame mechanism allows rear access to the equipment for service.
- **Floor standing** These stand-alone racks have mounting rails that hold the AV equipment in place and are used typically at fixed installations. They usually have enclosed sides and may also have doors at the front and/or rear. They may be fitted with rear mounting rails to carry deep and heavy equipment and to support cable management and power distribution equipment.

Figure 8-1
The swing
frame in a wall-
mounted rack
allows rear access
to the equipment
for servicing

- **Open frame**   As shown in Figure 8-2, these fixed racks have an open mounting-frame and can be combined side by side to accommodate more devices. Open-frame racks may be fitted with side panels and front and/or rear doors. Like floor-standing enclosed racks, they may be fitted with rear mounting rails to carry deep and heavy equipment, and to support cable management and power distribution equipment.

Figure 8-2
Open-frame
racks can be
combined to
accommodate
more devices

- **Two post**   Also known as relay racks or telco racks, two-post racks are used where floor space is tight and where the equipment mounted is not very deep or heavy at the rear. They are primarily used for mounting patch panels and network devices.

- **Slide-out**   These racks are used where there is only limited access to service the installed equipment. As shown in Figure 8-3, the enclosed equipment rack can slide forward out of its housing for installation or service. This type of rack may also have the ability to rotate or spin in place for ease of access during installation, wiring, and service.

- **Mobile/portable**   These racks could be metal, wood, or plastic with casters and handles for portability. A large selection is available for components such as mixing consoles, communications systems, radio microphones, amplifiers, outside broadcast systems, media servers, and lighting control. Portable racks may take the form of trunks, cases, or custom solutions for AV equipment and maintenance gear that requires protection and storage during transport.

- **Portable shock-mounted**   These portable racks for delicate equipment are vibration isolated in dense foam or suspended by shock-absorbing mounts. They are designed to protect the enclosed equipment from rough handling and transport by road, rail, or air.

Figure 8-3
Slide-out racks
may also rotate by
45 or 90 degrees

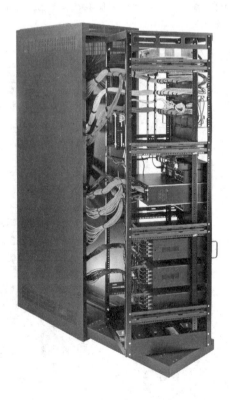

Rack Sizes

Racks vary in the width of equipment that can be mounted and are selected to match the width of specific components. The inside of the standard equipment racks used for audiovisual equipment is 19 inches wide (482.6 millimeters).

Many of the technical specifications for a rack, including size and equipment height, are determined by standards that have been established by numerous standards-setting organizations. The external width of the rack varies from 533 to 635 millimeters (21 to 25 inches). Racks are also classified by their vertical mounting height. The height of the usable mounting space in a rack is measured in *rack units*. One rack unit (RU) is equal to 1.75 inches (44.45 millimeters) in height. A 300-millimeter (1-foot)–high rack would be considered to be 6 RU in height, while a 2.1-meter (7-foot)–high rack would be considered 48 RUs high.

Like racks themselves, equipment height is measured in RUs, with most rack-mountable equipment designed to fit into a whole number of RUs. A simple audio mixer might be 1 RU high, while an amplifier might be 2 RUs high. Some equipment, like switchers, can be 10 RUs high or even taller. While 48 RUs is a widely used rack height, and all that rack space is handy, some taller racks may not fit through standard doorways or be transportable in a passenger elevator/lift.

Hole spacing on standard rack units is designed to match RU dimensions and provide stability for the mounted equipment. On each side of the mounting rail, there are typically three screw holes or cage nut slots for each RU. Figure 8-4 shows the standard spacing for mounting holes; notice that the holes are not evenly spaced. Equipment manufactured for a rack has "ears," or extensions, on both sides of the faceplate that align with the holes in the mounting rail. Sometimes the ears are ordered separately and attached to the sides of equipment before mounting. Rack screws through the ear holes secure the equipment in the rack.

Among other considerations noted earlier, the selection of a rack's size is based on the size of the space into which the rack will be installed. The rack must be able to fit into the space and able to pass through the entries and building passages to reach the installation site. Available options vary in depth from 300 to 800 millimeters (12 to 32 inches) or more.

Figure 8-4

Typical screw and cage nut holes with standard spacing on the mounting rail

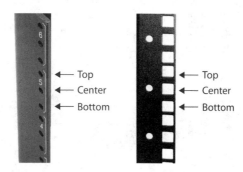

← Top
← Center
← Bottom

← Top
← Center
← Bottom

Figure 8-5
Flanged and
flat, non-vented
blank rack panels
in two sizes

Rack Accessories

A variety of accessories are available to help make rack space suitable for the items it will hold. These rack-mountable accessories include the following:

- Drawers, for storing such items as patch leads, adapters, documentation, memory media, microphones, and cables
- Keyboard drawers for pull-out computer keyboards
- Video monitor drawers for pull-out, pop-up video displays
- Rear mounting rails for supporting deep and heavy equipment and to provide mounting for cable management and power distribution systems
- Shelves, good for fractional-rack-width devices and consumer gear that is not rack mountable
- Sliding shelves, providing easier access
- Media holders, such as slots and dividers, for removable optical and memory media
- Lighting, helpful in a dimly lit presentation space or dimmed control area

Rack blank panels, shown in Figure 8-5, are metal strips that can be mounted in empty spaces between pieces of equipment. They can be flat or flanged with vent holes, as shown in Figure 8-6. Flat blanks are not as sturdy as their flanged counterpart, but both are more appealing than an empty space in a rack. They are available in different colors and finishes; choose ones that match the rack. Printing your logo and support information on them may be a good use of the blank space.

Figure 8-6 Flanged vented rack panels that allow air circulation and cooling

In addition to accessories, racks may require other installed components, such as the following:

- Casters (should be installed prior to mounting equipment on the rack)
- Rear lacing strips used for cable management
- Front cable management strips used for managing patch leads
- Power strips
- Security doors and lock boxes, good for securing the entire rack or just one storage bin

Additional mounting rails can be used for lacing bars in the rear, and universal mounting panels can be used for the additional support required for heavier equipment.

Rack Elevation Diagrams

It is easier to plan the rack-mounted equipment configuration on paper than it is to physically move the gear around to decide where the components should be located. The two types of diagrams that you will need to refer to when building an AV rack are rack elevation diagrams, which indicate where each component is located on the rack, and block diagrams, which illustrate the signal path of the various components in the rack (see details in the section "Rack Wiring" later in this chapter). The AV designer typically provides both diagrams, but in some cases, it may be necessary for you to create them.

A rack elevation diagram is a pictorial representation (see Figure 8-7) of the front of a rack. It shows the location of each piece of equipment within that rack and the number of RUs of each piece. The rack and equipment are drawn to scale, so all the components should fit as shown on the rack elevation diagram.

An additional side elevation or section diagram of the rack will simplify the construction process and clarify the location and orientation of drawers, cable management devices, and shelves.

When working with rack elevation diagrams, best practice is to first check the scale. Although equipment is sized in RUs, in reality, a device will usually be slightly smaller to ensure it fits in the rack and can be mounted and removed without binding or rubbing against other equipment.

Ergonomics

When building a rack, it is important to understand ergonomics and weight distribution. It is also important to note any special requirements the equipment to be mounted may have, such as air flow. When designing a rack layout, consider how the end users will interact with specific components; what do they need to see, use, or adjust? Also, consider how the field technician servicing the rack will access various components. Is it difficult to remove a piece of equipment for repair? Can the technician get to terminals for troubleshooting? Should you add test points to allow for quick and easy calibration and recalibration? Is there space for additional equipment to be added to the system?

Figure 8-7

Rack elevation diagram

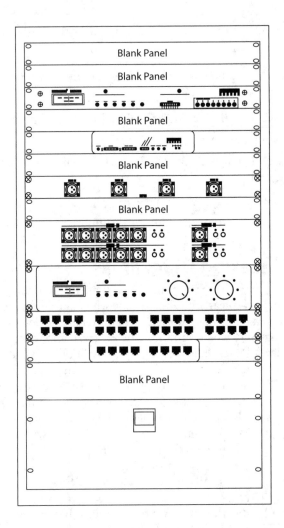

When you are building a rack, it is important to keep in mind the user interface components. Reference ergonomic best practices to ensure that these are within easy reach for everyone using the rack.

Equipment with user interface components such as recorders or media players should be placed within easy reach. For example, an optical disc player should be placed at a convenient height for the user to insert the media. Figure 8-8 illustrates user line of sight at both standing and seated positions. Devices that are not necessary for a user to interact with, such as a wireless microphone receiver, can be located at the top or bottom of the rack.

Different jurisdictions have their own accessibility regulations for mounting equipment. Consult your authority having jurisdiction (AHJ) to ensure your rack meets with all applicable accessibility requirements. In general, you should strive to provide all users

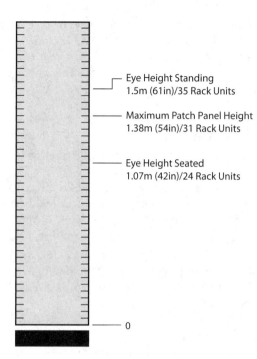

Figure 8-8
User line of sight
at standing and
seated positions

Eye Height Standing
1.5m (61in)/35 Rack Units

Maximum Patch Panel Height
1.38m (54in)/31 Rack Units

Eye Height Seated
1.07m (42in)/24 Rack Units

0

with an equal experience. For example, when positioning equipment with user interfaces in a rack, make sure it is at a height that allows a person in a wheelchair to insert media or operate controls easily.

Weight Distribution

Another consideration for equipment placement in a rack is weight distribution. If a freestanding rack has heavy equipment near the top, the whole unit could become unstable and tip over. To avoid this, put the heaviest equipment in the bottom of the rack. For example, a local uninterrupted power supply (UPS) or a powerful analog audio amplifier may be the heaviest devices that you would put in an equipment rack. UPSs may weigh more than 30 kilograms (66 pounds). When equipment racks are mounted to a wall, or even bolted to the floor, the added stability may make it acceptable to install some heavier components at the top of a rack.

For the stability of racks that are not fixed to the structure of a building, the AVIXA Rack Design standard (F502.02:2020) recommends that no less than half of the weight of the equipment in the rack should be contained in the bottom third of the rack.

Most equipment is supported entirely by the rack screws in the front mounting rail. Heavier equipment typically occupies two or more RUs, providing more support. Some equipment may need the additional support of a rear mounting rail.

Professional equipment will have integral rack ears for mounting or rack ears that may be removable. Either way, the equipment will be supported by mounting the rack ears to the rack mounting rails, using either rack screws or cage nuts and screws. Heavier equipment, such as UPSs and power amplifiers, will include additional mounting provisions at the rear of the equipment and need to be secured to additional rack mounting rails located in the rear of the rack. Always check the manufacturer's equipment manual for mounting and spacing requirements.

Cooling a Rack

Electronic devices use electricity, and some of that electrical energy is inevitably radiated as heat.

The amount of heat the equipment generates is directly related to its power consumption. For example, a 20-watt equalizer produces less heat than a 40-watt equalizer. If electronic equipment gets too hot, it may cease to operate. Therefore, manufacturers incorporate methods to circulate cooler air in and hot air out of electronic equipment.

TIP   The energy dissipated by equipment in an AV rack is measured in kilojoules (kJ). Strangely enough, the U.S. customary unit is the British thermal unit (BTU): 1kJ = 0.95BTU.

With all the heat-generating equipment close together in a rack, it is up to the AV designers and technicians to direct a flow of cool air into the rack and hot air out.

There are two main methods of cooling a rack:

- **Evacuation**   The evacuation cooling method uses fans to draw air *out* of the rack, usually through the top vents (see the top left of Figure 8-9). This is much like sucking on a straw. Cool air is drawn into the rack from the side and bottom vents and through vent plates. Since there may not be adequate filtering over every vent or slot, dirt can infiltrate the equipment rack. It is important to keep the equipment rack dust and dirt free with periodic maintenance.

- **Pressurization**   Cooler air is blown *into* the rack by fans in the bottom of the rack (see the lower right of Figure 8-9). The fans may have a filter to prevent dust and dirt from entering the rack. Vents on the vent plates and sides and top of the rack provide an escape for the hot air. The positive air pressure in the rack reduces the ingress of dust. In larger equipment rooms, pressurized, chilled air may be ducted into the bottom of the racks, often from under the floor.

Most electronic devices have vents through which the hot air can escape, typically in the back or top of the unit, and slots or vents where cool air can enter. Pay attention to the heat flow between the vents on adjacent equipment. Beware of mounting equipment with opposite heat flow immediately adjacent to each other because it will cause circulation of only the hot air.

Separate pieces of equipment that produce a lot of heat by placing blank or vented panels between them. Vent plates in the front of the rack may allow cool air to escape before reaching the equipment farther up the rack and may need to be replaced with blank plates. Equipment producing a greater amount of heat should go near the top of the rack.

Figure 8-9
Evacuation
(left) and
pressurization
(right) methods
to cool an AV rack

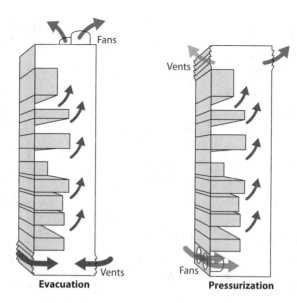

Evacuation

Pressurization

The installation of heavy analog audio amplifiers in a rack can create a dilemma. They can cause the rack to tip over if placed higher up in the rack, and they also create a lot of heat, which is best vented if it is placed near the top of the rack. Only after the rack is securely mounted in place to the wall or floor should such an amp be installed at a higher rack elevation.

TIP   To avoid inadvertently changing the thermal transfer configuration of a rack, always consult with the AV system designer or AV engineer before making major changes to the rack layout.

Building a Rack

As noted earlier, it is better to build the rack in the workshop rather than building it on the job site. At the workshop, you will have all your tools; other employees who could help; and more time to build the rack, assemble the components, and test the system before shipping it to the job site.

When preparing to build an AV rack, you should have the rack elevation drawings and a complete equipment list on hand. The equipment list should have notations on what is in inventory and what is in transit. If components are in transit, you will have to decide whether to begin rack building before they arrive.

As equipment to be mounted in the rack is removed from the shipping containers, gather all instruction manuals and materials that came with the products and store them in a container labeled with the job number to ensure that they are not discarded accidentally. Label power supplies, adapters, remote controllers, ancillary components, and cables shipped with the equipment to indicate the device they are intended for. Record equipment model numbers, serial numbers, firmware versions, and other potentially useful details such as frequencies and MAC addresses before mounting because they may not be visible after they are placed in the rack. Digital photos of the rear, top, and underside of each device can save a lot of heartache.

Rack Assembly

On some projects, the racks may require assembly. If this is the case, first complete these tasks and then build the rack following the manufacturer's instructions:

- Verify that the equipment rack is the intended size and design.
- Verify that the rack size matches the measurement indicated on the rack elevation drawing.
- Check whether the rack will have a door because doors require attention to additional factors.
- Ensure that you have all the required parts and rack accessories.
- Label the racks with the job name and number using a removable label.
- Verify that you will be able to move the rack to its final destination. Large rack assemblies can be particularly difficult to move.

Racks with doors require special attention during rack build. Some rack manufacturers require the door to be mounted or the door hinge hardware to be installed before mounting equipment into the rack. When assembling racks with doors, make sure that cables do not get pinched and that when the door is closed, there is space for protruding equipment elements such as user control knobs and antennas. This may require recessed mounting rails.

Rack mechanical accessories should also be attached before equipment is mounted. These may include power and grounding strips, casters, and fans. Use a level to ensure all equipment is mounted properly. Some accessories are difficult or impossible to include after a routine assembly.

Vertical lacing strips are a rack accessory that is typically installed after all the equipment is mounted. These will be used to tie groups of cables. They should be mounted toward the rear of the equipment.

Mounting Equipment

The installation of components in a rack requires care and precision. Keep the following guidelines in mind when mounting AV components into a rack:

- **Level**   Ensure correct sets of mounting holes.
- **Stability**   If you press a button on a device in the rack, the device and rack should not move.
- **Care**   Equipment and rack should be handled with care. Make sure the rack and the installed equipment are not scratched when you use tools.

Figure 8-10
Rack rail with a
clip nut installed

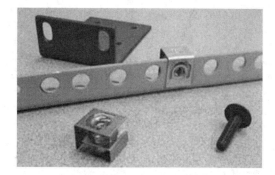

To ensure rack-mounted items will be level, verify that mounting rails are straight, stable, and aligned properly. Since the hole patterns are specific, you need to be sure that the mounting rails (including rear rails, if used) are perfectly parallel and that complementary holes align.

There are three mounting holes for each RU. Use masking tape alongside the mounting rail to mark the individual RUs. This will help you tally the equipment rack assembly to the rack elevation drawing. Some rack manufacturers include the rack units marked off on the rails for convenience.

There are three types of mounting rails:

- **Drilled and tapped** Rack screw holes are drilled and the holes are tapped or threaded so the mounting rail is ready to receive rack screws.

- **Drilled** Rack screw holes are drilled, but there is nothing to screw into. Drilled mounting rails require you to attach a nut to screw into, such as the clip nut shown in Figure 8-10.

- **Punched** Rectangular holes designed to accept cage nuts are placed in the standard positions on the rail within each RU. Cage nuts are only inserted in the positions actually required for mounting the equipment for the job, as shown in Figure 8-11. Cage nut insertion tools can be used to save wear and tear on your fingers.

There are different kinds of nuts for different applications, including clip nuts, spring nuts, and caged nuts. Be sure you have chosen the proper hardware for your rack. To mount the equipment, use a cordless drill with a clutch so you do not overtighten the rack screws and snap off their heads. The drill should be set at a low torque rating.

Figure 8-11 Punched rail with cage nuts installed

Use nylon washers between the screw and the equipment to prevent it from marking. This is an aesthetic decision; these washers do not isolate the equipment from ground loops.

Note that some of the equipment you are installing in the rack could also be designed to sit directly onto a solid surface, without requiring additional mounting procedures. These devices will commonly have surface-protecting rubber feet. In many cases, these rubber feet must be removed before rack mounting because they cause the equipment to be too large to fit in the rack space.

Some devices, especially consumer devices, do not have available direct mounting hardware. You must find a way to install this equipment securely, especially if the equipment requires user interaction. For example, you would not want a media player to shift when the user presses Play. You have two options:

- Provide a custom-made racking assembly that allows you to secure the device to the rack with brackets. This solution is the most secure, provides the best air flow, and looks the best.

- Place the component on a rack shelf. Minimally, use adhesive-backed hook and loop materials to secure the equipment to the shelf.

Do not remove the feet of equipment that will sit on a shelf. Often, there are ventilation slots at the bottom of the piece of equipment that should not be blocked.

Before you start mounting equipment on a rack, lay the rack on its back on top of a blanket or other padding so the rack is not scratched or damaged. This will make it easier to insert the equipment and slide it on the rack rails. Let gravity work for you, not against you. If you must insert equipment in an upright rack, get help. It is difficult to hold equipment steady while also securing it to the rack.

Fill all rack spaces with equipment, blanks, vents, or power strips so that it exactly matches your rack elevation drawing. After all the equipment is in place and you have verified that it matches the drawing, you can fasten equipment to the rack rail.

Machine screws are used to fasten the mounting ears to the rack rail. Make sure you use the proper size and type for your rail. For security purposes, you may install fasteners that require a special tool for insertion and removal. This reduces the risk of unauthorized tampering or theft. However, you must be sure that any future service technicians will have access to the required special tools.

Industry best practice is to insert the screw and washer into each mounting hole and turn them by hand to get them started. After they are smoothly started by hand, you may use a tool to tighten them further.

Mount power strips into the rack. Power strips that are generally 1 RU or 19 inches (482.6 millimeters) wide are available with the outlets easily accessible from the back of the rack. These are popular for smaller systems. As shown in Figure 8-12, vertical power strips mounted in the back of the rack are popular for larger systems.

Position vertical lacing strips in the rack to handle cable dressing (shown on the right in Figure 8-12); multiple lacing strips should be used to maintain signal separation.

Figure 8-12   Rack with vertical power strip (left) and vertical lacing strip installed (right)

Horizontal lacing bars can be attached to the rear mounting rail to support the weight of the cables as they reach for their termination points.

Mounting Wireless Equipment

Mounting a wireless interface device such as a wireless microphone receiver, network wireless access point, wireless talkback transceiver, hearing assistance transmitter, DMX transceiver, or in-ear monitor transmitter on a rack requires special attention. Make sure the antennas are located so that they can transmit and receive RF signals from the remote devices. Do not place RF antennas toward the back of the rack. If an antenna is located at the back of the rack, the equipment rack itself will act as a shield, greatly attenuating the RF energy that reaches the antennas. Some manufacturers design rack-mounted equipment with front antennas to solve this problem.

Special considerations must also be given if a rack has a door. If the door is metal, then it will interfere with the RF or IR signal reception if those receivers are rack mounted. A transparent rack door will not interfere with these signals. An RF antenna may require recessing the receiver in the rack so the rack door can shut. In some circumstances the antennas may need to be located outside the rack for proper transmission and reception and fed via the appropriate RF cables (usually coax).

Rack Wiring

After you have mounted all the components in a rack, you will need to refer to the block diagram to connect the inputs and outputs for signal flow to the devices. You will have to ensure proper power distribution, signal separation, and RF and IR reception for wireless devices.

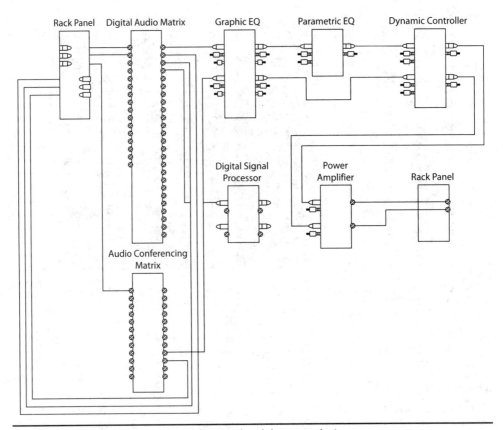

Figure 8-13 A block diagram shows the signal path between devices

Block Diagrams

A block diagram illustrates the signal path through a system. It is drawn with simple icons (triangles, circles, or boxes) for the devices the signal will pass through. A good diagram will have well-marked symbol labels because there are several common variations of the symbols used in drawing block diagrams. For example, an amplifier is usually, but not always, symbolized as a triangle. To help you visualize the signal flow, interconnecting lines (sometimes with arrows) indicate the signal path between devices, as shown in Figure 8-13. The diagram is read from left to right and from top to bottom.

Analyze two connected devices at a time to ensure they will work when connected. Then work your way into greater detail from there. This process will reveal any potential problem areas.

Connecting Devices

When connecting devices on an AV rack, you will need to select cables and connectors and then route and connect them to the appropriate devices.

Here is a process for connecting devices:

1. Select the proper cable and connectors.

2. Locate the two connection points.

3. Determine the cable path between these points.

4. Terminate one end of the cable, label it appropriately, and connect to the first piece of equipment.

5. Neatly run the cable to the second piece of equipment using the lacing bars.

6. Use removable tie strips to hold cables. After all cables are installed, you may replace the removable strips with permanent ones if allowed.

7. Estimate the length of cable. Do not forget to leave some for a service loop.

8. Cut the cable to length.

9. Terminate the end of the cable, label it appropriately, and connect it to the second piece of equipment.

Connecting devices on an AV rack is a simple task, but ensuring it is done correctly requires time and attention to detail. Do not rush the process, because you might end up with a substandard installation.

Power Distribution Within a Rack

Electricians will supply the needed power for the AV system, sometimes directly to the rack. However, installers have a responsibility to protect the integrity of the supplied power and earthing/grounding scheme. You also have a responsibility to provide clear power identification for future installers accessing the system.

You should attach an AC power wiring diagram on the inside of the rear door of a rack.

The AC rack wiring diagram should include the following:

- It should include the number of AC circuits, their location, and the number of outlets fed by each circuit; typically, rack power is supplied by one circuit.

- If more than one AC circuit is used, some jurisdictions require that the rack be labeled with a warning to this effect. Even if it is not mandated by the local AHJ, it is a good safety practice to keep all AC power feeds to a rack on the same power phase.

- Note in the rack the current capacity of each circuit, the location of the breaker panel, and the breaker or circuit numbers.

- Note in the rack whether an isolated ground system, separate from the equipment earth/ground system, is used.

- Pay attention to local, state, and national electrical codes. Always follow the most restrictive code for the region in which you are working. If applicable codes and standards appear to be in conflict, follow the requirements of the local AHJ.

PART III

Remember to keep power feeds separate from signal cables. Run each power lead into an outlet on the vertical power strip directly behind it. Consider using shorter power leads to reduce the risk of interference problems.

Power leads that are permanently attached to equipment should be bundled separately. However, detachable power leads may be bundled together. If the piece of equipment needs servicing, the technician can simply detach the power cable and leave it in place until the item is returned. Likewise, specialty power supplies or transformers should be separate from the power bundle for the same reason. The power supply unit would need to return with a service technician to the workshop or to the manufacturer if a problem is suspected.

Rack Grounding

Grounding, also known as *earthing*, provides a pathway for all unintended or undesired electric current, like that introduced by lightning, power disturbances, electromagnetic interference (EMI), or component failure, away from people and electric circuits and toward the earth or other absorptive body. Grounding/earthing protects the integrity of the power supply system and all of the electrical and electronic systems it supports. Most importantly, it protects people from electric shock, which is why it is mandated in most jurisdictions.

To provide an escape route for this charge, ground wires connect the metal element to the ground wires in the infrastructure, which are connected to the system ground. These ground wires serve as a backup only and under usual circumstances carry no current.

Powering a rack using a power distribution unit (PDU) or power strip does not necessarily provide adequate grounding of the rack itself. However, the use of an auxiliary earth bonding conductor connected to the rack's grounding stud does.

Racks cannot be effectively bonded through PDU rack ears or through power strip mounting clips. Rack power should be bonded to the grounding stud on the rack. Bonding ungrounded rack components, such as side panels, tops, and doors, provides a degree of noise immunity to sensitive analog equipment housed within. Additionally, the high microprocessor clock rates in digital equipment can give rise to wide-band EMI radiation, which is attenuated by the grounding of these loose sheet-metal parts to form a Faraday cage around the enclosed components.

Let's take a closer look at grounding the rack. There are some practical things you can do to assure that the racks installed in your system do not suffer from grounding-related problems.

Ideally, and in the view of most electrical authorities, there should be only one path to ground from the rack.

The rack should not be connected to the building's metal structure. Maintain electrical isolation from any metal conduits by placing plastic conduit couplings, cable troughs, and junction boxes outside of the equipment rack system to prevent corruption of the technical (signal) ground.

Figure 8-14
IEC Class II
protection
symbol

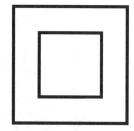

IEC Protection Class II
Double Insulated

Each copper ground busbar within the equipment rack should be bonded to the non-painted metal of the equipment rack with an appropriate antioxidant. Where required by the AHJ, utilize the ground busbar within each rack to ground the chassis of electronics that are not grounded by the third pin on a power plug.

Grounding the chassis of equipment rated as IEC Class II protection (double-insulated) is not required in most jurisdictions and is strictly forbidden as an additional risk factor in some others. IEC Class II protection is identified by the square-within-a-square symbol, as shown in Figure 8-14, usually located by the power input point on a piece of equipment.

Do not assume that bolting an equipment chassis to a rack will ground it; the paint or finish from the rack can electrically isolate it. This may result in hum in the system.

When using multiple adjacent racks, scrape paint off adjacent racks and use a copper jumper with a lug to connect racks to the same ground; simply bolting racks together will not ensure that the grounds are connected. For ungrounded shelf equipment that is not double insulated, you may need to run a grounding wire from metal on the shelved equipment to the rack, and possibly from any signal ground connector on the rear.

Signal Separation Within a Rack

In rack layout design, it is common to group equipment according to their function. For instance, you may want to install and connect all video components in one area of the rack to avoid electromagnetic interference, resulting in noise.

Signal separation protects against crosstalk between cables. As shown in Figure 8-15, cables carrying mic level signals (at about 1 to 10mV) are kept separate from cables carrying power (100 to 240V). The difference in voltage between a mic level signal and a power cable can be seen as a ratio of about 50,000 to 1 (47dB). Keeping the cables separate helps to maintain the integrity of the signals.

Signal separation allows audio wiring to remain shorter because most of that cable will be between the components themselves. Shorter cable runs mean less of an opportunity for the cables to pick up induced noise.

Figure 8-15
Signal separation
in an equipment
rack

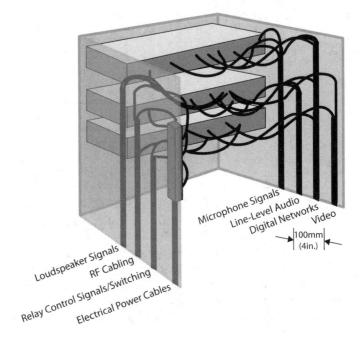

Review the best practices for signal separation in a rack, but note that for every "rule" regarding signal separation, there will be exceptions because no one scheme will work in 100 percent of the systems. Table 8-1 shows the recommended minimum cable separations listed in the AVIXA rack-related standards F502.01:2018 and F502.02:2020.

Signal Type	Mic Level	Audio Line Level	Video	Data Twisted Pair	RF Coax	Speaker	AC Power
Mic Level	–	Separate Bundles	Separate Bundles	100mm (≈4in.)	100mm (≈4in.)	100mm (≈4in.)	300mm (≈12in.)
Audio Line Level	Separate Bundles	–	Separate Bundles	Separate Bundles	50mm (≈2in.)	50mm (≈2in.)	100mm (≈4in.)
Video	Separate Bundles	Separate Bundles	–	Separate Bundles	50mm (≈2in.)	50mm (≈2in.)	50mm (≈2in.)
Data Twisted-Pair	100mm (≈4in.)	Separate Bundles	Separate Bundles	–	Separate Bundles	Separate Bundles	50mm (≈2in.)
RF Coax	100mm (≈4in.)	50mm (≈2in.)	50mm (≈2in.)	Separate Bundles	–	Separate Bundles	50mm (≈2in.)
Speaker	100mm (≈4in.)	50mm (≈2in.)	50mm (≈2in.)	Separate Bundles	Separate Bundles	–	50mm (≈2in.)
AC Power	300mm (≈12in.)	50mm (≈2in.)	50mm (≈2in.)	50mm (≈2in.)	50mm (≈2in.)	50mm (≈2in.)	–

Table 8-1 Recommended Minimum Separation of Cables from AVIXA Rack Standards

Best Practices for Signal Separation in a Rack

The following are best practices for signal separation in a rack (see Figure 8-16):

- Cables should be bundled into different signal types.
- Signals with the greatest level differences should be kept on separate sides of a rack.
- Looking at the rear of a rack, the following groups of cables should be in separate bundles on one side of the rack, with a minimum of 100mm (4in.) between bundles:
  - Loudspeaker level
  - RF
  - Power
  - Ground
- Looking at the rear of a rack, the following groups of cables should be in separate bundles on the opposite side of the rack, with a minimum of 100mm (4in.) between bundles:
  - Microphone level
  - Line-level audio (including SMPTE time code)
  - Video
  - Control
  - Digital networks
- When cables must cross, try to make them do so at 90 degrees to minimize the possible transfer of electromagnetic energy.
- Use a lacing rail mounted inside the rack to maintain better signal separation and reduce strain on the connection points.

PART III

Figure 8-16
Cable bundles separated by signal type (Image: alacatr / Getty Images)

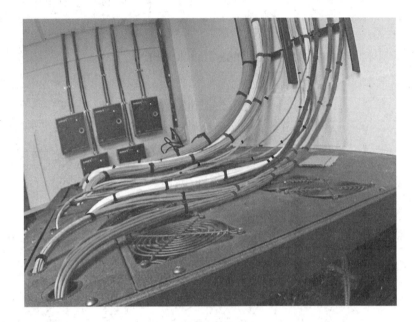

Rack Dressing

Rack dressing is one of the finishing touches of an installation. It involves making the rack look neat but also maintaining the functionality of the system. For example, when bundling cables, it is essential to consider the weight of the bundle because a heavy bundle can create tension on the connection points and result in system failure. Large cable bundles can also impede the flow of air in the cooling system. Secure the bundle with ties at regular intervals, as shown in Figure 8-17.

Figure 8-17
A heavy cable bundle weighs down on the connectors. Securing the cable with wide ties at regular intervals reduces tension. (Image: alacatr / Getty Images)

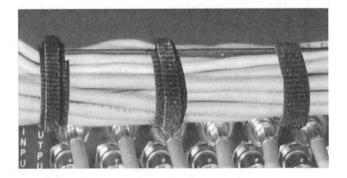

To help maintain the integrity of the connections within the rack, follow these rack dressing tips:

- Ensure cable bundles are straight and neat and then secure them with removable cable ties.

- Do not use narrow ties or overtighten them, because that may distort or damage the internal cable structure and compromise the performance of the cable. UTP, coaxial, and fiber-optic cables are particularly susceptible to damage from narrow, overtightened ties.

- Place cable ties around bundles at even intervals. The interval may range between 150 millimeters (6 inches) and 750 millimeters (18 inches), depending on cable type and bundle size.

- When you add a cable to a bundle, the existing cable ties should be removed and the entire assembly should be re-bundled.

- As you add cables to a bundle, they should flow naturally and not twist around each other.

Best Practice for Rack Dressing

Secure cable bundles with a hook and loop (Velcro) or other removable tie to the lacing bar before the point of connection, thereby reducing stress on the connector. The weight of a bundle can create heavy tension on the connection points and cause failure.

Avoid using narrow nylon cable ties (zip-ties) for bundling cable because these can crush or distort cables, especially when they are overtightened. Instead, use removable cable ties such as hook and loop (Velcro) cable strips whenever possible. This is especially true for fiber-optic cables, which have a glass core and can be damaged by tight, narrow ties.

In wall-mounted racks with swing-out doors, keep cables away from the edges so they are not pinched. Also, check the bend radius limits for the cable and recognize that cables flowing between the back wall and the rack when it is swung out will bend further when the rack is shut. Cable bends should never be less than the specified minimum radius. Be aware of possible abrasion when the cable passes through a hole in metalwork, crosses a hinge, or passes over a metal edge. You can file or soften the metal edge or use tubing to lessen the chance of abrasion.

When dressing a rack, you should also leave service loops. As shown in Figure 8-18, service loops are simply loops of extra cable length that permit a single repair of a termination without replacing the entire cable. Service loops provide flexibility in the rack wiring to move equipment within the rack and to ensure easier, faster maintenance. They also make it easier to move a device, especially if the device is on a sliding shelf.

Figure 8-18
Service loops
in equipment
cabling

Rack Build Checklist

Table 8-2 provides a checklist of rack build items. You can use it to make sure you have completed all the tasks. This is not intended to be an exhaustive verification checklist, but it will help you ensure that you have fulfilled your rack build requirements.

Category	Complete	Item
General	☐	The rack was completed in the time allotted.
	☐	Overall, the rack is neat and orderly.
	☐	There is no rubbish or trash left in the rack such as wire strippings and cable tie clippings.
	☐	Cable ties and straps are used to organize cables. Wide straps are used for coaxial, UTP, and fiber cables.
	☐	The appropriate cables were used for all interconnections.
	☐	The cables are all properly labeled.
	☐	The "as-built" drawing has all of the cables labeled, and the cables match the labels in the rack.
	☐	The labels on the "as-built" drawing match the cables in the rack.
	☐	The cables are well dressed with smooth transitions to their destinations.
	☐	For racks built by a team, cooperation and teamwork were exhibited.
	☐	The rack was assembled in a safe manner (nobody got hurt).

Table 8-2 A Checklist of Rack Build Items *(continued)*

Category	Complete	Item
Mounting	☐	Equipment is located in the correct rack spaces as instructed.
	☐	The proper equipment is installed as instructed.
	☐	Equipment is mounted level and square.
	☐	There are no loose equipment pieces in the rack.
	☐	The equipment was not marred or damaged during the exercise.
Power	☐	Permanently attached power leads can be easily removed from the rack.
	☐	Specialty power supplies such as transformers and plug packs/wall warts can also be easily removed in case of service.
	☐	There is proper signal separation.
	☐	No dissimilar signals are bundled together (for example, video with audio or network with power).
Electronic	☐	All cables have been routed to their proper places according to the diagram.
	☐	All connections have been properly accomplished (pin-to-pin wiring is correct).
	☐	A consistent ground scheme has been followed.
	☐	All soldering shows good flow and wetting with no cold solder joints.
	☐	The wires on the captive screw terminations were not tinned with solder.
	☐	The captive screw connectors are not clamping onto insulation of the wires.
	☐	Associated DIP switches, jumpers, and mode switches are in the correct positions as per manufacturer manuals.

Table 8-2 A Checklist of Rack Build Items

Load-In

Once you have completed the workshop build, you must transport racks and other equipment to the client site. As with everything in installation, loading and unloading a vehicle requires a plan. Although this plan can vary according to company or venue rules, keep these tips in mind:

- Identify the area where the vehicle will be loaded. Make sure you have approval to load the vehicle in this area. If you are unsure of where you should load equipment, talk to the lead AV technician.
- Park the vehicle so that it can be safely accessed from the loading area. Use a loading area that is in good repair and clear of obstructions. If at all possible, park the vehicle on level ground.
- Put all loading equipment such as ramps and lifts safely in place.
- Plan how to pack the vehicle prior to placing items on the vehicle.

- Load the items into the vehicle according to a plan.

- Weigh the vehicle after it is loaded to ensure that the proper distribution and weight limits are met.

- Upon arrival at the venue, follow local rules and regulations for loading and unloading procedures. It is your responsibility to find out what the rules are and to follow them.

After you arrive at the job site, the lead AV technician should brief the labor crew on the company's case-and-label system. This will help ensure that the cases get delivered to the correct places in the venue.

As the equipment is being unloaded from the vehicle, enter the case numbers or bar codes into the log book or inventory tracking system. This list can be compared to the packing list or pull list to confirm that all of the equipment has arrived at the site.

Chapter Review

Upon completion of this chapter, you should be able to select the style, size, and accessories of the rack to match your client's needs and project requirements. Given an elevation diagram, you should also be able to build a rack, applying the ergonomic, weight distribution, mounting, and cooling factors you studied.

Review Questions

The following questions are based on the content covered in this chapter and are intended to help reinforce the knowledge you have assimilated. These questions are similar to the questions presented on the CTS-I exam. See Appendix E for more information on how to access the free online sample questions.

1. A key objective for mounting AV equipment in a rack is to _____.

   A. hide it discretely in a closet

   B. protect the integrity of the system

   C. ensure only technical personnel can access it

   D. roll the gear in and out easily

2. What type of rack would you use for an AV event at a remote location?

   A. Stand-alone rack

   B. Open-frame (gangable) rack

   C. Portable rack with casters

   D. Portable shock-mounted rack with casters

3. What type of rack would you use for a room with limited space for servicing equipment?

   A. Fixed stand-alone

   B. Portable

   C. Wall-mounted

   D. Slide out

4. Your client has asked you to mount a consumer media and gaming console in the rack. The device measures 340 × 80 × 265 millimeters (W × H × D) (13.4 × 3.1 × 10.4 inches). How many RUs of rack space will this device require?

   A. 1 RU

   B. 2 RUs

   C. 6 RUs

   D. 8 RUs

5. Where should an optical disk player be placed in a rack?

   A. At the bottom of the rack

   B. At the top of the rack

   C. Next to an audio speaker

   D. In the user's direct line of sight

6. Which diagram is used for positioning mounted equipment within a rack?

   A. A rack building diagram

   B. A rack elevation diagram

   C. A rack mounting diagram

   D. A rack system diagram

7. What are the methods used to cool an AV equipment rack? (Choose all that apply.)

   A. Pressurization

   B. Decompression

   C. Refrigeration

   D. Evacuation

8. What is the problem with the cable bundle dressing in the picture?

   A. The cables are not coax.

   B. The cable ties are too close.

   C. The ties are overtightened.

   D. The cable insulation is too soft.

9. Why do wall-mounted racks with swing-out glass doors require special attention during installation? (Choose all that apply.)

   A. The swing-out door could jam in the carpet.

   B. The door may require installation before mounting the equipment.

   C. The door could pinch the cables.

   D. The door gets in the way of mounting the equipment.

10. Which are best practice methods for dressing a rack? (Choose all that apply.)

   A. Bundle cables by signal type.

   B. Use removable cable ties to secure a cable bundle.

   C. Secure heavy cable bundles to lacing bars.

   D. Keep power cables separate from signal cables.

Answers

1. **B.** Protecting the integrity of the system is a primary objective of rack-mounting AV equipment.

2. **D.** Portable racks with shock mounts protect AV components during transport to an event location.

3. **D.** Slide-out racks are used where there is only a limited service area.

4. **B.** A device that is 80 millimeters (3.1 inches) high would require 2 RUs (89 millimeters or 3.5 inches) of rack space.

5. **D.** The user's line of sight would be the ideal height at which an optical disk player should be mounted in a rack.

6. **B.** Rack elevation diagrams are used for locating equipment within a rack.

7. **A, D.** Pressurization and evacuation are methods used to cool AV equipment racks.

8. **C.** The ties are overtightened on the cable bundle shown in the picture.

9. **B, C.** When working on wall-mounted racks with swing-out glass doors, check whether the manufacturer's instructions require door installation before mounting the equipment. Also, make sure that the door, when opened, does not pinch the cables.

10. **A, B, C, D.** All items identified are best practices for dressing a rack.

PART IV

AV Installation

Audio Systems Installation

In this chapter, you will learn about
- Audio signals
- Audio system components
- Loudspeaker impedance
- Loudspeaker placement and installation
- Microphone installation

In its most basic form, an audio system has a sound source at one end and a destination for that sound at the other. In most situations, there are multiple and varied sources of sound. The sources could be several vocalists with instruments at a concert; playback devices such as an optical disc, a personal media player, or a computer; multiple panelists at a conference; or several performers in a stage production. Common applications of audio systems include sound reinforcement, intercom and paging, audio conferencing, broadcasting and streaming, program replay, background audio, sound masking, and audio for video conferencing.

Typical audio systems capture sound from its source and then process, amplify, and route it so listeners can hear the sound or it can be properly recorded. Sound systems include everything from microphones to signal converters, from processors to loudspeakers.

Before installing an audio system, you need to think about the basic principles involved in sound amplification, such as sound wave frequency and wavelengths, harmonics, and decibels. Your knowledge of sound waves, the electrical audio signal path, and audio components will be utilized during the installation, setup, and verification stages of the project. In this chapter, you will learn about audio signal types and levels and about microphone and loudspeaker installation. In Part V of this book, you will learn about the setup and verification methods to ensure audio signal integrity along its path from source to loudspeaker.

Decibels

When discussing audio, you need a way to talk about how people experience the sound the system produces. The most common unit of measurement for sound is the decibel (dB), which is one-tenth of a bel. The bel (B) is the name given to the base-ten logarithmic ratio between two numbers (e.g., a 10 to 1 ratio equals 1 bel, and 1,000 to 1 ratio equals 3 bel). However, as the bel is a relatively large unit, the decibel has been adopted for most applications.

The advantage of a logarithmic ratio for comparisons is that a comparatively small range of ratios can describe variations over a range of values, varying over many orders of magnitude. For example: a 2 to 1 ratio is 3 decibels, while a 1 million (1,000,000 or 10^6) to 1 ratio is just 60 decibels, and a 1 trillion (1,000,000,000 or 10^9) to 1 ratio is only 90 decibels, thus saving on the writing and misreading of countless strings of zeros.

 VIDEO Check Appendix D for a link to *Logarithms*.

A decibel does not really measure sound pressure levels (SPLs); it measures the difference between two SPLs where one of those levels is the reference point. Because the human ear's response to sound is logarithmic rather than linear, expressing SPLs in decibels allows for a more linear representation of changes to sound levels.

Here are some accepted generalities in relation to human hearing:

- A "just noticeable" change—either louder or softer—requires a 3dB (2 times) change in SPL. For example, an increase from 85dB SPL to 88dB SPL would just be noticeable to most listeners.
- A 10dB (10 times) change in SPL is required for you to subjectively perceive either a change twice as loud as before or a change one-half as loud as before. For example, an increase from 85dB SPL to 95dB SPL is perceived by most listeners to be twice as loud as before.

Calculating Decibel Changes

Since a decibel is a comparison of two values, you need a starting point and an end point to compare. Those values will be real physical measurements in linear units, such as pascals (pressure), volts (potential difference/EMF), watts (power), or amps (current). The pascal (Pa) is the international unit for pressure, and 1 pascal = 0.00015 pounds per square inch.

Calculating the decibel change from one voltage, or power measurement, to another will tell you how much louder or softer the output of a sound system will seem after that change.

To calculate the change in decibels between two power measurements, use the 10 log equation. For example, the equation

$$dB = 10 \times \log_{10} (P_1 \div P_2)$$

would give you the difference in decibels if you compared one power in watts (P_1) against another power in watts (P_2).

Because power, and hence SPL, from a source varies as the square of the voltage and current in a circuit ($P = V^2 / R$ and $P = I^2 \times R$), and because field strength decreases as the inverse square of the distance from a source ($1 / d^2$) due to the inverse-square law, a different equation is required to calculate ratios with these measurements. This quadratic (square) relationship requires a using 20 log formula for the difference in decibels when comparing two voltages, currents, or distances. For example, the equation

$$dB = 20 \times \log (V_1 \div V_2)$$

would give you the difference in decibels if you compared one voltage (V_1) to another voltage (V_2).

This equation:

$$dB = 20 \times \log (D_1 \div D_2)$$

would give you the difference in decibels if you compared the sound pressure level at one distance from the source (D_1) to the sound pressure level at some other distance away from the source (D_2).

Now that you know how to perceive differences in sound levels logarithmically and that you use the decibel for a logarithmic scale, let's explore further how the decibel is used.

Decibel Reference Levels

A decibel can be a comparison between two arbitrary values, or it can be a comparison of a value to a predetermined starting point, known as a *reference level*. This is also sometimes referred to as a *zero reference*. For example, you could compare two voltages using the 20 log formula to calculate the decibel ratio between them. Or you could find an absolute voltage measurement in decibels by comparing a measured voltage to its reference voltage level.

Quantity	Decibel Abbreviation	Reference Level
Sound pressure	dB SPL	20µPa at 1kHz
Voltage (often used in consumer electronics)	dBV	1V
Voltage (mostly used with professional AV equipment)	dBu (dBv)	775mV
Power	dBW	1W
Power	dBm	1mW

Table 9-1 Common Reference Levels Used in Audiovisual Applications

For sound pressure, the reference level is the threshold of human hearing at 1 kilohertz, which is 20 micropascals (µPa). Human beings perceive that sound pressure level as silence. Any unit you might quantify in decibels has its own reference level.

As shown in Table 9-1, decibels will be abbreviated differently to indicate the reference level used. For example, dB SPL indicates that the reference level is a sound pressure level of 20µPa at 1kHz, whereas dBV indicates that the reference level is a voltage of 1V.

Some units, such as volts and watts, have more than one zero reference, as shown in Table 9-1. Use the one that makes sense for the application. If you are taking power measurements at a radio station, use dBW. If you are measuring the power of wireless microphones, use dBm.

Perceived Sound Pressure Level

Not all sounds with the same sound pressure level seem equally loud. Very low- and high-frequency sounds are harder to hear than sounds in the middle of the audible frequency range. In addition, the quieter a sound at the fringes of human perception is, the harder it is to hear.

Figure 9-1 shows a graph of the equal loudness curves. The threshold of human hearing is 0dB SPL at 1kHz. This graph shows how loud different frequencies must be for the human ear to perceive them as equally loud as another. For instance, what sound pressure level would a 40Hz tone have to be to be perceived equally as loud as a 1kHz tone? The answer depends on the tone's actual sound pressure level.

The dotted curve on Figure 9-1 represents the threshold of human hearing, which is what the human ear perceives using a 1kHz reference. The x-axis of the graph shows actual frequency, and the y-axis shows actual dB SPL. At the threshold of human hearing, the 40Hz tone must be about 50dB SPL louder before the human ear can perceive it equally as loud as a 1kHz tone. A 200Hz tone, however, would have to be only 15dB SPL louder to be perceived equally as loud as a 1kHz tone.

Notice that at overall louder listening levels, the hearing response curve begins to flatten. A 90dB, 40Hz tone seems as loud as a 70dB, 1kHz tone. Human perception of the energy across the audible spectrum is more even at overall louder listening levels. Also note that your ears are more sensitive in the normal speech frequency range, 500Hz to 4kHz. And your ears are most sensitive to higher-pitched sounds, such as a crying baby.

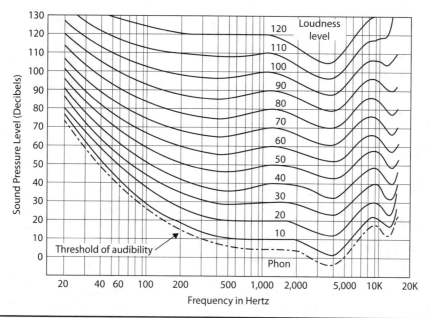

Figure 9-1   Equal loudness curve

You will use the decibel to measure many aspects of audio system performance: ambient noise, audio signal level, gain structure, loudspeaker performance, and so on. As you build and verify an audio system, it is important to keep in mind that humans and machines respond differently. An audio system may have a "flat" response, reproducing all frequencies equally. The human audio system, our ears, does not.

VIDEO   Check Appendix D for a link to *Equal Loudness*.

You will learn more about sound pressure levels and how to measure them using an SPL meter in Chapter 14.

Audio Signal Pathway

An audio signal is an electrical representation of sound. The audio signal path starts at the sound source, takes acoustic energy, and converts it into electrical energy so it can be routed, processed, further amplified, possibly recorded, and eventually converted back into acoustical energy.

Here is the sequence of an audio signal pathway, as shown in Figure 9-2:

1. The sound source creates sound waves in the air.

2. These sound waves are picked up by a microphone.

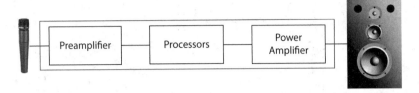

Figure 9-2 Components in the audio signal pathway

3. The microphone converts the sound wave energy into an electrical signal.

4. The electrical signal is processed.

5. Eventually, the signal ends up in an output device (such as an earphone or loudspeaker).

6. The output device converts the electrical signal back into acoustic energy (sound waves).

The process of conversion of energy from one form to another is called *transduction*. A microphone is a transducer that converts acoustic energy into electrical energy. A loudspeaker is also a transducer: one that converts the electrical energy into acoustic energy. This means you have transducers on either end of an electrical audio path.

Balanced and Unbalanced Circuits

All electrical circuits—and the cabling used to connect them—generate electromagnetic energy fields that interact with other electrical circuits and cabling, including AV circuitry and cabling. This interference and noise degrade the quality of the signal and, in an audio system, may introduce hum, buzz, crackles, and splats.

One way to reduce the noise in a circuit is to employ an actively balanced electrical circuit, as shown in Figure 9-3, which offers a defense mechanism against noise. In a balanced circuit, the signal is transmitted over two identical parallel wires with the signal on one of the wires inverted. These signals are said to be *in balance*. At the receiver, the signals are combined, with the inverted signal reversed. This produces a higher-strength

Figure 9-3
A balanced circuit.
The signal spikes
are cancelled out
by inversion.

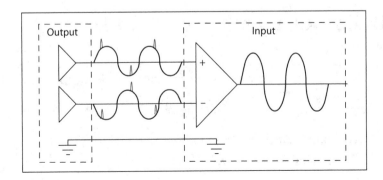

Figure 9-4
Balanced shielded
twisted-pair cable

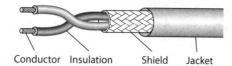

Conductor   Insulation   Shield   Jacket

signal but, more importantly, cancels out any external signals picked up by both wires. This cancellation process is known as *common mode rejection.* As a general rule, balanced circuits are used wherever possible in audio, despite the cost premium. In audio cables, the two signal conductors are usually twisted around each other and surrounded by a shield that is tied to electrical ground. This type of cable is known as *shielded twisted-pair cable,* as shown in Figure 9-4.

In the lower-cost *unbalanced* circuit, shown in Figure 9-5, the signal is transmitted down a single signal wire. Without the second, inverted signal to provide common mode rejection, all external signals picked up by the signal wire would be present in the system output. As shown in Figure 9-6, the single conductor in an unbalanced cable is usually surrounded by a cable shield that also acts as the return electrical path for the circuit. Unbalanced circuits are also known as *single-ended circuits.*

With either a balanced or unbalanced design, the longer the cable run, the more noise the cabling is subjected to. Therefore, unbalanced lines are extremely limited in the distance they can cover and their ability to transfer a usable signal.

A quick way to determine whether a piece of equipment is balanced or unbalanced is to look at its outputs or inputs. Unbalanced audio equipment and cables typically have only one pin or segment per signal, plus one return connector. The most common of these are the RCA (phono) plug and the 3.5mm (1/8in.) jack plug. While a three-section 3.5mm (TRS) jack plug is in wide use on portable equipment and computers, this connecter is used only in unbalanced circuits. It actually carries two separate unbalanced signals: the left and right channels of a stereo signal, plus a common signal return wire.

Figure 9-5
An unbalanced
circuit

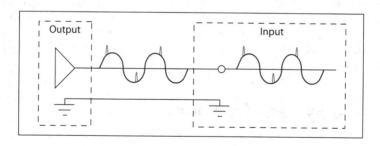

Figure 9-6
Unbalanced
shielded cable

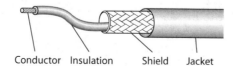

Conductor   Insulation   Shield   Jacket

PART IV

By preference all mic- and line-level audio cable runs should be over balanced circuits, but unbalanced runs should not exceed 5 meters (16 feet).

Audio Signal Levels

The voltage of the audio signal will vary depending on the signal source. The types and voltage levels of common audio signals are

- **Mic level**   Regardless of the type, microphones produce a signal with a voltage in the region of just a few millivolts (mV). This *mic-level* signal typically ranges from −60 to −50dBu.

- **Line level, professional**   *Line level* is the level where all audio signal routing and processing are performed. The analog signal outputs from most digital audio systems are usually at line level. In professional audio systems line level ranges from 775mV to 1.23V (0dBu to +4dBu).

- **Line level, consumer**   In consumer-grade devices, line level is 316 mV (−10dBV). One indicator of a consumer-level signal is the use of RCA (phono) or 3.5mm (1/8in.) jack connectors.

- **Loudspeaker level**   After the signal is routed and processed, it is sent to a power amplifier for signal amplification to drive a loudspeaker. Amplifier output can vary from a few volts up to about 100 volts. As noted earlier, the loudspeaker takes the amplified electrical signal and transduces the electrical energy into acoustical energy. In active loudspeakers, as the amplifier is inside the loudspeaker cabinet, the input to the cabinet will be at line level.

As you install your audio system, pay attention to the audio signal levels at each point in your audio signal chain. The audio-level inputs and outputs will help you know whether the components are compatible with one another. For example, while you could plug a microphone directly into the input of a power amplifier, which is designed to accept a much higher line-level signal, you would not be able to amplify the sound to the desired dB SPL. Conversely, it would be most unfortunate if you were to connect the output of a power amplifier into a device expecting either a mic- or line-level signal because you would almost certainly damage components, and possibly even produce some impressive puffs of smoke.

Audio System Components

The components of an audio system capture, store, process, and output the audio signals. Your job is to make sure that these components are compatible with one another and installed correctly. Let us take a tour through some of the components of an audio system that you may have to install.

Microphones

Microphones vary by type of transducer and directional characteristics. The selection of a microphone (mic) for any particular application is based on several factors, including directional or pickup patterns, sensitivity, impedance, and frequency response. Installers should also be aware of the mic's physical design and how the mounting surface will affect the overall audio performance. Here is how the basic types of microphones work.

Dynamic Microphones

In a dynamic microphone, you will find a coil of wire (a *conductor*) attached to a diaphragm and placed in a permanent magnetic field. Sound pressure waves cause the diaphragm to move back and forth, thus moving the coil of wire attached to it.

As the diaphragm-and-coil assembly moves, it cuts across the lines of flux of the magnetic field, inducing a voltage into the coil of wire, as shown in Figure 9-7. The voltage induced into the coil is proportional to the sound pressure and produces an electrical audio signal. The strength of this signal is very small and is called a *mic-level signal.*

Dynamic microphones are used in many situations because they are economical and durable, and they will handle high sound pressure levels. Dynamic microphones are very versatile because they do not require a power source.

Condenser Microphones

In the study of electricity, you will find that if you have two oppositely charged (*polarized*) conductors separated by an insulator, an electric field exists between the two conductors. The amount of potential charge (*voltage*) that is stored between the conductors will change depending on the distance between the conductors, the surface area of the conductors, and the dielectric strength of the insulating material between the two conductors. An electronic component that uses this principle is called a *capacitor.*

A condenser microphone contains a conductive diaphragm and a conductive backplate, as shown in Figure 9-8. Air is used as the insulator to separate the diaphragm and backplate. Electrical energy is required to polarize, or apply, the positive and negative charges to create the electric field between the diaphragm and backplate.

Sound pressure waves cause the diaphragm to move back and forth, subsequently changing the distance (spacing) between the diaphragm and backplate. As the distance changes, the amount of charge, or *capacitance,* stored between the diaphragm and backplate changes. This change in capacitance produces an electrical signal.

Figure 9-7

The workings of a dynamic microphone

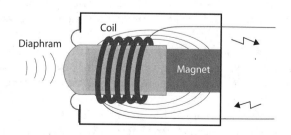

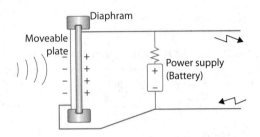

Figure 9-8
The workings of a condenser microphone

The strength of the signal from a condenser microphone is not as strong as the mic-level signal from a typical dynamic microphone. To increase the signal, a condenser microphone includes a preamplifier powered by the same power supply used to charge the plates in the microphone. This preamplifier amplifies the signal in the condenser microphone to a mic-level signal but is not to be confused with a microphone preamplifier found in a mixing console.

The power supply to charge the capacitor elements and drive the preamplifier may be a battery included in the microphone body, an external mains-powered supply device, or come from an external *phantom* power supply system.

As the diaphragm used in a condenser microphone is usually lower in mass than other microphone types, the condenser microphone tends to be more sensitive than other types and responds better to higher frequencies, with a wider overall frequency response.

Electret Microphones

An *electret microphone* is a type of condenser microphone. The electret microphone gets its name from the prepolarized material, or the *electret,* applied to the microphone's diaphragm or backplate.

The electret provides a permanent, fixed charge for one side of the capacitor configuration. This permanent charge eliminates the need for the higher voltage required for powering the typical condenser microphone. This allows the electret microphone to be powered using small batteries, or normal phantom power. Electret microphones are physically small, lending themselves to a variety of applications and quality levels such as in lavalier and head-worn microphones.

MEMS Microphones

A *microelectromechanical system* (MEMS) microphone is a member of a group of microscopic mechanical devices that are built directly onto silicon chips using the same deposition and etching processes used to construct microprocessors and memory systems. They are minute mechanical devices that can be directly integrated with the pure electronic circuitry of the chip. The best known of these in the AV world are the digital micromirror devices (DMDs) used for light switching in DLP projectors. Many other MEMS devices are in wide commercial use as accelerometers, air pressure sensors, gyroscopes, optical switches, inkjet pumps, and even microphones.

Microphones built using MEMS technologies are generally variations on either condenser microphones or piezo-electric (electrical signals produced by the mechanical movement of a crystal) microphones. Although their tiny size means that MEMS microphones are not particularly sensitive, especially at bass frequencies, the huge advantage of MEMS microphones is that they can be constructed in arrays to increase sensitivity and they can be placed on the same chip as the amplifiers and signal processors that manage both gain and frequency compensation. The on-chip processing can include analog-to-digital conversion, resulting in a microphone with a direct digital output. In addition to dedicated MEMS microphones, smartphones, smart speakers, tablet computers, virtual reality headsets, and wearable devices can incorporate chips that include a range of MEMS devices, including microphones, accelerometers, and gyroscopes and all of their associated processing circuitry.

Defining Phantom Power

Phantom power is the remote power used to power a range of audio devices, including condenser microphones. The supply voltage typically ranges from 12V to 48V DC, with 48V being the most common. Positive voltage is applied equally to the two signal conductors of a balanced audio circuit, with the power circuit being completed by current returning through the cable's shield. Because the voltage is applied equally on both signal conductors, it has no impact on the audio signal being carried and does not cause damage to dynamic microphones.

Phantom power is frequently available from audio mixers. It may be switched on or off at each individual microphone input, enabled for groups of microphone channels, or enabled from a single switch on the audio mixer that makes phantom power available on all the microphone inputs at once. If phantom power is not available from an audio mixer, a separate phantom power supply that sits in line with the microphone may be used.

Microphone Physical Design and Placement

Whether dynamic, condenser, electret, or otherwise, microphones come in an assortment of configurations to meet a variety of applications. The following are some common microphone configurations:

- **Handheld** This type is used mainly for speech or singing. Because it is constantly moved around, a handheld microphone includes internal shock mounting to reduce handling noise. Handheld mics are available in both wired and wireless configurations.

- **Surface mount or boundary** This type of microphone is designed to be mounted directly against a hard boundary or surface, such as a conference table, a stage floor, a wall, or sometimes a ceiling. The acoustically reflective properties of the mounting surface affect the microphone's performance. Mounting a microphone on the ceiling typically yields the poorest performance because the sound source is much farther away from the intended source (for example, conference participants) and much closer to other noise sources, such as ceiling-mounted projectors and HVAC diffusers.

- **Gooseneck**   Used most often on lecterns and sometimes conference tables, this type of microphone is attached to a flexible or bendable stem. The stem comes in varying lengths. Shock mounts are available to isolate the microphone from table or lectern vibrations.

- **Shotgun**   Named for its physical shape and long and narrow polar pattern, this type of microphone is most often used in film, television, and field-production work. You can attach a shotgun microphone to a long boom pole (fishpole), to a studio boom used by a boom operator, or to the top of a camera.

- **Instrument**   This family of microphones is designed to pick up the sounds of musical instruments, either directly from acoustic instruments or from the loudspeaker cabinet of an amplified instrument. This type of microphone is usually either a condenser or a dynamic device, depending on the loudness, dynamic range, and frequencies to be picked up from the instrument. Some specialized instrument transducers use direct mechanical or magnetic pickups.

- **Lavalier and headmic**   These microphones are worn by a user, often in television and theater productions. A lavalier (also called a *lav* or *lapel mic*) is most often attached directly to clothing, such as a necktie or lapel. In the case of a headmic, the microphone is attached to a small, thin boom and fitted around the ear. As size, appearance, and color are critical, lavaliers and headmics are most often electret microphones.

- **Beamforming array**   Beamforming arrays have multiple microphone elements, usually condenser microphone capsules or MEMS microphones. These elements are configured in arrays of varying shapes and linked together through a digital signal processing system to form narrow beam patterns that can be electronically steered to pick up the desired sounds while rejecting ambient noise. Array microphones can be placed in convenient locations such as on a wall, on a tabletop, or on the ceiling, as shown in Figure 9-9. Microphone array devices are primarily used in meeting rooms, in conferencing spaces, on desks, and on lecterns.

Figure 9-9   A ceiling-mounted multimicrophone array used for meeting rooms (courtesy Sennheiser)

Microphone Polar Patterns

One of the characteristics to look for when selecting a microphone is its *polar pattern*. The polar pattern describes the microphone's directional capabilities—in other words, the microphone's ability to pick up the desired sound in a certain direction while rejecting unwanted sounds from other directions.

Polar patterns are defined by the directions from which the microphone is optimally sensitive. These polar patterns help you determine which microphone type you should use for a given purpose. There will be occasions when you want a microphone to pick up sound from all directions (like an interview), and there will be occasions when you do not want to pick up sounds from sources surrounding the microphone (like people talking or someone rustling papers). The polar pattern is also known as the *pickup pattern* or a microphone's *directionality*.

As a microphone rejects sounds from undesired directions, it also helps to reduce potential feedback through the sound system. The following polar patterns are available:

- **Omnidirectional**   Sound pickup is uniform in all directions.

- **Cardioid (unidirectional)**   Pickup is primarily from the front of the microphone (one direction) in a cardioid pattern. It rejects sounds coming from the side, but the most rejection is at the rear of the microphone. The term *cardioid* refers to the heart-shaped polar patterns.

- **Hypercardioid**   A variant of the cardioid, this type is more directional than the regular cardioid because it rejects more sound from the side. The trade-off is that some sound will be picked up directly at the rear of the microphone.

- **Supercardioid**   This type provides better directionality than the hypercardioid, as its rejection from the side is better. It also has more rear pickup than the hypercardioid.

- **Bidirectional**   Pickup is equal in opposite directions, with little or no pickup from the sides. This is sometimes also referred to as a *figure-eight pattern* because of the shape of its polar patterns.

Figure 9-10 shows various microphone polar patterns.

When selecting a microphone for a particular use, consider where you can place the microphones in the space. Microphone placement affects system performance. Make sure the microphone is not too far from or too close to the presenter, loudspeakers, or obtrusive furniture.

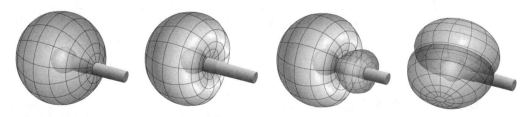

Figure 9-10   Microphone polar patterns: (L to R) omnidirectional, cardioid, supercardioid, and bidirectional

Preamplifiers

The first step in the audio chain after the microphone will usually be a preamplifier. A microphone preamplifier, often known as *preamp* or *mic pre*, takes the low mic-level signal and amplifies it to line level to make it compatible with all the subsequent signal processing devices in a system. A preamp may also include some filtering or tone adjustments.

Audio Mixers

In its most basic form, an audio system has a sound source at one end and a destination for that sound at the other. In almost all situations, there is more than one source.

Audio systems deal with multiple and varied sources of sound. The sources could be several vocalists with instruments at a concert, portable playback devices or media servers, multiple participants in a conference, or several actors in a theater performance. All of these signals come together in the *audio mixer*.

All audio mixers serve the same purpose: to combine, control, route, and possibly process audio signals from a number of inputs to a number of outputs. Usually, the number of inputs will be larger than the number of outputs.

Audio mixers are often identified by the number of available inputs and outputs. For example, an 8-by-2 mixer would have eight inputs and two outputs. Each incoming mic- or line-level signal goes into its own channel. Many mixers provide individual channel equalization adjustments, as well as multiple signal-routing capabilities, via *main* or *auxiliary buses*.

An audio mixer is often also known as a *mixing console*, an *audio console*, or a *mixing desk*.

Regardless of the size and complexity, any mixer that accepts mic-level inputs will have microphone preamps. Once the mic level is amplified to line level by the preamp, it can be processed by the rest of the mixer.

Between the inputs and outputs, the typical audio mixer provides multiple gain stages for making adjustments. These adjustments allow the console operator to balance or blend the audio sources together to create the appropriate sound balance for the listening audience.

Some audio mixers will open and close microphone channels automatically, like an on/off switch as levels cross preset thresholds. These are called *gated automatic mixers*. Others will turn up microphone channels that are in use and attenuate (or *mute*) the microphone channels not in use, like a volume knob. These are called *gain-sharing automatic mixers*. The channels set for automatic mixing should be used only for speech. Other sound sources, such as music, should not be set for automatic mixing. Most applications, especially music, require live operator intervention to achieve an acceptable sound mix.

 NOTE *Automatic mixers* should not be confused with *automated* mixers, which are automated by computer and store presets, control settings, and various mixing moves.

Audio Processors

Numerous types of processors can refine and modify audio signals. The intended use and listening environment will determine which type is right for each application.

Some common processors include compressors, limiters, expanders, gates, filters, and delays. Sometimes multiple audio-processing functions are combined in a single device such as a digital signal processor (DSP). Specialized processors, such as acoustic echo cancellers, are used in conferencing systems.

Dynamic Range Processing

Compressors, limiters, and expanders are processors that affect the dynamic range of audio signals. The term *dynamic range* refers to the difference between the loudest and quietest levels of a signal. A signal level that varies greatly between the loudest and quietest parts is said to have a *wide* dynamic range. Compressors and limiters operate in the same way but are different in their uses.

Compressors A compressor controls the dynamic range of a signal by reducing the part of the signal that exceeds the user-set threshold. When the signal exceeds the threshold, the overall amplitude is reduced by a user-defined ratio, thus reducing the overall dynamic range, as shown in Figure 9-11. In other words, they keep loud signals from being too loud.

The amount of reduction above the threshold is determined by an adjustable ratio. The reduction reduces the variation between highest and lowest signal levels, resulting in a compressed (reduced) dynamic range. Compressors can be used to prevent a signal from driving into system distortion. They are useful when reinforcing energetic presenters, who may occasionally raise their voice for emphasis.

The *compression ratio* is the amount of actual level increase above the threshold that will yield 1dB in gain change after compression. For example, a 3:1 ratio would mean that for every 3dB the gain increases above the threshold, the audience would hear only a 1dB difference after compression. Likewise, if the level were to jump by 9dB, the final level would jump only 3dB.

Figure 9-11
Audio signal
shown before
and after
compression

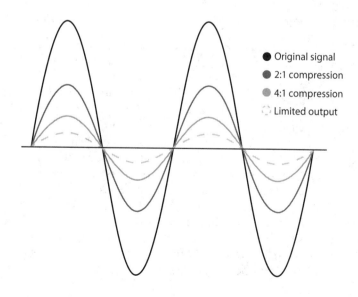

- ● Original signal
- ● 2:1 compression
- ● 4:1 compression
- ○ Limited output

PART IV

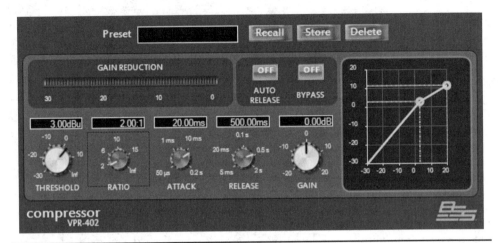

Figure 9-12 Software compressor interface showing sample threshold and ratio settings

Figure 9-12 shows some common settings for compressors. In this example, the compressor's threshold has been set to 3dBu, so any input signal above +3.0dB will be compressed. The compression ratio is set to 2:1, which means that for every 10dB the sound increases above the threshold, the sound will be compressed by 5dB, as shown on the display graph.

You can also set the attack and release times on a compressor. The attack time is how long it takes for the compressor to react after the input level exceeds the threshold. The release time determines when the compressor lets go after the level settles below the threshold. Both of these functions are measured in milliseconds. Compression, when used too enthusiastically, can reduce the dynamic range of material until it lacks color and brightness.

Extreme compression of high-amplitude signals is called *limiting*.

Limiters   A limiter is an audio signal processor that functions like a compressor except that signals exceeding the threshold level are reduced at ratios of 10:1 or greater. Limiters limit the level of all signals above an adjustable threshold. In other words, they prevent high-amplitude signals from getting through. Limiting is used to prevent damage to components such as loudspeakers and to prevent signal clipping in analog-to-digital conversion. It is triggered by peaks or spikes in the audio signal (like a dropped microphone or a hit from a drum stick), and it reacts quickly to cut them off or reduce them before they exceed a certain point. The amount of limiting above the threshold is determined by a more aggressive ratio than a typical compressor reduction ratio. The reduction limits the variation between the highest and lowest signal levels, resulting in a limited dynamic range.

In Figure 9-13, you can see some sample limiter settings. In this example the threshold is set to +10dB, so the limiter will activate when the input signal is +10dB. The attack time is set to 1ms, so the limiting will begin quite quickly as the input signal level increases, but the release is set to 200ms, so the limiting will decrease more slowly as the input signal level drops below the threshold.

Figure 9-13
A software
limiter interface
showing sample
limiter settings

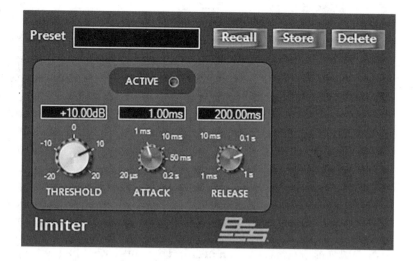

Expanders   Expanders are audio processors that are more properly thought of as *downward expanders*. Expanders typically reduce the level of all signals below an adjustable threshold. The signal level reduction increases the variation between the highest and lowest signal levels, which results in an increased dynamic range. It is used for reducing unwanted background noise. The amount of reduction below the threshold is determined by an adjustable ratio.

As you can see in Figure 9-14, downward expanders have many of the same settings as compressors and limiters. In this example, the threshold is set to −40dB. When the input signal falls below this threshold, it will reduce at a 4:1 ratio. This allows you to eliminate low-level noise from your audio signal.

Figure 9-14
An example
of settings for
a downward
expander

Gates A gate is an audio processor that allows signals to pass only above a certain setting or threshold. Gates mute the level of all signals below an adjustable threshold. Figure 9-15 shows a software gate. Notice the change in the output signal amplitude when a gate is applied to the input signal. This means that the signal levels must exceed the threshold setting before they are allowed to pass. This can be used to turn off unused microphones automatically. Figure 9-15 shows a software gate with a threshold of 60dB. You can control how the gate activates by setting the gate's attack and hold times.

Filters A filter attenuates or passes certain frequencies from a signal. High- and low-pass filters attenuate or pass certain frequencies from a signal. Notch filters will "notch out" a specific frequency or band of frequencies, while low or high filters will attenuate the upper or lower range of frequency content. Graphic and parametric equalizers include a variety of filters and amplifiers to enable the frequency characteristics of a signal to be customized for a particular venue, system response, or application.

Filters are classified by the rate of attenuation on the signal. This is shown in terms of decibels per octave, where an octave represents a doubling of frequency. A first-order filter attenuates at a rate of 6dB per octave (see Figure 9-16), while a second-order filter attenuates at a rate of 12dB per octave (see Figure 9-17). A third-order filter attenuates at 18dB per octave, a fourth-order filter attenuates at 24dB per octave, and so on. Each order has 6dB more roll-off than the one before it. To show how dramatic this roll-off can become, Figure 9-18 shows an eighth-order filter, which attenuates at 48dB per octave.

Equalizers

Equalizers (EQs) are frequency controls that allow you to boost (add gain) or cut (attenuate) a specific range of frequencies. The equalizer found on the input channel of a basic audio mixer may provide simple high-, mid-, and low-frequency controls.

Figure 9-15
A software gate with a threshold of –60dB

Figure 9-16
First-order filter centered on 180Hz

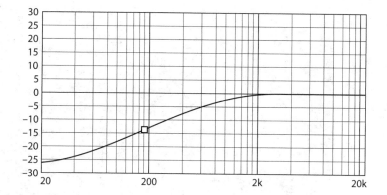

Figure 9-17
Second-order filter centered on 180Hz

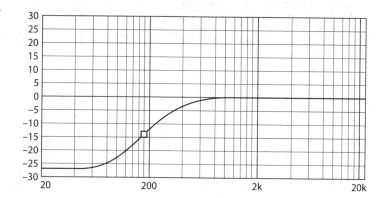

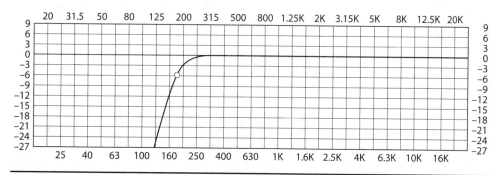

Figure 9-18 Eighth-order filter centered on 180Hz

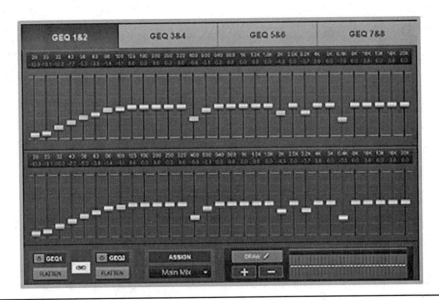

Figure 9-19   A graphic equalizer screen

In the audiovisual world, two types of sound-system equalizers are common:

- **Graphic equalizer**   A common graphic equalizer is the one-third-octave equalizer, which provides 30 or 31 slider adjustments corresponding to specific fixed frequencies with fixed bandwidths. The frequencies are centered at every one-third of an octave. The numerous adjustment points allow for shaping the overall frequency response of the system to produce the required effect. The graphic equalizer is so named because the adjustment controls provide a rough visual, or graphic, representation of the frequency adjustments, as shown in Figure 9-19.

- **Parametric equalizer**   A parametric equalizer, as shown in Figure 9-20, offers greater flexibility than a graphic equalizer. Not only does the parametric equalizer provide boost and cut capabilities, as does the graphic equalizer, but it also allows center frequency and filter bandwidth (often called the filter's *Q*) adjustments. A simple *parametric equalizer* may be found on the channel inputs of many audio mixers.

There are many different types of equalizers, ranging from simple tone controls to fully parametric equalizers. Some mixing consoles offer a combination of fixed and semi-parametric controls. Graphic and parametric can be separate devices, built in to audio mixers, or included as a function in a DSP unit.

Figure 9-20
A parametric
equalizer screen

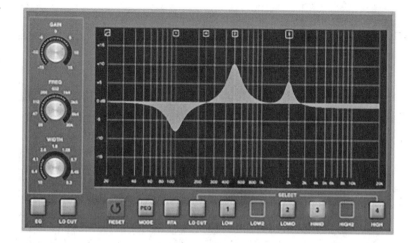

Delays

A *delay* is an adjustment of the time in which a signal is sent to a destination, often used to compensate for the distance between loudspeakers or the differential in processing required between multiple signals. If the delay is an unintended by-product of signal processing, it is usually referred to as latency.

Delays are used in sound systems for loudspeaker alignment, either to time-align components within a loudspeaker enclosure or loudspeaker array or to time-align the loudspeakers distributed in a system.

Within a given loudspeaker enclosure, the individual components may be physically offset, causing differences in the arrival time of the wave fronts from those components. This issue can be corrected either physically or by using an electronic delay device to provide proper alignment.

Electronic delay is commonly used in sound-reinforcement applications. For example, consider an auditorium with an under-balcony area. The audience seated directly underneath the balcony may not be covered very well by the main loudspeakers. In this case, supplemental loudspeakers are installed to cover the portion of the auditorium beneath the balcony.

While the electronic audio signal arrives at both the main and under-balcony loudspeakers virtually simultaneously, the sound coming from these two separate loudspeaker locations will arrive at the audience underneath the balcony at different times and sound like an echo. This is because sound travels at about 343 meters per second (1,125 feet per second), which is much slower than the speed of the electronic audio signal, which travels at approximately 150,000 kilometers per second (90,000 miles per second).

In this example, an electronic delay would be used on the audio signal going to the under-balcony loudspeakers. The amount of delay would be adjusted so that the sound from both the main loudspeakers and the under-balcony loudspeakers arrive at the audience at the same time. Similarly, delays can be used to time-align the audio from a video replay to be in synch with the vision throughout a viewing area served by multiple loudspeakers.

PART IV

Echo Cancellers

In audioconferencing and videoconferencing applications, two distinct types of echo need to be minimized:

- **Electronic echo**   This can occur on a line used for bidirectional communication. An electronic echo canceler, DSP, or hybrid will attempt to discern which audio came from which direction and cancel it out to avoid feedback.

- **Acoustic echo**   This refers to the environmental echoes created by the far-site sound bouncing around walls and furniture and returning to the microphones. Acoustic echo cancelers are used in conferencing systems. Acoustic echo cancellation is often one function of a conferencing device, DSP, or other audio component; it's not usually handled by a stand-alone "acoustic echo canceler." It may be integrated with other functions, such as microphone mixing or amplification.

The echo-canceling function is rated by the echo's tail length, or the amount of reverberation memory the device can handle. Tail lengths may vary from 40ms to as much as 270ms.

Echo cancellation may be implemented in a hardware echo-cancelling device or as a function of a software unified communications system such as a web-conferencing application.

 TIP   Avoid using both hardware and software echo cancellation in the same setup, as the results can be unpredictable and difficult to troubleshoot.

Figure 9-21 depicts an echo-cancellation system configuration.

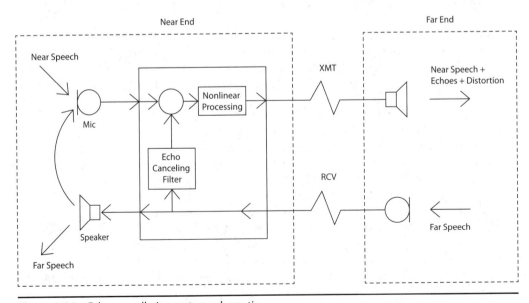

Figure 9-21   Echo cancellation system schematic

Digital Signal Processors

An audio DSP is a microprocessor-based device that analyzes the incoming digital audio streams and performs mathematical manipulations and transformations on the signals to produce a range of processing functions, which may include the following:

- Mixing
- Automated level control
- Filtering
- Equalization
- Limiting
- Reverberation
- Echo and chorus
- Compression and/or expansion
- Time delay
- Pitch shifting
- Echo cancellation
- Acoustic echo cancellation
- Feedback suppression
- Temporal manipulation
- Matrix routing and mixing
- Loudspeaker processing

A DSP may be incorporated into a mixer or may be a stand-alone processing device. In addition to digital audio inputs and outputs, stand-alone DSPs may have analog inputs and outputs to interface with analog equipment. A single DSP device can be used to replace many other pieces of processing equipment. Some DSPs have just a few dedicated functions and may be configured from simple front-panel controls, but the majority are multifunction devices that must be connected to an external system for configuration and programming.

Programming and configuring a DSP may require an external computer system running proprietary software to be connected to the DSP via a serial, USB, or Ethernet interface, although it is common for DSPs to include a web interface that allows configuration via a web browser from any device on the Internet.

As signal processing takes time for digitizing, signal analysis, processing, and conversion back to analog for output, a DSP introduces a delay (latency) into the signal path. Latency is expressed in milliseconds between a signal's input and its output. The selection of an appropriate DSP for a task should include consideration of the likely signal latency that may be introduced.

Assistive Listening Systems

Every jurisdiction has specific regulations regarding supporting people with disabilities. One aid for people with hearing disabilities is an assistive listening system, which transmits the audio content of the program to hearing-impaired attendees. These technologies may also be extended for use in simultaneous-translation, tour-guide systems, or multilingual environments.

Assistive listening systems are usually connected into the audio system via an auxiliary feed of the main system output, receiving the same material as any recording or streaming system, and the amplifiers which drive the program loudspeakers. These systems provide a wireless connection to devices worn or carried by listeners. The technologies used for transmission are generally either wireless radio, infrared (IR) light, or audio frequency induction loop signals.

Wireless RF technologies used may include low-powered commercial-radio FM radio transmitters, transmissions in the unlicensed industrial/scientific/medical (ISM) bands, DECT cordless-phone technology systems, digital streaming over Wi-Fi, and even Bluetooth systems. As with any RF systems in an installation, the frequencies to be used should be included in the spectrum allocation plan for the project, and care should be taken that there is no interference between the rooms and spaces in the project that have assistive listening systems.

In the case of IR-based systems, there must be a clear line of sight between the IR transmitter and the listeners' receivers throughout the space. This may require the use of multiple transmitters.

Induction loop systems take advantage of the "T" or *telecoil* induction coil included in most hearing aids, enabling anyone with a suitably equipped hearing aid to pick up the program audio feed without requiring any additional equipment such as the receivers and headsets used in IR and RF systems. The telecoil acts as the secondary coil in a transformer, picking up the magnetic field from a "primary coil" in the listening space that is fed by a powerful audio amplifier. An induction loop system requires that a heavy wire "antenna" be included in the areas of a space where the assistive hearing system is to be used. The induction loop cable is usually placed on the floor, often using specialty cables designed to be installed under floor coverings.

Some jurisdictions require that a minimum area or number of seats in each space must be equipped with such a loop. In many jurisdictions there are requirements that ticket booths, service desks, and reception desks be fitted with induction loop hearing assistance devices.

Power Amplifiers

The power amplifier is the last device used before the signal reaches the loudspeakers. Power amplifiers *amplify* electronic audio signals sufficiently to drive the loudspeaker. They do this by increasing the gain (voltage and power) of the signal from line level to loudspeaker level. Some basic amplifiers have only a power switch and input sensitivity controls. Many include digital signal processing, speaker-protection logic, and network monitoring and control. Powered, or *active,* loudspeakers include built-in power amplifiers.

Potentially, the more powerful the amplifier is, the greater the amplification of the signal it can provide, and the louder the sound that the loudspeaker can achieve.

Power amplifiers are typically connected to loudspeakers with heavier-gauge stranded wire than used at mic or line level. The size of wire will depend on the distance between the power amplifier and the loudspeaker, and the current required. Loudspeaker cabling will be unshielded and may or may not be twisted.

A common connector used for loudspeaker cabling is the Speakon connector. Speakons are widely used for professional loudspeaker connections because they are rugged, durable, locking, and almost idiot-proof.

Loudspeakers

Loudspeakers are used for sound reinforcement, communications, sound masking, or sound reproduction in an audio system and are generally found at the end of the audio signal path. The acoustic energy that was transduced into electrical energy by the microphone and processed through the audio system is transduced back into acoustic energy by the loudspeaker. Most loudspeakers share a common characteristic: they have acoustic drivers mounted in an enclosure. Loudspeakers vary based on the type of drivers used, system design, and performance specifications, including sensitivity, impedance, and frequency response.

 VIDEO   Check Appendix D for a link to *Loudspeaker Impedance*.

Powered loudspeakers can simplify setup, and some designs provide for easy portability. Many of the signal processing requirements may be built into portable powered loudspeakers. Typically they will have a line-level input, but some are also equipped with mic-level input and a microphone preamplifier. Some also include some basic filtering or equalization.

Crossovers

The audio spectrum includes a wide range of frequencies. It's not physically possible to create a single acoustic driver that can reproduce the entire range either accurately or efficiently. This is resolved by constructing loudspeaker systems with multiple drivers, as shown in Figure 9-22. A loudspeaker enclosure containing more than one frequency range of drivers is known by the different frequency ranges it covers.

So that each driver is sent only those frequencies that it will transduce accurately, an electrical filtering device known as a *crossover* is used. The crossover device may be placed before the amplification stage, which then supplies only the appropriate signal for its intended driver speaker, or it may be placed on the input to a multidriver loudspeaker cabinet. Many low-power-handling loudspeaker cabinets incorporate an unpowered (*passive*) crossover network to divide the frequencies of the incoming signal and send them to the appropriate driver.

Figure 9-22
Multidriver
loudspeaker

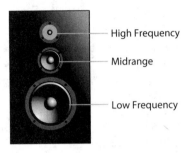

High Frequency

Midrange

Low Frequency

Loudspeaker Systems

All loudspeakers have coverage patterns in which they project sound waves at specific frequencies. Therefore, different loudspeaker types and configurations are needed for various audio applications. There are two categories of loudspeaker systems.

Distributed Systems

A distributed system delivers even sound coverage of the same program material throughout a space. The loudspeakers are distributed at regular intervals. Wherever you are in the space, the distributed system should sound the same. Distributed systems are used where even coverage is more important than drawing people's attention to the directional source of the sound. For example, the paging system in an airport would be a distributed system.

Ceiling-mounted loudspeakers are commonly used in distributed systems where the sound source should not be visually intrusive. For ceiling-mounted loudspeakers in a distributed audio system, directionality is not a major concern, as long as you place them where the designer intended.

Point-Source Systems

Point-source systems are used where the directionality of the sound source is important. If the sound should appear to originate from a particular place in space, one or more loudspeakers, or clusters of loudspeakers, are positioned to give the illusion of sound coming from that place. Sometimes the system consists of a single cluster of loudspeakers directly above or behind the source point, but more frequently the system consists of multiple loudspeaker clusters with their signals phase-, time-, and amplitude-aligned to give the illusion of point-source directionality.

The basic two-source stereo speaker arrangement for the audio component of a video replay system may extend out to become dozens of loudspeaker clusters, distributed throughout a space, for the replay of highly spatialized, surround-sound content for cinematic screenings and visualization spaces.

For many productions in large spaces, phase-aligned arrays of loudspeakers are placed above, below, and to the sides of the stage to generate a coherent wave-front for the entire audience. These are often supplemented with additional time-aligned speaker arrays located deeper into the audience space to augment the sound levels for more distant listeners.

Safety becomes critical when an array of loudspeakers is mounted overhead with people underneath. All rigging and suspension hardware should meet local safety standards and be installed by qualified riggers.

When installing point-source systems, each loudspeaker cabinet is aimed in a certain direction and angle to cover the listening area as intended.

Installing Audio Components

All but the simplest of systems will have a system signal flow or block diagram drawings. These show how specific outputs are routed or wired to specific inputs. The system flow drawings are also known as the *pin-to-pin* drawings, and they are part of the integrator's design package. You will know where and how to install audio components by reading your system flow diagrams. Notice how the signal flow diagram shown in Figure 9-23 includes equipment such as graphic equalizers, power amplifiers, a digital signal processor, and an audio conference mixer.

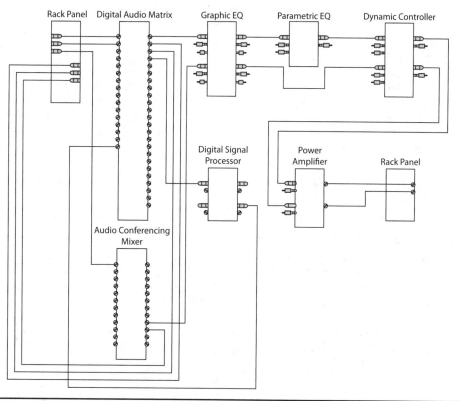

Figure 9-23 Signal flow diagram

Before you begin to connect an audio system, make sure the components are compatible with one another. Always read the operating manual for the device to verify correct signal levels and compatibility with other components in the system.

It is always good practice to test cable assemblies before installing them. It is quicker, easier, and always more efficient to test before the installation than to troubleshoot after the fact. Not only does this help you check the connections, but it also reveals problems (such as short circuits) or incorrectly wired pinouts that may originate in the connectors.

Microphone Installation

Although microphones are the first component in an audio signal chain, they may be the last to be installed; ceiling-mounted microphones may be installed earlier in the process.

The type of microphones for the project will be selected by the system designer, but to make sure that the microphones installed work the way the designers intended them to, you need to be aware of the characteristics of each type of microphone and how microphone placement affects audio system performance. Microphone input adjustments, such as setting gain, will be discussed in Part V of this book. In this section, you will learn about key considerations when installing microphones.

One of the characteristics to pay attention to when installing a microphone is its *polar pattern*, or pickup pattern. The pickup pattern describes the microphone's ability to pick up the desired sound in a certain direction while rejecting unwanted sounds from other directions. You need to make sure the mics are installed so the pickup patterns are oriented correctly.

With regard to microphone placement, consider these factors:

- **Distance**   If microphones are placed too far away or too close to the presenters, the presenters will not be comfortable using the system. If microphones are too close to the source, they may pick up unwanted sounds such as pops from breathing or be driven into distortion by SPLs that exceed the dynamic range of the microphone. However, if the microphone is placed too far away from the source, it may pick up unwanted reflections of the source, or it may need to be increased in gain to the point of picking up its own signal coming back from loudspeakers and so become part of a feedback loop.

- **Proximity**   If the microphone's pickup pattern intersects with a reinforcement loudspeaker's coverage area, feedback becomes an issue.

- **Aesthetics**   Many architects and interior designers take the position that microphones should be heard but not seen. Microphones and microphone arrays that are suspended from the ceiling can be obtrusive. Table microphones placed with their cables trailing can be seen as objectionable as well.

If situations similar to these are discovered, or if the location of microphone placement is unclear, carefully examine the drawings for microphone installation information and contact the designer for guidance.

Best Practice for Microphone Placement

Here are some best practices for microphone placement:

- Place the microphone as close as you can to the desired sound source, without creating sightline problems for the audience or any camera coverage.

- Keep the microphone directed away from loudspeakers and other undesired sources.

- Avoid ambient and undesired noise by being mindful of the microphone's polar pattern.

- Aim the microphones toward the area where the sound source (presenter or musician) will be and away from the places where they will not be.

- Isolate the microphones from mechanical noise sources such as objects being tapped on tables or lecterns, furniture being moved, or feet walking on floors.

- Keep the system gain to the minimum value needed to achieve the desired sound reinforcement (you will learn about gain control and adjustments in Chapter 14).

Microphone Feedback Avoidance

Feedback is unwanted noise caused by an inadvertent loop of an audio output back to its input. Acoustic feedback is generated between a microphone and a loudspeaker when the same signal is caught in a loop and repeatedly amplified. This happens when there is too much gain in the signal path; the audio system may be too loud, or the source (presenter or musician) may be too far away from the microphone.

Use these guidelines for mic placement to avoid feedback:

- Place mics near the person and physically as far from and behind the loudspeakers as possible, as shown in Figure 9-24. Never aim the microphone toward the direction of the loudspeaker, as shown in Figure 9-25.

- Avoid placing microphones directly underneath ceiling-mounted loudspeakers. Also, alert presenters to potential problems so that they avoid walking under overhead loudspeakers.

- The presenter should stand at an appropriate distance from the microphone. The appropriate distance is determined by the sensitivity of the microphone. If the presenter is too far away, you may need to increase the gain of the microphone, which may cause feedback.

- Avoid having too many microphones open at any time. Utilize the mute buttons or gating/automatic gain sharing features on a mixer to manage the microphones. You can also decrease the gain on unused microphones.

Figure 9-24
Correct
placement
of microphone

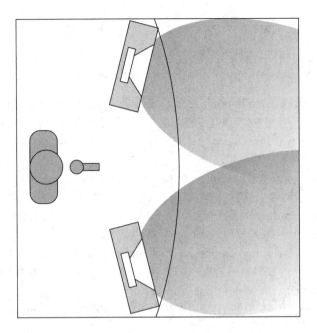

- Avoid placing many microphones with wide pickup patterns close to each other.
- Proper placement of a lavalier microphone on the presenter is critical to good microphone performance.

You can eliminate feedback with proper microphone placement and by using a microphone with the appropriate pickup pattern for the application. The best way is to place

Figure 9-25
Incorrect
placement of
microphone
that will result in
feedback

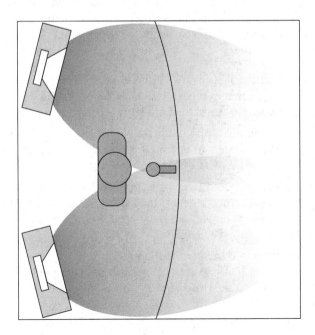

the microphone near the source of the sound that is to be reinforced. Avoid microphones aimed at loudspeakers, and should feedback begin, reduce the gain on the microphone until the feedback goes down.

In dire circumstances, feedback elimination devices can sometimes be of assistance. Depending on their sophistication, feedback eliminators employ a range of processing tricks, including phase manipulation, automated gain control, narrow-band dynamic filtering, and digital signal cancellation, to help solve the problem, but the best solution by far is to keep the signals from the loudspeakers from getting into the microphone in the first place.

Microphone Mounting

Many types of microphones and mountings are used in staging and live events. Placement and mounting of these microphones will vary with the requirements of any particular event and with the types of sound sources to be captured. These may vary from a single presenter or a seated panel to a choir, an orchestra, a brass band, or a musical presentation with a mix of vocalists, acoustic, electronic, and amplified instruments.

In the permanent or fixed installations found in boardrooms, meeting rooms, conference rooms, or lecture halls, holes are drilled, mounts and microphones are installed, and cables are run and hidden for an integrated look. In some instances of custom furniture, drawings and templates may be sent to the furniture manufacturer for drilling holes for microphones and other audiovisual connection boxes. In other instances, the installer is responsible for drilling and mounting microphones. In these cases, it is critical to verify that the furniture, in terms of number of units and locations, matches the drawings. Check for obstructions, such as furniture framing and legs, before performing any drilling. If you need to install a metal microphone plate, a metal punch tool will help you puncture holes into the plate and secure it to the wall. Consider that conference tables and other custom furniture can be very expensive, so protect all surfaces from any potential marring or damage.

Wireless Microphone Considerations

Sometimes called *radio mics,* wireless microphones use radio frequency (RF) transmission instead of a cable to carry the signal from the microphone to the rest of the audio system. Many wireless microphone systems use analog frequency modulation (FM) transmission, similar to analog FM radio broadcasts. However, digital wireless systems are increasingly being used due to their wide dynamic range, low distortion, and capability for secure communication via encryption. Even the best available wireless microphone link is less reliable and less noise immune than a standard microphone cable.

There are two form-factors for transmitters: body pack (or belt pack) and handheld.

- **Body/belt pack** The body/belt pack can be clipped to a belt, placed in a pocket, secreted in a costume, or attached in some other way to the user. The microphone is then connected to the transmitter unit with a short, very flexible cable. Virtually any type of microphone, including instrument pick-ups, can be connected to a transmitter pack. If the microphone used is a condenser or electret, microphone power is usually supplied from the batteries in the transmitter pack.

- **Handheld**   In a handheld microphone, a standard microphone mechanism is often integrated onto the top of a transmitter, making the microphone and transmitter a single unit. In some systems, a small "plug-on" transmitter is attached to the output connector on the base of a regular handheld microphone, although these may have limited output power and/or limited operational life between battery charges/replacements.

 TIP   When pairing transmitters and receivers in a multichannel wireless microphone system, make sure they share the same group and channel.

Wireless microphone systems often employ diversity receivers with more than one antenna. The receiver compares the signals from each antenna and uses the stronger one. The correct placement of antennas is crucial to success; pay special attention to this and other factors discussed in the "Mounting Wireless Equipment" section of Chapter 8. In addition, note that if you are mounting multiple antennas, they should be separated by at least 1 meter (3 feet). Typically, an antenna should be mounted at least 300 millimeters (1 foot) from any metallic surface.

Wireless microphone antenna systems employ 50-ohm coaxial cable. Check the cable impedance to ensure there is no significant loss of signal over distance. If you have a long run between the antenna and receiver, you will need to use low-loss coaxial cable. Consult with the system designer before selecting a cable. When connecting wireless microphones to the audio or AV system, pay attention to all special considerations, especially those identified by the manufacturer.

Every aspect of installing and configuring wireless microphone systems applies equally to wireless in-ear monitoring (IEM) systems.

Spectrum Management   The coordination of wireless frequencies allocated to wireless microphone systems, IEM systems, assistive listening systems, communications systems, remote control devices, and the multitude of other wireless technologies involved in audiovisual installations and operations is critical. A full-spectrum frequency allocation plan is an important component of any AV installation or production. The plan must take into account local frequency allocation regulations.

The Association of Professional Wireless Production Technologies (www.apwpt.org) is an international organization that attempts to keep track of the changing spectrum allocations and the sources for up-to-date regulatory information for event production. APWPT maintains information on at least 30 countries in the Americas, Europe, Asia, and Oceania. The manufacturers of your wireless equipment are usually a good source of local spectrum information.

Loudspeaker Installation

Loudspeakers are selected, installed, and integrated into the audio system to produce uniformity of coverage throughout the intended listening area. Designers take into account several factors when selecting the location, including the directivity and sensitivity of the loudspeakers, overall intelligibility of the system, avoidance of feedback, and amplifier

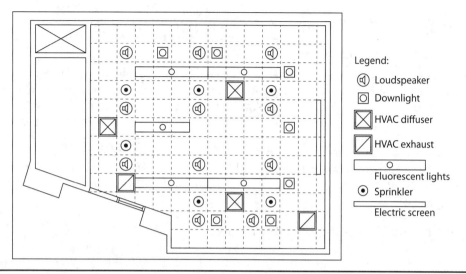

Figure 9-26   Reflected ceiling plan diagram indicating loudspeaker placement

power requirements. As with other components, your job is to make sure the loudspeakers are installed where indicated in system design diagrams.

The first step is to verify and inspect the specified location of each loudspeaker before any cable is pulled. If you identify a potential design flaw or if you think that there might be a problem during installation, contact the designer for a solution. Do not make any changes without getting prior approval. This is what the RFI-RFC-CO process is for.

Loudspeaker Placement

Architectural reflected ceiling plan (RCP) diagrams indicate where the loudspeakers should be placed. The two categories of loudspeakers, distributed and point-source, each have special characteristics, applications, and placement requirements.

Typical of many AV installations, the RCP in Figure 9-26 shows loudspeakers sharing ceiling space with lighting, an electrically lowered screen, HVAC units, and sprinklers. Refer to your RCP diagrams frequently to make sure the loudspeakers you are installing do not interfere with the spaces planned for these other necessary items.

Pulling Cable to Each Loudspeaker

Once you have verified the loudspeaker locations, you may begin to pull cable. You may elect to wire the loudspeakers on a workbench or table instead of directly on the ceiling from a ladder or work platform. Follow these guidelines when pulling cable to loudspeakers:

1. Verify the correct gauge of cable to be used by referencing the technical drawings or cable legend sheets. Also verify whether the cable needs to be plenum rated.

2. Be aware of the "deck," the structure above a drop ceiling. This could be a concrete floor or could be structural support beams above where you are working.

3. All cabling, including service loops, should be suspended from infrastructure above a drop ceiling, using hooks or other appropriate hardware. This protects the cabling from damage.

4. Carefully remove ceiling tiles from the ceiling and place them aside. Clean gloves can help to ensure that fingerprints do not appear on ceiling tiles (do not use the same work gloves that you use to pull cables or handle steelwork).

Pushing to finish the job on time puts the installer under a lot pressure. However, nothing is more important than the safety of everyone on the job site. Safety should always be the top priority when pulling cable. Use appropriate ladders, work platforms, and protective equipment such as harnesses, gloves, masks, kneepads, headgear, and safety goggles.

Ceiling-Mounted Loudspeaker Installation

In many installations, some loudspeakers need to be installed in a ceiling. In Chapter 7, you learned the basic principles, process, and safety measures for ceiling-mounted AV equipment. When mounting loudspeakers in the ceiling, you need to know how to support the weight of the loudspeaker. Follow the set of instructions that have been approved by a structural engineer. Use the exact part numbers defined in the instructions, or if these are not available, seek guidance from the engineer via the RFI-RFC-CO process.

For a distributed audio system, it is common to install loudspeakers into plaster sheeting or suspended ceiling tiles. The weight of the loudspeaker components can be substantial, so use a tile bridge for support to keep ceiling tiles from sagging after the loudspeaker is installed. Heavier loudspeaker components may need to be supported independently from the ceiling, with connections back into the building structure.

As you install the loudspeaker and connect the appropriate wiring, there can sometimes be additional wires that are not used. These wires should be insulated from each other and tied off to prevent them from touching the moving elements of the loudspeaker. Make sure no debris or foreign matter will cause vibration during operation.

Loudspeakers are designed to vibrate, and whenever you suspend something, you are working against gravity. Vibration can cause mounts and attachments to come loose, and gravity pulls them out.

If present, line up screw-head slots so they create a pattern with the grid. Tighten loudspeaker and pre-painted grill screws; do not overtighten, or they may crush the ceiling material or strip the screws. Repaint chipped screw heads.

TIP To remove a ceiling tile from a grid support structure gently, lift the ceiling tile out of the grid and then turn it until it slides through the support structure.

When suspending loudspeakers overhead, use mounting brackets that have been designed for that type of loudspeaker. Certain loudspeaker enclosures have been built by a manufacturer who has designed the use of the loudspeaker for overhead suspension and employs a minimum of a 5:1 design factor. This means it can withstand at least five times the load being applied to it.

Loudspeakers for suspension should come with manufacturer-installed mounting points already on the loudspeaker cabinet. Read the system manufacturer's instructions to make sure you have the correct brackets and follow the proper procedures. Consult the project manager or a structural engineer if you have any doubts or questions.

Best Practice for Loudspeaker Installation

Wash your hands or put on clean gloves before handling ceiling tiles so as not to leave handprints. Also, handle ceiling tiles gently because they break easily. If you ruin the tile in any way, most likely your company will have to replace it. This can be a real problem in an existing facility because matching replacement tiles may no longer be available.

Most loudspeaker manufacturers provide a template for the proper cutout of the ceiling tile. Always follow manufacturer guidelines.

Back Boxes

While a finished installation will leave only the loudspeaker grill or baffle cover visible, a back box or back can (see Figure 9-27) is hidden inside the ceiling cavity. Back boxes help extend the low-frequency response of the loudspeaker. They keep sound contained by preventing or reducing the amount of sound that is transmitted through the ceiling space to an adjoining room. Back boxes can provide the loudspeaker with physical security from dust and damage. Some back boxes are plenum rated. Verify that you are using the correct type of back box in your installations.

Figure 9-27
In-ceiling
loudspeaker
with a back box

PART IV

Loudspeaker Installation Guidelines

Although it is not possible to predict the local conditions that you may encounter in an installation environment, it is a good idea to follow these guidelines that apply to any installation site:

- Never attach or suspend heavy loads to/from a wall or ceiling surface without a second person to help.

- Always make a secure attachment to the structure of the building. When two flat surfaces are mated, one takes on the integrity of the other, which is why they have to be mated together flat and tight. (This is why you generally want to avoid wall mounting on plaster board.)

- Be absolutely certain of the structural integrity of anything that is to be used to support external loads. Hidden structures can have hidden weaknesses. Do not rely on nails to support overhead loads. Nails can pull out over time and stress. They are not dependable. Verify the quality of suspension points supplied by another contractor. Seek help from the project manager or the structural engineer when uncertain or in doubt about the load limits of the building.

- Document and archive all discussions or problems related to safety. If you work with a structural engineer, follow instructions to the letter. If you modify another contractor's work, you take responsibility.

- Know the load ratings on the product being used. Working load limit (WLL) and safe working load (SWL) mean the same thing. The WLL of a part is the load rating for which the part can be safely loaded. The load rating must not be exceeded, or the part may fail.

Following these guidelines will aid in creating a safe work environment. Ceiling-mounted loudspeakers especially pose safety risks, so follow all possible safety precautions.

Direct-Coupled and Constant-Voltage Loudspeaker Wiring

There are two common configurations for wiring loudspeakers: direct coupled and constant voltage. Direct-coupled configurations are usually connected directly to the loudspeaker, while constant-voltage configurations employ a transformer at the power amplifier output and transformers at each individual loudspeaker.

Direct-Coupled Loudspeakers

In a direct-coupled (low-impedance) configuration, the amplifier is connected directly to the terminals of the loudspeaker. If multiple loudspeakers are installed onto one amplifier channel, they can be wired in series, parallel, or series/parallel combination. The loudspeakers must be wired in such a way as to ensure that the load impedance of the loudspeaker wiring configuration is within the impedance range that the amplifier can drive.

A direct-coupled system typically provides better low-frequency response and improved fidelity in comparison to many constant-voltage systems. Direct-coupled systems are best used for program audio and live performance work.

Direct-coupled systems have limitations. The loudspeaker connections are low imped-ance, and the resistance of the wires will dissipate some of the energy on the way to the loudspeakers. Therefore, larger-gauge wire is required to minimize the resistance. As cable loss is a significant factor in system efficiency, the designer may need to locate the amplifier close to the loudspeakers. Active/powered loudspeakers are usually directly coupled to their internal amplifiers.

Constant-Voltage Loudspeakers

Constant-voltage is the name given to a higher-voltage system of distributing signals to loudspeakers over a large area, with lower signal losses than for a typical direct-connection system. Despite the confusing name, the voltage in a constant-voltage audio circuit varies with the amplitude of the signal, just like any other audio signal.

In a constant-voltage system the audio amplifier uses a step-up transformer to produce its output at a higher voltage than in direct-coupled systems. From the power equation ($P = I \times V$) we know that at a higher voltage, less current is required to transmit the same amount of power. And as we know from Ohm's law that resistance losses in a cable are lower if the current is lower, it becomes apparent why the constant-voltage system is more efficient at transferring power and can be run over longer distances or with smaller-gauge wire. Because of the low cable losses, the voltage is nearly constant across all of the speakers in the network. Typical output voltages for constant-voltage-capable amplifiers are 25V, 50V, 70.7V, and 100V. Many commercial AV amplifiers have both low-impedance outputs for direct-coupled loudspeakers and transformer-driven, high-impedance outputs for constant-voltage networks.

As most loudspeakers cannot operate at the high voltages present in constant-voltage systems, they require a step-down transformer to provide signals at an appropriate voltage for the speaker drivers. As a side effect, the transformers used to step up and step down the voltages for constant-voltage networks act as filters on the audio program material, causing low-frequency attenuation and some distortion at high frequencies. There are also some power losses, as transformers are not completely efficient at power transfer between their windings.

Some loudspeakers for commercial AV applications come fitted with a multi-tap step-down transformer specifically for use in constant-voltage systems, while other loudspeakers require that an external multi-tap transformer be connected before the constant-voltage feed. Constant-voltage loudspeakers are generally referred to by their intended operational voltage (25V, 50V, 70.7V, and 100V). In constant-voltage systems all loudspeakers are connected in parallel with the amplifier.

Typical applications of constant-voltage loudspeaker systems include the following:

- 100V systems are widely used throughout the world.
- 70.7V systems are still used in North America due to historical electrical regulations.
- 25V and 50V systems may still be used in some smaller-scale applications.

Figure 9-28   The transformer on a loudspeaker

It is most common to install loudspeakers that have been fitted with a transformer, as shown in Figure 9-28. However, if the loudspeaker that was purchased does not have the transformer connected, you will have to connect it yourself. You will need to locate the secondary side of the transformer and then solder those two wires to the voicecoil tabs on the loudspeaker. Refer to the manual to ensure that these wires are connected properly.

A constant-voltage system allows you to connect many loudspeakers to one amplifier or to connect loudspeakers to the amplifier from a considerable distance. For example, an airport paging system's loudspeakers tend to be located a long way from the audio source, but each transformer receives the same level of signal. Even though the line is common to all loudspeaker points, the transformer winding length (tap) determines how much of the available energy is coupled to the loudspeaker itself. The AV designer must calculate and specify the tap value for the transformer that will be used for each connection.

Loudspeaker Transformers

Voltage can be manipulated in an electrical circuit with the use of transformers. Transformers are common electrical devices that are used in power supplies, in audio and video circuits, and, particularly for audiovisual use, in loudspeaker systems. Transformers transfer energy from one circuit to another without physical connection, using the principle of magnetic induction between two windings, or coils of wire. One winding is connected to the power source (primary winding), and the other winding is connected to the load (secondary winding). The ratio of the number of coils (or turns) of wire in the primary and secondary sides determines the impedance and/or voltage ratio. The transformer shown in Figure 9-29 has three input taps and outputs for both 4-ohm and 8-ohm loudspeakers.

Many constant-voltage loudspeakers come with preassembled loudspeaker transformer kits. However, if you must assemble the transformer to the loudspeaker, make sure you do not wire it backward. Connect the appropriate pair of wires on the secondary side of the transformer to the voicecoil tabs on the loudspeaker.

Figure 9-29
Multi-tap 70-volt-line transformer

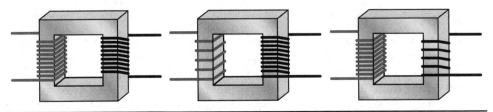

Figure 9-30   L to R: 1:1, step-up, and step-down transformer windings

Types of Transformers

Transformers have the ability to increase or decrease the voltage in a circuit or keep it the same.

The following types of transformers are available (see Figure 9-30):

- **1:1 transformer**   This type of transformer has an equal number of primary and secondary windings. It is used for electrical (galvanic) circuit isolation to solve problems such as ground loops causing audible hum, buzz, or rolling hum bars on a display.
- **Step-up transformer**   This type has more windings on the secondary side than on the primary side. Voltage and impedance increase, while available current decreases.
- **Step-down transformer**   This type has fewer windings on the secondary side than on the primary side. Available current increases, while voltage and impedance decrease.

Loudspeaker Taps

Most speaker transformers for use in constant-voltage systems have multiple wires on the primary side that allow you to adjust the voltage level to each loudspeaker. These wires are commonly referred to as *taps*. Taps are intermediate connections to the transformer windings that allow you to select different power levels from the transformer (see Figure 9-31).

Figure 9-31
Multi-tap
step-down
transformer

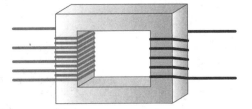

Either the transformer manufacturer will identify these wires in some way and provide a chart for their values or you can write the values on the wires themselves. The taps can be selected for the appropriate amount of power in watts or impedance steps that they will deliver to the loudspeaker and must comply with the designer's intentions for the desired performance.

Many multi-tap speaker transformers are packaged with a switching device for the simplified selection of the required tap point.

VIDEO   Check Appendix D for a link to *Setting Loudspeaker Taps.*

Many transformer manufacturers pre-strip the tap wires for termination. When terminating, be sure that the wires from the amplifier are connected to the tap points specified on the designer's drawings. The consequences of connecting the wrong wire include exceeding the capability of the amplifier or decreasing the signal to a barely audible level. You will connect only one tap value and the common connection on the transformer to the loudspeaker.

Securing Unused Tap Wires

The unused tap wires should be isolated and secured to prevent a short circuit caused by wires touching each other or the metal chassis of the loudspeaker. It is important to leave the unused wires for possible future adjustment.

Follow these steps to secure unused tap wires:

1. Wrap the end of each unused wire individually with a piece of electrical tape.

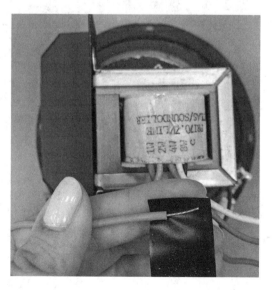

2. Group all the unused wires together.

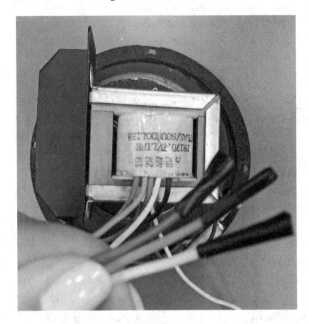

3. Wire-tie the unused wires together in a loop for possible future use.

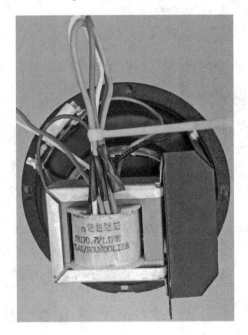

Loudspeakers are the final step in the audio signal pathway, but are not necessarily the last component you will install in the audio system.

Loudspeaker Wiring

As an installer, you will be responsible for wiring the loudspeakers so they work the way the system designer intended them to. Wiring diagrams are similar to block diagrams, in that they show the way the signal is supposed to flow through the circuit. In analog systems, you will also be testing the loudspeaker system to make sure that it has the correct impedance. Impedance depends upon the wiring scheme used in your loudspeaker installation, so you first need to identify how your specific system is wired.

Networked loudspeakers are generally fed with a power supply and a looped (or occasionally star topology) data network connection. Some lower-powered networked loudspeakers are powered via a power over Ethernet (PoE) system, which is similar in concept to audio phantom power.

Analog loudspeaker systems are typically wired either in series or in parallel, although sometimes a combination of both may be used. When connecting analog loudspeakers, each loudspeaker location should be fed with two readily identifiable cables (usually colored either red and black or white and black). The black wire is always the negative (–), and the red or white wire is the positive (+).

Series Loudspeaker Circuit

In a series loudspeaker circuit, each loudspeaker's coil is connected to the next loudspeaker in the series in sequence, as shown in Figure 9-32. The power amplifier's positive output terminal connects to the positive terminal of the first loudspeaker. The first loudspeaker's negative terminal connects to the second loudspeaker's positive terminal. The second loudspeaker's negative terminal connects to the third loudspeaker's positive terminal, and so on. The last loudspeaker's negative terminal completes the circuit by connecting to the amplifier's negative terminal.

Series circuits are wired plus to minus to plus to minus. Therefore, when wiring a series circuit, you will connect the red/white wires to black wires and connect the black wires to red/white wires.

Figure 9-32
A series loudspeaker coil is connected to the next loudspeaker's coil in sequence

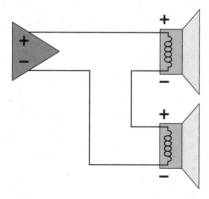

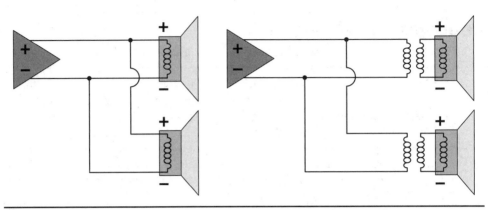

Figure 9-33 In a parallel loudspeaker circuit, the positive output of the amplifier connects to every loudspeaker's positive terminal

Parallel Loudspeaker Circuit

In a parallel loudspeaker circuit, the positive output of the amplifier connects to every loudspeaker's positive terminal, and each loudspeaker's negative terminal connects to the amplifier's negative terminal, as shown in Figure 9-33.

In a parallel wiring scheme, you will connect a positive wire from an amplifier to another positive terminal on each of the loudspeakers in the circuit. Likewise, the amplifier's negative wire is connected to the negative terminal on each of the loudspeakers in the circuit. So, you will connect red to red, or white to white, and black to black.

Connecting Loudspeakers

Loudspeakers have either a single connection point for each of the positive and negative terminals or a pair of bridged connection points (the input and output, or "loop out") for each of the positive and negative terminals. If there is only a single connection point per terminal, then both the input and the output connection to the next loudspeaker need to be connected to that point. The presence of an output or "loop out" connection point, as shown in Figure 9-34, simplifies the connection to the next loudspeaker in the line.

With any loudspeaker wiring scheme, be consistent with which wire you use for the positive terminal and the negative terminal. The termination block is usually a euroblock-style screw-down compression connector.

Regardless of the type of connection points, you will need the same tools to connect the wires. Before you start connecting the equipment, make sure you have the following tools: wire cutters, cable/jacket stripper, crimp tool, spools of two-conductor cable, and possibly crimp-type connectors (closed end or butt splice).

 TIP To connect loudspeakers that have only one connection point, use butt-splice connectors or closed-end connectors.

Figure 9-34
Loudspeaker
with two pairs
of connection
points; from
the left:
positive input,
negative input,
positive output,
negative output

 VIDEO Check Appendix D for a link to *Connecting Loudspeakers*.

Loudspeaker Impedance

Impedance (Z), measured in ohms, is the total opposition to current flow in a circuit. It includes the resistance (R) of the materials in the circuit plus their *reactance (X)*. Reactance is the opposition to current flow in the circuit caused by changes in the voltage and current, and is directly related to the frequency of the current. Reactance has two components: *capacitive reactance (X_C)* due to the capacitance in the circuit and *inductive reactance (X_L)* due to the inductance in the circuit.

On the back of almost any loudspeaker, you will see a nominal impedance rating, in ohms, for that loudspeaker. The impedance of a loudspeaker is related to many factors, including the construction of the speaker, the circuitry included in the speaker enclosure, the signal frequency range, and the wiring configuration of the total speaker system.

Most common loudspeakers will have a nominal impedance rating of 4, 8, or even 16 ohms. You may even see some that have an impedance of 6 ohms. You will need to determine the resulting impedance of the entire loudspeaker line when you connect and wire together a loudspeaker system.

Best Practice for Amplifiers

How you wire the loudspeakers together determines the circuit's impedance, and you need to be certain that the power amplifier is rated for that load.

If you are just wiring up a couple of loudspeakers, determining the circuit's impedance and checking the power amplifier's rating is a simple process.

Amplifiers have specified impedance that is expected to be connected to their output terminals. Matching this specified impedance with the loudspeaker load maximizes the energy transfer from amplifier to loudspeaker to acoustic energy and reduces the possibility of the amplifier being overdriven.

PART IV

Impedance Meter

Measuring impedance allows you to make decisions that can prevent damage to the circuit components. Impedance meter readings can also indicate short circuits, open loudspeaker lines, transformers installed backward, low-impedance loudspeakers on a high-impedance system, and the total impedance or load of the loudspeaker system.

An impedance meter will output a test tone in the audible frequency range. This tone is used by the meter to detect how the circuit reacts to that frequency. It will give an indication in ohms.

Here is the procedure for using an impedance meter:

1. The loudspeaker lines are disconnected from the amplifier.
2. The technician or engineer will calculate what the impedance of the line should be.
3. The appropriate range is selected on the meter.
4. The meter is calibrated to that range (or scale).
5. The meter leads are connected to the line to be tested.
6. The meter is read.
7. The reading is compared to the anticipated reading.
8. Only after being satisfied that the reading obtained is within an acceptable tolerance range should the loudspeakers be connected to the amplifier.

If the measured impedance is within 10 percent of the anticipated reading, the system is probably correctly wired. If the measured impedance is significantly higher than anticipated, it probably means the system is wired incorrectly; perhaps some transformer taps were set at lower output levels than specified. This will not damage the system; it just will not work as intended. If the measured impedance is significantly lower than anticipated, you must reduce the load presented to the amplifier.

Series Circuit Impedance Formula

Use the series circuit impedance formula to calculate impedance in loudspeakers wired in series.

Series Circuit Impedance Formula

The formula for calculating the total impedance of a series loudspeaker circuit is the simple arithmetic sum

$$Z_T = Z_1 + Z_2 + Z_3 \ldots Z_N$$

where

Z_T = The total impedance of the loudspeaker circuit
$Z_1 \ldots Z_N$ = The impedance of each loudspeaker

Figure 9-35
Three loudspeakers wired in series

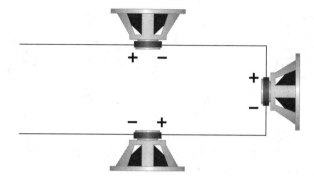

Figure 9-35 shows three loudspeakers wired in series (that is, plus to minus to plus to minus). If you want to calculate the total impedance of this loudspeaker line, simply add the impedances of the individual loudspeakers as you would any series circuit.

If each of these loudspeakers has a nominal impedance of 8 ohms, the total impedance would be 24 ohms. You simply add the impedances together in a series circuit: 8 ohms each times 3 equals 24 ohms.

If, for example, you had four loudspeakers wired in series, each with a nominal impedance of 4 ohms, again you would add the impedances together and find a total impedance of 16 ohms. That's four loudspeakers at 4 ohms each.

Parallel Circuit, Same Impedance Formula

Another method of connecting loudspeakers in a circuit is to wire them in parallel. This means the positive output of the amplifier connects to every loudspeaker's positive terminal, and each loudspeaker's negative terminal connects to the amplifier's negative terminal. Loudspeakers wired in parallel may have the same impedance or differing impedances.

Parallel Circuit Impedance Formula:
Loudspeakers with the Same Impedance

This is the formula to calculate the circuit impedance for loudspeakers of the same impedance, wired in parallel:

$$Z_T = \frac{Z_1}{N}$$

where

Z_T = The total impedance of the loudspeaker system
Z_1 = The impedance of each loudspeaker
N = The number of loudspeakers in the circuit

Figure 9-36
Three
loudspeakers
wired in parallel

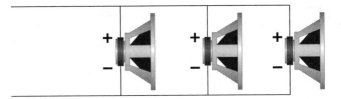

If all of the loudspeakers are of the same impedance (as is often the case in loudspeaker systems), the impedance of the loudspeaker divided by the number of loudspeakers wired in parallel equals the impedance of the circuit.

Figure 9-36 shows three loudspeakers wired in parallel. If each loudspeaker is rated at 4 ohms, the circuit's impedance is 1.33 ohms: $Z_T = 4 \div 3 = 1.33\Omega$.

Designers usually specify identical values. It is then easy to take the number of loudspeakers and divide into representative values. For example, three 8-ohm speakers will result in $8 \div 3 = 2.67\Omega$.

Parallel Circuit, Differing Impedance Formula

Use the parallel circuit impedance formula to calculate the impedance of loudspeakers of differing impedance wired in parallel.

Parallel Circuit Impedance Formula: Loudspeakers with Differing Impedance

This is the formula to calculate the circuit impedance for loudspeakers of differing impedance wired in parallel:

$$Z_T = \frac{1}{\dfrac{1}{Z_1}+\dfrac{1}{Z_2}+\dfrac{1}{Z_3}\cdots\dfrac{1}{Z_N}}$$

where

$Z_1 \ldots Z_N$ = The impedance of each individual loudspeaker
Z_T = The total impedance of the loudspeaker circuit

If you have three loudspeakers wired in parallel, with the first rated at 4 ohms, the second rated at 8 ohms, and the third rated at 16 ohms, the circuit's impedance is 2.29 ohms: $Z_T = 1 \div [(1/4)+(1/8)+(1/16)]$.

Series/Parallel Combination Formula

In a series/parallel loudspeaker circuit, groups of loudspeakers called *branches* are wired together in series. Typically, loudspeakers in the same branch have the same impedance and are wired together in parallel, as shown in Figure 9-37.

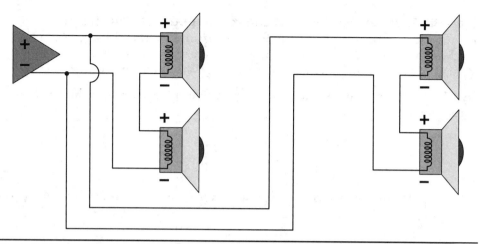

Figure 9-37   Loudspeakers wired in a combination circuit

Loudspeakers may be wired in a combination circuit, with several series branches wired together in parallel. This is also known as a *series/parallel circuit*.

To calculate the total impedance of a series/parallel circuit, calculate the total impedance of each branch using the series circuit impedance formula. Then calculate the total circuit impedance of the circuit using the parallel circuit impedance formula.

Series/Parallel Circuit Impedance Formula

The formula to calculate the expected total impedance of a parallel circuit may also be used to calculate the total impedance of a series/parallel circuit:

$$Z_T = \frac{1}{\dfrac{1}{Z_1} + \dfrac{1}{Z_2} + \dfrac{1}{Z_3} \cdots \dfrac{1}{Z_N}}$$

where

$Z_1 \dots Z_N$ = The total impedance of each circuit branch
Z_T = The total impedance of the loudspeaker circuit

This formula is used *after* the series portions of the circuit have been calculated.

It is rare to encounter loudspeakers wired in a series/parallel combination in the field. Although the idea of implementing a series/parallel combination would be to present a proper load to the output of the power amplifier, such systems may be complex to troubleshoot, requiring separate isolation and testing of each branch.

Example 1: Series/Parallel Loudspeaker Impedance Calculation

You want to verify that the audio amplifier's impedance matches the specifications indicated by the system designer. In the meeting room there are six 8-ohm loudspeakers wired in two branches of three loudspeakers each. What is the impedance of the circuit?

Step 1 Calculate the impedance of each of the similar branches using the series circuit impedance formula.

$$Z_T = Z_1 + Z_2 + Z_3 \dots Z_N$$
$$Z_T = 8 + 8 + 8$$
$$Z_T = 24 \text{ ohms for each branch}$$

Step 2 Calculate the total circuit impedance using the parallel circuit impedance formula for the two similar branches.

$$Z_T = Z_1 \div N$$
$$Z_T = 24 \div 2$$
$$Z_T = 12 \text{ ohms}$$

Answer The total impedance of this loudspeaker circuit is 12 ohms.

Example 2: Series/Parallel Loudspeaker Impedance Calculation

What is the impedance of a series/parallel circuit with two branches of three 8-ohm loudspeakers and one branch of four 16-ohm loudspeakers?

Step 1 Calculate the impedance of each branch using the series circuit impedance formula.

$$Z_T = Z_1 + Z_2 + Z_3 \dots Z_N$$
$$Z_T = 8 + 8 + 8$$
$$Z_T = 24 \text{ ohms}$$

Each branch of 8-ohm loudspeakers has a total impedance of 24 ohms.

$$Z_T = Z_1 + Z_2 + Z_3 \dots Z_N$$
$$Z_T = 16 + 16 + 16 + 16$$
$$Z_T = 64 \text{ ohms}$$

The branch of 16-ohm loudspeakers has a total impedance of 64 ohms.

Step 2 Calculate the total circuit impedance using the parallel circuit impedance formula.

$$Z_T = 1/[(1/Z_1) + (1/Z_2) + (1/Z_3)]$$
$$Z_T = 1/[(1/24) + (1/24) + (1/64)]$$
$$Z_T = 1/0.09895833\dots$$
$$Z_T = 10.10526\dots \text{ ohms}$$

Answer Rounded to one decimal place, the total impedance of this loudspeaker circuit is 10.1 ohms.

Chapter Review

In this chapter, you studied the basics of audio signal routing and components used in an audio system. You learned about the key factors involved in audio system installation and how to install and wire microphones and loudspeakers.

Upon completion of this chapter, you should be able to do the following:

- Identify a series or parallel wiring scheme in a loudspeaker arrangement
- Identify the causes of feedback in an audio system
- Compare and contrast direct-coupled and constant-voltage loudspeaker systems
- Apply loudspeaker taps and transformers to a loudspeaker
- Calculate loudspeaker impedance for loudspeakers wired in a parallel circuit with no errors

You have learned about the installation of audio components. Equally important are the setup and verification processes discussed in Part V.

Review Questions

The following questions are based on the content covered in this chapter and are intended to help reinforce the knowledge you have assimilated. These questions are similar to the questions presented on the CTS-I exam. See Appendix E for more information on how to access the free online sample questions.

1. What unit is used to express decibel changes in voltage in professional audio equipment?

   A. dB SPL

   B. dBu

   C. dBV

   D. dBm

2. Noise caused by circuit interference can be reduced by ___.

   A. using balanced shielded twisted-pair cable

   B. muting the microphone

   C. distancing the microphone from the loudspeaker

   D. extending the cable run to another room

3. The difference between the loudest and quietest levels of an audio signal is the ___.

   A. volume level

   B. stereo level

   C. dynamic range

   D. frequency range

4. Which of the following is mic-level voltage?

   A. 316 millivolts (–10dBV)

   B. 1.23 volts (+4dBu)

   C. Up to about 100 volts

   D. 0.3 to 1 millivolts (–60 to –50dBu)

5. What does a microphone's polar pattern indicate?

   A. Sound sensitivity in different directions

   B. Performance in cold temperatures

   C. Peak noise level

   D. 6dB drop-off at 270 degrees

6. Which of the following are transducers?

   A. Expanders

   B. Filters

   C. Gates

   D. Loudspeakers

7. What are typical applications of a distributed loudspeaker system? (Choose all that apply.)

   A. Paging system at an airport

   B. Background music in a restaurant

   C. Public address system in an office building

   D. Subwoofers in a concert hall

8. The red wire in a loudspeaker cable carries a ___ signal.

   A. ground

   B. negative

   C. positive

   D. neutral

9. Which loudspeaker system is represented in the following picture? The loudspeakers are shown as circles.

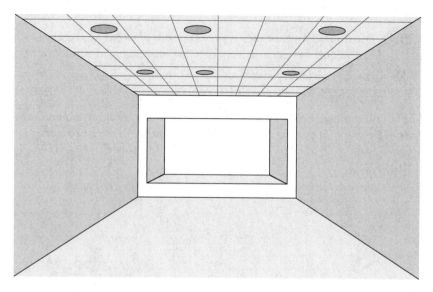

    A. Distributed system

    B. Center-cluster system

    C. Point-source system

    D. Portable system

10. What is the expected impedance of the loudspeaker circuit shown in the picture?

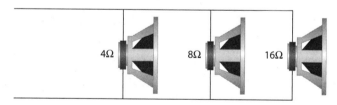

Answers

    1. B. The unit dBu is used to express decibel changes in voltage in professional audio systems.

    2. A. Using balanced shielded twisted-pair cable, you can reduce the noise caused by circuit interference.

    3. C. Dynamic range is the difference between the loudest and quietest levels of an audio signal.

    4. D. Mic-level voltage is 0.3 to 1 millivolt (–60 to –50dBu).

5. **A.** The polar pattern of a microphone indicates its sound sensitivity in different directions.

6. **D.** Loudspeakers are transducers that convert electrical to acoustic energy.

7. **A, B, C.** Distributed loudspeaker systems are not ideal for applications where extended low-frequency response, very high power, and low distortion are priorities.

8. **C.** The red wire in a loudspeaker cable is positive.

9. **A.** The picture shows a distributed loudspeaker system.

10. The circuit's impedance is 2.29 ohms. The three loudspeakers shown in the picture are wired in parallel, with the first rated at 4 ohms, the second rated at 8 ohms, and the third rated at 16 ohms. Use the formula $Z = 1 \div [(1/4) + (1/8) + (1/16)] = 2.29$.

Video Systems Installation

10

In this chapter, you will learn about
- Digital and analog video signals
- Digital video requirements
- Video components
- Video wiring schemes
- Installing video components

The transition from standard-definition (SD) to high-definition (HD) and ultra-high definition (UHD) video and the convergence of digital video with information technology (IT) have increased the use of video across all industries. The demand for higher-quality video and easier-to-use systems has also grown, resulting in more UHD, high frame rate (HFR), high dynamic range (HDR), and wide color gamut (WCG) system installations.

Digital video has forged a closer working relationship between audiovisual (AV) and IT professionals. Therefore, you will need to be more mindful of terms that may sound similar but have different meanings. For example, the acronym MAC is used for *media access control,* so if you hear "MAC," do not assume it refers to a Macintosh computer or macaroni.

The convergence of video, audio, and data in AV-IT applications also introduces challenges regarding the interoperability of HD video components and computers. Although interoperability issues are usually addressed in the design phase, installers are sometimes called on to troubleshoot computer-video problems. For this and other reasons, it is a good idea to know as much as possible about the differences and similarities in the ways HD and UHD video and computer systems handle signals, control, and connectivity.

Additionally, even though the AV industry has largely transitioned to digital formats, installers should be able to integrate legacy analog systems with newer digital components. This is necessary because the components in large systems are not always replaced or upgraded simultaneously. Also, most audio transducers such as microphones and speakers are analog devices and still have analog connections.

Ensuring appropriate signal levels across all AV equipment, properly adjusted cameras and displays, and correctly positioned projectors and display screens are all part of the responsibilities of the AV installation team.

In this chapter, you will learn the basics of video signals and systems, with a focus on digital video. You will also learn how to install video source components such as cameras, projectors, projection screens, and flat-panel displays. In Part V of this book, you will learn about the setup and verification methods to ensure AV signal integrity.

Duty Check

This chapter relates directly to the following tasks on the CTS-I Exam Content Outline:

- Duty A, Task 5: Conduct On-Site Preparations for Installation
- Duty B, Task 2: Mount Substructure
- Duty C, Task 4: Mount Audiovisual Equipment

These tasks comprise 12 percent of the exam (about 12 questions). This chapter may also relate to other tasks.

Video Signals

A video clip or sequence is really a series of still images or frames. A video signal contains the information needed to display a complete video image. When a camera captures an image, the light focused on the camera's imager is captured as separate red, green, and blue (RGB) signals. These color signals must be combined with synchronizing information called *sync*. Sync preserves the time relationship between video *frames* and correctly positions the image horizontally and vertically. This information is called *horizontal sync* and *vertical sync* (H and V). Horizontal sync determines the point where the line of pixels end on the right edge of the image. Vertical sync pulses are used to determine the top and bottom of the screen. Hence, a video signal consists of five basic kinds of information: R, G, B, H, and V.

Video Terms and Concepts

When installing AV devices, you will need a clear understanding of the video specifications that are listed with each product. These specifications will help you ensure that products you will be installing are compatible with each other. For instance, two pieces of equipment that specify 1080/60p video with HDMI 2.0 connectivity will generally work well together. Format names and related specifications like these are useful for determining compatibility and can also give a general idea as to picture or sound quality.

One common mistake, however, is to strictly equate picture quality with format specifications. For example, a 1080/60p professional AV monitor may well have superior picture quality when compared with a 4K/60p display intended for desktop computer use.

Any one of several factors, including the compression ratio of the signal along the transmission chain, can also affect picture quality. For example, a 16Mbps network video streaming service signal at 4K/60p will almost certainly have less horizontal resolution, more compression artifacts (often visible as distortion in fast-moving objects), and less color detail and depth than a 1080/60p signal played back from a local media server with a bandwidth of 100Mbps or more.

Aspect Ratio

The aspect ratio of an image describes its width and height proportions, stated as the ratio of width to height (W:H). The aspect ratio of all standards-based HD and UHD video is 16:9 (1.77:1), although some computer video outputs are 16:10 (1.6:1). Digital Cinema (DCI) cinemascope images have an aspect ratio of 2.39:1, while many personal digital devices tend toward a portrait format aspect ratio of 1:2. Legacy CRT television images and computer displays mostly have a "square" aspect ratio of 4:3 (1.33:1).

Frame Rate

Every video signal for display, whether it is analog or digital, is sent as a series of still images called *frames*. The way those frames are reproduced varies. The number of frames sent per second is referred to as *frame rate*. Frame rates are stated in frames per second (fps) or as a *frame frequency* in hertz (Hz). For instance, a 1080p HD video stream may have a frame rate of 60fps or 60Hz. That means it will contain 60 frames of video per second.

The standard frame rate for sound films was established at 24fps and that remained the commercial cinema standard until cinema projection went digital. For simplicity of sync-signal generation, the frame rate of television was adopted at 25fps in countries with 50Hz mains power and 30fps in countries with 60Hz mains power. Some scanning gymnastics are required to scan 24fps cinema film for screening on 25fps and 30fps video systems.

While cinema experimented with 48fps for features such as Peter Jackson's *The Hobbit Series*, higher frame rates for smoother motion stability are mostly found in digital video, where frame rates of 48, 60, 96, and 120fps are common. Frame rates above 24/25/30fps are generally referred to as HFR video.

Each picture in Figure 10-1 represents a frame. If the images were played in quick succession, they would show a skilled AV tech rolling up a kink-free cable. The frame rate tells you how many of these frames would play during a specified time interval (usually one second). The more frames per second you display, the smoother the video will appear. Frame rate is not the same thing as refresh rate.

Figure 10-1   Image frames in a video sequence

Refresh Rate

The number of times per second a display will draw the image sent to it is the *refresh rate*. Displays often use scaling circuitry to match the frame rate of a source. A display's refresh rate should be greater than or equal to the frame rate of the signals sent to it. The specifications for a display device should include a detailed list of the source refresh rates that it supports.

NOTE Frame rate is the number of frames per second sent from a display source. Refresh rate is the number of times per second a display will draw the image sent to it.

Interlaced Scanning

Interlaced scanning combines odd and even lines of video to produce a full frame of video. As shown in Figure 10-2, this process draws all the odd lines of a video signal from top to bottom and then draws all the even lines. Each set of lines is called a *field*. Two fields create a *frame*. If you watch an interlaced video image closely, you may be able to see a motion artifact called *interlace flicker* as the fields are scanned. Interlaced scanning is mostly used where video bandwidth is limited.

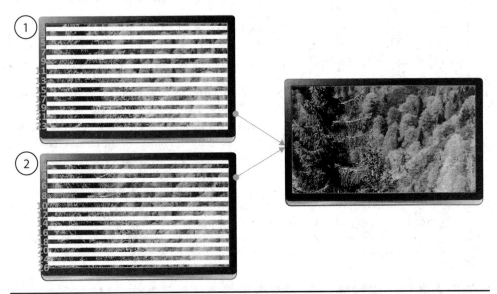

Figure 10-2 Interlaced scanning traces all the odd lines first and then all the even lines to create a video frame

Figure 10-3
Progressive
scanning draws
the image's scan
lines sequentially
from top to
bottom

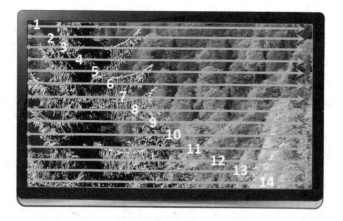

Progressive Scanning

Progressive scanning draws all the image lines in order, from top to bottom, as shown in Figure 10-3. The progressive scanning method eliminates artifacts or interlace flicker commonly seen in interlaced scanning. However, because it transports the whole frame at once, it requires more bandwidth than interlaced scanning.

As an installer, why should you care about scanning and frame rates? You need to be able to identify the frame rate, refresh rate, and scan type of sources and displays used on the project. Audiovisual equipment uses a range of different high-definition video format standards. These different standards have compatibility considerations. Therefore, you must take care to use the frame, refresh, and scan rates specified by the AV designer.

High-Definition Video Formats

The image resolution of a video signal is based on the type of scanning, the number of lines, and other factors. The video format generally refers to screen or display resolution and frame rate, but also applies (with some differences) to cameras and other components such as graphics generators and video-processing equipment.

Here are the most common high-definition video signal and display formats available today:

- **720p** This is an HDTV signal format. The number 720 represents 720 horizontal lines, and the *p* indicates that it uses progressive scanning. The aspect ratio is 16:9, and the resolution is 1280×720. It typically delivers approximately 60 fields per second (720p60 or 720/60p) in North America and 50 fields per second (720/50p) in many other regions. A 30 field per second rate (720/30p) is found in some video conferencing systems.

- **1080i** This has a resolution of 1920×1080 and uses interlaced scanning at 25 or 30 frames per second. A single frame of 1080i video has two sequential fields of 540 lines of 1,920 horizontal pixels. The first field contains all the odd-numbered lines of pixels, and the second field contains all the even-numbered lines. In areas such as Europe, Australia, much of Asia, Africa, and part of Latin America, 1080i television signals are broadcast with 50 fields (1080/50i). In North America, Japan, and most of Latin America, this format is broadcast with 60 fields per second (1080i60 or 1080/60i).

- **1080p** This has a resolution of 1920×1080, and it uses progressive scanning at 60 frames per second. Typically, everything you install, from displays to switching equipment, should be at least 1080p compatible to simplify image setup and maximize display quality. Many sources, such as Blu-ray Disc players and computer systems, output at 1080p. It is sometimes referred to as *full HD*.

- **Ultra HD 4K** This has a minimum resolution of 3840×2160 pixels in a 16:9 (1.77:1) aspect ratio. Originally used in digital cinema, DCI 4K video resolution is 4096×2160 in a 1.9:1 full frame format, 3996×2160 for 1.85:1 conventional wide screen, and 4096×1716 for 23.9:1 CinemaScope screenings. The related video resolution consumer television format is known as 2160p, *4K Ultra HD*, and *UHDTV-1*, which specifies 3840×2160 pixels, at a 16:9 aspect ratio.

- **Ultra HD 8K** This has a minimum resolution of 7680×4320 pixels in a 16:9 aspect ratio, exactly double the width and double the height of 4K. The related video resolution consumer television format is known as 4320p, *8K Ultra HD*, and *UHDTV-2*.

Video formats usually include rounded frame numbers such as 60p. But in many cases, the precise frame rate is 59.94 (or 29.97) per second. In practice, this usually makes no difference because most common displays and other equipment tolerantly accept all typical rates. However, occasionally the transfer of video is interrupted because of small frame rate or related differences between equipment.

Analog Video

Analog is a method of transmitting information by a continuous but varying signal. Video was transmitted as an analog signal for many decades, but has now been replaced by digital transmission. Analog-capable equipment remaining in use will most likely be connected to a digital-to-analog conversion device that accepts digital video inputs or will be a transition device with both analog and digital inputs.

Analog Video Signal Types

The video connectors shown in Figure 10-4 are those still in current use in some applications. The signals they carry are

- **RGBHV** This is considered the highest-quality analog signal. These signals are typically used for computer and signal processors. They are referred to as "full-bandwidth RGB with separate sync," and they include red, green, blue, horizontal, and vertical sync. The RGBHV signals are carried on separate pins on the HD15 (VGA) connector and the dedicated analog pins on the DVI-I and DVI-A connectors.

Figure 10-4
Typical connectors used for analog video

- **Composite video** These video signals combine all aspects of a video signal on a single cable, and some compromises are necessitated by this process. This composite analog signal is available only in standard definition, and the quality is substantially lower than RGBHV. Consumer and some low-end professional equipment use a single RCA or phono connector for composite video connections, while some professional video equipment uses a BNC connector this purpose.

As you probably noticed, Figure 10-4 also shows a radio frequency (RF) connector that is used in the distribution of combined video and audio signals. This connector is used for video distribution to multiple TVs and other devices. It typically uses an *F*-type screw-on coaxial connection. Some RF distribution systems deliver HDTV content to appropriately equipped television sets using terrestrial digital video broadcast transmission standards.

As shown in Figure 10-5, a custom input faceplate enables the display of either analog or digital signal.

Although most devices will have exclusively digital outputs, you need to be familiar with legacy analog devices. The use of owner-furnished legacy equipment will ensure that a diminishing amount of existing analog equipment may remain in use in the field. You should be prepared to work with analog signals alongside digital.

Figure 10-5
A custom display faceplate with analog and digital inputs

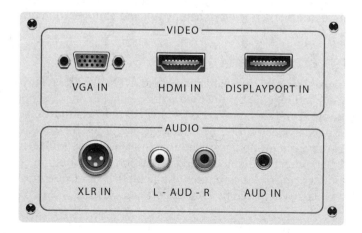

Digital Video

In the digital video world, varying signals are represented by a series of numbers based on the signal level at particular times. Those numbers are stored and transmitted as streams of binary digits (1s and 0s), where the signal is either a 1 (on) or a 0 (off).

As digital video formats are merely strings of numbers, they are capable of carrying more than just the video signals. Some formats include embedded audio, time code, and other data. Some common digital AV transport mechanisms also carry control signals, general-purpose data streams, and DC power. These aspects of digital video are discussed later in this chapter.

Digital Video Signal Types

Uncompressed digital video signals exist in several forms; these forms can contain information in any one of a variety of formats (for example, 1080/60i or 2160/30p) and can be delivered via any one of several transport carriers such as HDMI, DVI, DisplayPort, HDBaseT, AVoIP, SDVoE, or SDI.

Serial Digital Interface (SDI) is a set of serial data standards developed by the Society of Motion Picture and Television Engineers (SMPTE) to transport digital video data over the BNC-terminated 75Ω coaxial cable originally used for professional analog video, although it is also frequently carried over optical fiber. SDI data includes audio and control information, but not content protection.

AV over IP can be either a full-bandwidth or compressed transport stream, depending on the codec (encoder/decoder) employed. As TCP/IP is a serial data stream, all video must at the very least be serialized from the original RGB and sync channels before transmission. Many hardware and software systems are available to convert an uncompressed video format such as HDMI for transmission over IP, to be decoded into the original uncompressed format for display or recording. The resolution and frame rate of the original video, together with the available bandwidth of the IP stream, will dictate how much (if any) compression is required by the codecs at the endpoints of the IP network. Some systems such as software-defined video over Ethernet (SDVoE) allow on-demand reservation of Ethernet network bandwidth and configuration of the transport codecs.

Video Transport

Video signal transport mechanisms such as HDMI, DisplayPort, and Thunderbolt are capable of handling much more than a video signal in one direction. They may also carry audio, control, and DC power. DVI is a video-only transport mechanism.

Most digital video transport technologies are licensed technologies and belong to different commercial groupings or industry organizations. HDMI is widely used in both home entertainment equipment and professional AV equipment, while DisplayPort is more commonly used for computer displays. DVI was frequently used in computers, but it has been largely replaced by DisplayPort and HDMI.

Here are some common video signal carriers you will encounter during AV installations:

- **HDMI**   High Definition Multimedia Interface is an advanced method for interconnecting digital sources to digital displays with the addition of audio, data, and control functions. HDMI 2.1 has a maximum bandwidth of 48Gbps. Since it can be used in a wide variety of AV applications, it has become the most often used connection type for direct video signal transmission over a distance up to 5 meters (16 feet). Technologies such as HDBaseT, AVX, and AVoIP are frequently used to extend the reach of HDMI.

- **DVI**   Digital Video Interface is a connection method between a source device (typically a computer) and a display device. It is a direct method for transferring data. However, the digital signal can travel only up to 5 meters (16 feet). As shown in Figure 10-6, there are many different connectors available for DVI signals, including DVI, DVI-D, DVI-I, and mini DVI. The two multipin connectors for DVI signal transport are DVI-D for only digital information (no analog video information can be sent) and DVI-I for digital or analog information. As DVI and HDMI share the transition minimized differential signaling (TMDS) link protocol, DVI-D interfaces are compatible with HDMI interfaces through the use of simple passive adapters. DVI usage is on the decline.

- **DisplayPort**   Many computers and display devices are equipped with a DisplayPort connection. This Video Electronics Standards Association (VESA)–developed connection method utilizes a 20-pin, high-resolution, high-speed digital display, audio, and data interface with up to 77.4Gbps effective bandwidth in version 2.0. It is typically used to connect a computer to one or two displays and is backward compatible with DVI, HDMI, and other current

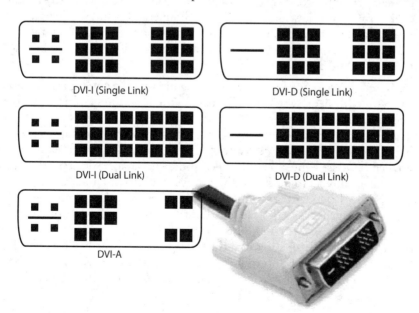

DVI-I (Single Link)

DVI-D (Single Link)

DVI-I (Dual Link)

DVI-D (Dual Link)

DVI-A

Figure 10-6   Many different connectors are available for DVI signals

PART IV

Figure 10-7
DisplayPort and
Mini DisplayPort/
Thunderbolt
2 connectors
(Images: *Left*,
Gudella / Getty
Images; *right*,
Beeldbewerking /
Getty Images)

display interfaces through the use of appropriate passive or active adapters. The Mini DisplayPort, as shown in Figure 10-7, is a smaller variant of this interface, often found on thin form-factor portable devices. DisplayPort has native support for fiber-optic cables and support for standard cables up to 15 meters (50 feet) in length at lower resolutions. It also supports high-bandwidth digital content protection (HDCP) signals and bi-directional communication.

- **Thunderbolt** Thunderbolt is a technology developed by Intel and Apple that can transfer high-speed data, video, audio, and DC power over one cable in two directions. It also permits a user to connect up to six displays, hard drives, and other compatible devices via a daisy-chain connection or hub. Most often found on Apple computer equipment, Thunderbolt combines DisplayPort and PCI Express (high-speed serial computer expansion bus) capabilities into one serial data signal. Thunderbolt versions 1 and 2 use a connector electrically identical to Mini DisplayPort shown in Figure 10-7. Thunderbolt 3 and 4, which use a USB Type-C connector, have a maximum bandwidth of 40Gbps. The Thunderbolt 3 and 4 multistream transport (MST) capability allows USB 3 or 4 or DisplayPort signals to be embedded in its data stream.

- **HDBaseT** A variety of devices, including high-resolution displays, projectors, video source devices, switchers, and matrix switchers, can be connected to each other over distances of up to 100 meters (328 feet) via a single UTP cable (Cat 5e+) using HDBaseT devices and technology. HDBaseT is a proprietary protocol that has been licensed to a multitude of AV equipment manufacturers who have installed HDBaseT interfaces into their products. Adapter devices are readily available to connect to devices without built-in HDBaseT interfaces. HDBaseT carries HDMI-standard uncompressed 2K/4K video, as well as audio, USB, and control signals. It is also capable of carrying up to 100 watts of DC power, making it unnecessary to plug some HDBaseT displays and other end nodes into the wall. HDBaseT can also carry 100Mbps Ethernet, which allows HDBaseT devices to access content on connected computers. All devices in the HDBaseT signal chain must be capable of transmitting and/or receiving HDBaseT. It has been designated as the IEEE 1911 standard.

- **Ethernet**   Ethernet is an Institute of Electrical and Electronics Engineers (IEEE) standard that defines how hosts exchange information over a local area network (LAN). As Ethernet is a very low-level protocol, for machine-to-machine data delivery, there are many protocols that are built on top of Ethernet for the delivery of specific data types such as audio and video. Most data transport for video, audio, and control is via the TCP/IP protocol family, which uses the Ethernet LAN to deliver raw data. Ethernet can be operated over a variety of media, including UTP (Cat 5+), optical fiber, and wireless using the IEEE 802.11xx protocols (Wi-Fi). You will find additional information on Ethernet in Chapter 12.

- **SDVoE**   Software Defined Video over Ethernet is a TCP/IP-based protocol stack that provides low-latency transport of video, audio, control, and other AV signals over a standard 10Gbps Ethernet network using standard cabling, switching, and routing hardware. In addition to signal transport, the associated software can perform video scaling, image synchronization, and multi-image scaling and processing.

- **SDI**   Serial Digital Interface is a family of connection methods for local transport up to 100 meters (328 feet) of high-quality, uncompressed, unencrypted, and standardized digital video over a coaxial cable (75 ohm) with a BNC connector. It is used in live production, broadcast, image magnification (IMAG), staging, and video conferencing applications, although its lack of support for copy-protected content can be a significant problem. Many SDI video processors include the capability for audio channel swapping and signal control. SDI has low latency (delay) and is excellent where lip sync is an issue. SDI is often considered better suited than HDMI for connection directly to recording devices because it has a locking connector and will transport time code. It is often used for infrastructure wiring in professional production and presentation environments.

  The SMPTE has established standards for several SDI formats, as shown in Table 10-1.

- **Fiber-optic cabling**   Fiber-optic cabling offers high bandwidth and maintains total electrical isolation, making it virtually immune to EMI and RFI. Fiber optics are able to maintain signal integrity in electrically noisy environments, experiencing little signal degradation over long distances. It is relatively durable and resistant to the effects of aging and corrosion. Fiber-optic lines are also useful for security because they produce no stray electrical fields that can be tapped.

Format Name	Max Data Speed	SMPTE Standard	Application
SDI	270Mbps	SMPTE 259M	SD (480i, 576i)
HD-SDI	1.5Gbps	SMPTE 292M	HD (720p, 1080i)
3G-SDI	3Gbps	SMPTE 424M	HD (1080p60)
6G-SDI	6Gbps	SMPTE ST 2081	UHD (1080p120, 2160p30)
12G-SDI	12Gbps	SMPTE ST 2082	UHD (2160p60)
24G-SDI	24Gbps	SMPTE ST 2083	UHD (2160p120, 4320p30)

Table 10-1   Single-Link SDI Formats

PART IV

It is therefore much more difficult to covertly capture data from a fiber-optic line than from copper wires or radio waves.

A common misconception about fiber is that it has no distance limitation. There are distance limitations depending on which "mode" you are using, as well as the design and composition of the cable. Multimode fiber has multiple light paths, which limits its distance. Single-mode fiber has a single light path, which allows it to travel farther, yet still has a limit. Optical fibers are covered in more detail in Chapter 6.

- **USB** Universal Serial Bus is a general-purpose, bi-directional serial digital data and power bus that can carry any form of digital data, including serialized audio and video data streams, at rates up to 40Gbps. It is terminated in a range of standardized connectors (Types A, B, and C). Small form-factor versions (mini and micro) of the Type A and Type B connectors are also in use on some portable devices. Since USB version 3.2, the Type C USB connector can be switched into an Alternate (Alt) signaling mode that allows it to be configured to carry a variety of digital signals, including PCI Express and DisplayPort. While USB may be limited to transmission distances of less than 5 meters (16 feet), there are USB-to-Ethernet adapter systems that use TCP/IP to allow USB control to be extended to any point on the Internet.

Digital Video Bandwidth Requirements

Digital video bandwidth requirements have a direct impact on the final signal quality. Exact bandwidth requirements will depend on resolution, frame rate, bit depth, the sampling method, the type of compression employed, and other factors. Video can be sampled at different rates and compressed using different methods to reduce the overall amount of data to be stored or transmitted.

The sampling frequency of a digitized signal is expressed as a ratio of luminance (Y) to color (Cb and Cr) sampling (Y:Cb:Cr). For example, a 4:1:1 sampling describes luminance as being sampled four times at 3.37MHz and the color component as being sampled one time each at 3.37MHz. Other sampling ratios are 4:2:2 and 4:4:4. Common methods of compressing video include MPEG2, MPEG4, Motion JPEG (MJPEG), AOMedia Video 1 - AV1 (AOM AV1), Advanced Video Coding - AVC (H.264 or MPEG-4 Part 10), High Efficiency Video Coding - HEVC (H.265 or MPEG-H Part 2), Versatile Video Coding - VCC (H.266 or MPEG-I Part 3), and Essential Video Codec - EVC (MPEG-5 Part 1).

If the signal is too big for the available bandwidth, the system will not work. As an installer, you will be unlikely to change the bit depth, sampling method, or compression modality. You may be able to adjust the frame refresh rate, screen resolution, compression ratio, and related factors. As always, check any variation with the system designer/ engineer, and be sure to document your changes and the system settings.

Signal Loss over Distance

Attenuation of a signal is directly related to the length of the cable over which it is traveling. This attenuation must be taken into consideration if an install calls for high-bandwidth signal flow. If a digital signal is carried too far on a cable, the attenuation becomes so great that the signal becomes entirely unreadable. This is called the *cliff effect*.

The edge of the "cliff" will vary based on the quality of the cable and the rate of the data. Signals with a higher bandwidth requirement cannot run as far as signals needing a lower bandwidth. For example, a 1080p signal might run only 20 meters (66 feet), while a 720p or 1080i signal might run as far as 40 meters (132 feet) on the same cable.

When selecting a cable, you need to consider the total bandwidth of the data being transported. Some video will require a higher-bandwidth cable if the video has a high frame rate (HFR), a high dynamic range (HDR), or a wide color gamut (WCG). To future-proof an AV installation, you may want to choose a higher-quality cable than the minimum for the specified bandwidth requirement. The cable you use to transport a 1080p signal today will likely not work for the 2160p signal your customer may want to use in the future.

HDMI Repeaters

When an HDMI cable is not long enough to connect two devices, you could simply use an HDMI end-to-end coupler to connect two HDMI cables and extend the distance. While this may seem like an easy fix, and each of the joined cables on their own will not attenuate the signal beyond usability, together they may just have too much attenuation to support the signal. Therefore, instead of an HDMI coupler, you could install an HDMI repeater. As shown in Figure 10-8, the original digital signal is degraded along the path, and the signal is restored after a repeater is used.

Figure 10-8
An HDMI repeater restores an attenuated digital signal

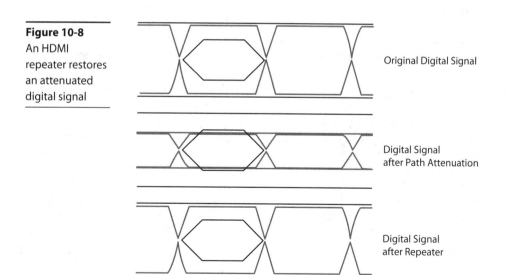

Original Digital Signal

Digital Signal after Path Attenuation

Digital Signal after Repeater

HDMI repeaters perform two functions:

- They join two different HDMI cables.
- They boost the signal to counteract signal degradation.

With an HDMI repeater, you can exceed 5-meter (16-foot) cable lengths without signal loss.

Consider using a repeater in these instances:

- **Single-input, single-output devices** These are used primarily for cable extension.
- **Multiple-input, single-output devices** These are used primarily to switch among multiple sources.
- **Single-input, multiple-output devices, where only one output is active** These are used to distribute among multiple displays or signal sinks (a device that displays the video).

In all cases, each HDMI input will fulfill all of the requirements of an HDMI destination (or sink) when it is connected with an active sink device.

An alternative to simple HDMI repeaters is to use repeaters, switch matrixes, and splitters based on HDBaseT, AVX, and SDVoE technologies, which allow for much longer runs over both cable and optical fiber and simpler distribution architectures.

Video System Components

Video components for installation range from image capture and processing to projection and display. Video source devices include cameras, media recorders/players, caption and graphics generators, media servers, and streaming or broadcast signal receivers. Video processors include a variety of devices such as switchers, scalers, effects generators, extenders, and converters. Video display components include projectors, projection screens, LED screen modules, and flat-panel displays.

Your installation may involve several of the video components discussed here as well as many others. The essential information for each product is contained in the operation or installation manuals and, to a lesser extent, in the specification sheets or brochures. If the required manuals are not included in the equipment delivered to the installation site, you can often find PDF versions on manufacturers' websites. Web searches are your friends. Your up-to-date knowledge of product capabilities will be helpful in problem-solving and troubleshooting during installation. For example, many lampless projectors can be side or vertically mounted where the installations call for portrait-aspect ratio or floor projection. In this case, you would want to check the manual and manufacturer's warranty for that specific model to see whether the warranty covers the required mounting angle. You should also be aware of what adjustments you will be able to make with certain components during setup and verification to achieve the desired results.

Cameras

Video cameras may be either a one-piece, integrated pan/tilt/zoom (PTZ) camera system with zoom lens and pan-tilt mechanism; a box-type camera with separate zoom lens and pan-tilt mechanism; an all-in-one camera unit; or a tripod-mount camera with either interchangeable or integrated zoom lens.

PTZ cameras are capable of remote directional, zoom, and picture control. They are used in meeting rooms and training centers for video conferences, in auditoriums for image magnification, and to record events. They are typically linked to control systems via serial connections or via a LAN. Figure 10-9 shows a mounted PTZ camera with cables for connection to the network. Many PTZ cameras incorporate auto-focus, auto-iris, auto-white balance, and other automatic adjustments. Some PTZ systems also incorporate automatic tracking, enabling the camera to follow the movement of a selected subject, such as the teacher in a classroom or the current speaker in a video conference.

High-end cameras typically have either three sensors, one each for RGB, or a single high-resolution sensor with a matrix of R, G, and B filters. They usually output either XX-SDI, video over IP, or HDMI. Utility or less expensive cameras usually have a single sensor and may output HDMI, video over IP, or video over USB. Two common measurements reflecting camera quality are horizontal resolution (for example, 720 lines) and *signal-to-noise ratio*. The latter indicates the ratio between the level of the captured image and the electronic noise introduced by the camera pick-up and processing system. It is expressed in decibels, for example, S/N 55dB, where higher numbers indicate a cleaner or better-quality signal.

Figure 10-9

Auto-tracking PTZ camera systems can follow a moving subject or adjust to focus on the current speaker in a meeting

Lighting

Lighting for video can be a challenge because the contrast in the image must not exceed the contrast ratio of the camera's imaging device. If the dynamic range of the device is exceeded, signal clipping can occur, which may cause objects to appear as a "white hole" in the image. If the light levels are insufficient for an adequate exposure, noise can be introduced into the image when the automatic gain system attempts to compensate. Light sensitivity and dynamic range tend to increase with succeeding generations of image sensors.

Color Temperature

White light is produced when a light source produces a mix of the main colors in the visible spectrum. All types of light source produce different distributions of the visible colors, so many types of white light have a different tint—some are actually reddish white, and others are bluish white. This difference in balance of the spectral colors is referred to as a light source's *color temperature*.

Color temperature is the scientific measurement for expressing the distribution of the spectral colors radiating from a light source, expressed in Kelvin (K). As the Kelvin measure is a reference to the colors emitted from a theoretical hot object under defined conditions, the measurement is expressed in Kelvin, never degrees. The higher the color temperature, the bluer the light. The lower the color temperature, the more orange the light.

Color temperature is important to visual display. Here are two examples of how you will see color temperature applied:

- Electronic cameras are substantially more sensitive to changes in light's color temperature than the human eye. Setting the correct white balance is a critical step in producing good-quality images from any electronic imaging device.
- Most monitors, flat-panel displays, and projectors have color-temperature selection and adjustment capabilities. These allow you to alter the white balance of an image for the lighting conditions in the environment or match projected images and flat-panel displays in a presentation.

Color Rendering Index

The *color rendering index* (CRI) of a light source indicates how accurately it shows the colors in the objects it illuminates. By definition, full-spectrum light sources have a perfect CRI of 100. Video cameras may have problems producing accurate images under light sources with a CRI under 60, even after a full white balance. The closer the CRI is to 100, the more accurate the color quality of the image. It can be important to render colors accurately when a client or a product has specific colors as part of their identity or brand.

Camera Controller

These may be joystick-type pan/tilt/zoom control units, a complete camera control unit (CCU) for exposure and image control, or a combination of these. In basic systems, a handheld wireless remote may be the primary means of moving and adjusting the camera.

Control connections to cameras can be via serial data (RS-422 or RS-232), USB, or IP network connections via wired or wireless Ethernet. Some camera controllers have a touch-screen user control interface. Simple video conferencing and conference room cameras may be equipped only with a handheld, wireless remote control.

Camera Mounts

These are used to mount the cameras to the tripod, dolly, wall, or ceiling and are often supplied by the camera manufacturer or a third-party vendor to match the size, weight, and color of the specific camera models. For some live production, event, or presentation work, cameras may be mounted on robotically-stabilized, handheld or drone mounts. Care should be taken to use the proper mount for each camera and application and to follow proper safety practices in securing the camera and mount.

Video Signal Processors

Video processors are common in professional AV installations because of the demand for high-quality, versatile, and cost-effective systems. Video processors have also become more powerful and reliable at the same time that their costs have continued to fall. Products generally regarded as video processors range from simple range extenders to powerful multifunction digital devices capable of 3D image mapping, standards conversion, temporal manipulation, edge blending, and multi-image display.

Video Switchers

Switchers provide a means to select or combine specific video source signals for display. Some switchers include audio switching, mixing, effects, or image processing functions.

Simple switchers can select one signal from multiple sources and send it to a specific destination. These might be used in a classroom or conference room to select inputs such as a media player, document camera, Internet stream, or computer output to a single projector or flat-panel display. Even simple switchers usually include a basic frame buffer on each input to allow synchronous (roll-free) switching between inputs. Simple switchers that do not switch synchronously between sources are often disparagingly referred to as *crash switchers*.

Matrix routing switchers can simultaneously send several different input signals to different outputs and often also separate (de-embed) audio signals from video signals. These sophisticated routers might be used to send various sources to many rooms within a building or used as a "master control" switcher in a large auditorium or production facility.

Presentation switchers or scaling switchers can adjust multiple input formats for display on various devices and may even combine multiple images for display with a single device or across multiple screens.

Production switchers share many of the characteristics of presentation switchers but are often intended primarily for delivery of one or more combined video streams for live production, IMAG, recording, or broadcast. Production switchers often include such capabilities as fade and wipe effects, image zoom, picture-in-picture, chroma key, luminance key, captioning, video effects processing, and still-frame stores.

Switchers are usually defined by the number of inputs and outputs they support, as well as capabilities for scaling, source synching, effects, signal processing, and workflow automation. Video switchers are also defined by the video formats they support. Some switchers also have separate signal buses for audio, video, RF, and other signals.

Some network-based AV distribution systems incorporate a range of functions into their nodes, including virtual switchers, image mappers, scalers, media servers, transcoders, keyers, effects generators, and titling.

Matrix Switchers

A *matrix switcher* is capable of selecting any of its multiple input sources to simultaneously go to one or more of its multiple outputs. Many matrix switchers also accommodate multiple signal types within a single unit. For the most part, one signal input type can be routed only to an output of the same signal type, but some matrix switchers incorporate scalers, audio embedders and de-embedders, and signal converters. Some matrix switchers can also handle EDID negotiations and spoofing on a port-by-port basis.

Many manufacturers produce HDBaseT matrix switchers to route HDMI and other input types with embedded audio, RS-232, infrared bi-directional remote control, and Ethernet.

Scalers and Scan Converters

Scalers are increasingly being integrated into inputs of switchers. They are available as small stand-alone devices, as hardware or software modules in larger video processors and switchers, and as built-in components in many modern projectors and displays.

Essentially, a scaler changes an image's format from one resolution to another, as shown in Figure 10-10. For example, a scaler can take an incoming signal at 1280×720, remap it, and output it at 1920×1080 to match the display's preferred resolution. These processors convert the vertical/horizontal scan frequency of a video signal "up" or "down" to another size or format for different applications.

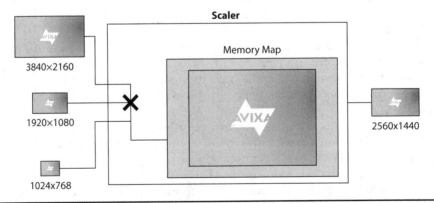

Figure 10-10 A simple scaler will take an incoming video signal and change its format for display on a device with a different resolution

Although the cost of these devices has fallen significantly, there are still substantial differences between low-end and higher-cost units, including picture quality, the smoothness and sharpness of the converted image, and the ease of adjustment and use.

Video wall controllers are essentially scalers that accept an input source and scale the output for display on multiple devices arranged in an image matrix. A single 2160p input signal may be scaled for display across a 6×9 matrix of 1.4m (55in.) 1080p display screens to form a large video wall display.

Media Servers

Media servers are a family of devices that record, store, process, and output media streams (video and often the associated audio). They often include the ability to simultaneously replay multiple media files while blending and processing them into a single output stream. The processing capabilities often include real-time image transformation, rescaling, temporal distortion, blending, warping, 3D mapping, and output of multiple streams at different refresh rates and resolutions. Media servers are used for live presentations, stage productions, visualization environments, television production, video walls, digital signage, theme park rides, museum displays, and trade exhibitions—to name but a few applications.

Media servers can be operated by any combination of live operator control; a programmed sequence; or respond to outside events such as the movement of objects, visual and electronic triggers, time code, safety interlocks, and push-buttons.

Media Players/Recorders

Media players store, replay, and in many cases also record media streams. The storage medium may be optical disc, magnetic hard drive, solid-state drive (SSD), flash media card, USB flash drive, or some combination of these. The device may take the format of a dedicated player with fixed or removable media, a personal media device, or a computer-based system. Input may be a digital video stream, the insertion of removable media, or by file-based transfer. Output may be in any digital or analog video and audio formats or via a network media stream.

Projectors

Projectors, such as the one shown in Figure 10-11, are used in a wide variety of applications where large images are necessary and the ambient lighting can be controlled, and they are often mounted overhead. Since they can be heavy, you will have to make some special considerations while you set them up.

Projected images use additive light mixing to achieve the full range of colors. The projection system's brightness is extremely important, as is the contrast between the bright and dark areas of the image. As black is the absence of any projected image, the screen can get no darker than when the projector is turned off.

A projection system includes the projector and its optics system, screen, and quality of setup.

Figure 10-11 A projector on a retractable ceiling mount

Projector Optics

Projector optics consist of a system of light sources, optical devices, and lenses that carry and focus an image on a screen for a viewer to see. A projection device operates at optimum performance only when used with the light source it was designed to use.

The lens that focuses the image onto the screen is referred to as the *primary optic*. Four factors related to primary optics influence the quality of the projected image: refraction, dispersion, spherical aberration, and curvature of field.

- **Refraction** *Refraction* is the bending or changing of the direction of a light ray when passing through the boundary between different materials, such as between glass and air. How much light refracts, meaning how great the angle of refraction, is called the *refractive index*. The refractive indexes of the materials used in a lens have an impact on the image quality from that lens.

- **Dispersion** *Dispersion* can be seen when a white light beam passes through a triangular prism. The different wavelengths of light refract at different angles, dispersing the light into its individual components. Lenses with strong dispersion can produce images with color fringes, known as *chromatic aberration*.

- **Spherical aberration** Light passing through the edge of a spherical lens leaves the lens at a different angle to the light passing through the center of the lens and so has a different focal length. The resulting distortion is called *spherical aberration*.

- **Curvature of field**   *Curvature of field* results in a blurry appearance around the edge of an otherwise in-focus object (or the reverse). This is the result of a curved lens that projects a curved image field onto a flat surface such as a projection screen. The more steeply curved the projection lens and the closer it is to the screen, the more noticeable is the effect. Very short focal length (wide angle) spherical lenses have steep curves.

Lamp-Based Light Sources

In lamp-based optical systems, the lamp works with a reflector to collect and direct the light. The lamp reflector may be internal or external. Figure 10-12 shows three lamp and reflector configurations.

Three light source technologies are used in lamp-based projectors:

- **Tungsten halogen**   Tungsten halogen, also known as *quartz halogen* or *QI lamps*, produce light by passing a current through a length of tungsten wire until it becomes white hot. To prevent the tungsten from boiling off the filament and blackening the lamp's quartz glass envelope, the envelope is filled with halogen gas. Although relatively inexpensive, these lamps are somewhat low in output, are energy inefficient, and have short lives. In some jurisdictions, tungsten halogen lamps are now considered too low in energy efficiency to be used.

- **Metal halide discharge**   Metal halide discharge lamps produce light by passing a current between a pair of tungsten electrodes immersed in a high-pressure mixture of metal-halide gases. This causes the gases to ionize and discharge light. A wide range of gas mixes, electrode configurations, and quartz envelope shapes are available in a wide range of outputs and physical sizes. Discharge lamps are significantly more efficient than tungsten halogen lamps and tend to have much longer lives.

- **Xenon arc**   Xenon arc lamps produce light by passing a current through a pair of tungsten electrodes immersed in an extremely high-pressure atmosphere of the inert gas xenon. The xenon plasma produced by the arc glows to produce light. Xenon arc lamps are very energy efficient and available in very high-output versions. The extreme gas pressure inside the quartz envelope makes them hazardous to handle, and high-output lamps can be difficult to change safely in situ.

Figure 10-12
Three types
of lamp and
reflector
configurations

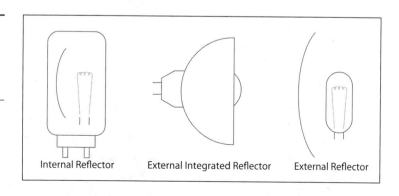

Internal Reflector   External Integrated Reflector   External Reflector

PART IV

Solid-State Light Sources

Solid-state light sources include a mix of LED sources, diode lasers (laser LEDS), and phosphor materials excited by LEDs and lasers. In some configurations the sources are a mix of red, green, and blue LEDs to produce white light. In others, some of the light comes directly from LEDs, and some comes from phosphor materials illuminated by LEDs or lasers. In RGB laser systems, the light comes from the diffused beams of red, green, and blue lasers. The configurations vary between manufacturers and models. Solid-state light sources are energy efficient and have operational lives in the tens of thousands of hours, which in some cases approach the total operational life of the projector.

Video Projection Technologies

Video projection technologies in current use include LCD, DLP, and direct laser.

LCD

The LCD technology used in projection works on the same light-blocking principle as LCD display panels. Light from the projector's light source is split into the three video primary colors—red, green, and blue. Each color of light is passed through a separate LCD panel, which creates the pixel pattern for that color component of the image. When all three of the color images are optically recombined, the additive color behavior of light results in full-color images that are passed through the projection lens to the projection surface. This class of projector is known as a *3LCD device*.

DLP

Digital Light Processing (DLP) projectors use an array of thousands of individually controlled microscopic mirrors on a digital micromirror device (DMD) chip to display images on screen. The individual mirrors, only microns in size, independently tilt back and forth over a small arc to reflect or block light, as shown in Figure 10-13. When a

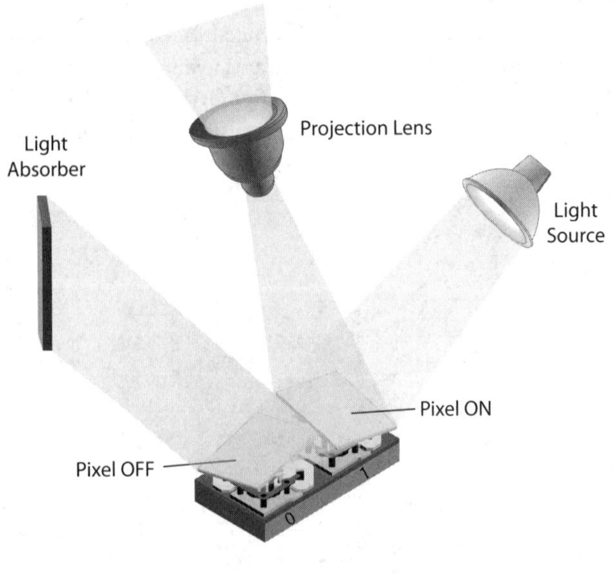

Figure 10-13
Microscopic mirrors in a digital micromirror device

mirror is tilted on, light reflects through the projection lens. When the mirror is tilted off, the light is reflected to a light absorber.

Because DMDs have only two states—on and off—pulse-width modulation (PWM) is used to create shades of gray in images. In a PWM system, the ratio of "on" cycles to "off" cycles within a specific time interval creates a particular luminance value. The on-off cycle is so fast in a DMD that our eyes don't perceive any flicker.

Color in DLP projection systems is created in one of two ways:

- **3DLP**   Similarly to a 3LCD system, light from the projector's light source is split into the three video primary colors—red, green, and blue. Each color of light is directed at a separate DMD chip, which creates the pixel pattern for that color component of the image. The images from the three DMD chips are optically combined to form an additive full-color image.

- **Single-chip DLP**   In these systems, a single DMD device produces the entire image. It is alternately illuminated at high speed by the component colors—usually, but not always, red, green, and blue—to form an additively mixed full-color image due to the viewers' persistence of vision. The alternating colors illuminating the DMD are usually generated by passing the white light from the projector's light source through a rapidly spinning filter wheel that has segments of each of the component colors. The color wheel is synchronized with the patterns being sent the DMD chip.

Direct Laser

Direct laser projection systems use a scanning mirror device to draw out the image directly on the projection surface using separate primary color beams from a collimated, narrow-beam laser source. Each frame of the image is scanned line by line, similarly to the raster pattern on CRT screens.

Projector brightness is measured in lumens and varies from less than 2,000 lumens for ultra-compact portable models up to more than 60,000 lumens for large venue installation units.

Projection Screens

Projection screens come in two general types: front or rear projection. Knowing the properties of each type will enable you to install them correctly and enhance the projection environment. Projectors produce light, but they do not deliver the light directly to the viewer's eye. Projection screens, however, give substance and quality to projected images.

Projection screens are passive devices. This means they cannot amplify or create light energy. They can reflect or transmit light rays at wide or narrow angles, thereby providing gain, for brighter images over a limited viewing area. Screen gain is the ability of a screen to redirect light energy to a narrower viewing area, making projected images appear brighter to viewers sitting near the center axis of the screen. The higher the "gain number" of a screen, the narrower the viewing angle over which it provides optimal brightness.

Front-Projection Screen

With front projection, the viewer and the projector are positioned on the same side of the screen. A front screen is simply a reflecting device. It reflects both the desired projected light and the unwanted ambient light back into the viewing area, so front-screen contrast is dependent on the control of ambient light.

Front-screen projection generally requires less space than rear projection. It offers simplified equipment placement. In front projection, the projector is in the same space as the viewer. This can result in problematic ambient noise from cooling fans in the projector. Another downside is that people, especially the presenter or instructor, can block all or part of the image by walking in front of the projected light. Discomfort or distraction can occur when the light from the projector shines directly in the presenter's eyes.

Ultra-short throw projectors have helped to address this problem by allowing substantial screen coverage from throws less than 500mm (18in.). Front projection may be used because of its lower cost or because of space limitations behind the screen.

Rear-Projection Screen

Rear projection is a transmissive system where light passes through a translucent screen toward the viewer. Rear projection needs more space for installation because it requires a projection room or cabinet behind the screen to accommodate the projector. Mirrors are commonly employed in rear projection applications to fold the projector's optical path, thereby saving depth. The ready availability of ultra-short-throw lenses has reduced the depth requirements for rear projection systems.

Many of the advantages of rear-projection systems are becoming less relevant with the wide availability and lower costs of employing high-brightness, high-resolution, direct view displays, which require little depth behind the screen.

Direct View Displays

Monitors, flat-panel displays, flat-screens, video walls—they are all terms for a common type of display technology. The prominent flat-panel technologies include LCD, OLED, LED, and microLED. The pixel resolution of a display is known as its *native resolution*.

The brightness of direct view displays is typically measured in nits or candelas per square meter (cd/m^2), with brightness levels ranging from around 350 nits for hospitality or meeting room use up to 6,500 nits for displays intended for outdoor signage or other applications with a lot of ambient light.

LCD

LCDs use a grid of pixels to create images on a screen. Each pixel on the screen is filled with a liquid crystal compound that changes its optical properties in the presence of an electric field.

To create an LCD pixel, light must first pass through a polarizer. The polarized light then travels through a sandwich of transparent switching transistors and liquid crystals.

The tiny liquid crystals within each pixel act like light shutters, passing or blocking varying amounts of light. As polarizing filters cannot block 100 percent of the incoming light, the "black" from LCD panels is actually dark gray.

A color LCD has three filters in each imaging pixel array, one for each primary color: red, green, and blue. The backlight can be LEDs, mini-LEDs, micro-LEDs, or compact fluorescent tubes. The backlight LEDs can be clustered in groups with red, green, and blue elements or only white LEDs. LEDs have lower power consumption than compact fluorescent tubes, higher brightness, wider color range, better white balance, and longer life, and they can include local area dimming for higher contrast. LED backlights are also more compact than compact fluorescent tubes, allowing for much slimmer panel profiles.

LED

Light-emitting diode (LED) displays are a matrix of red, green, and blue LEDs, grouped to form the pixels of the display. LEDs are emissive devices, meaning that they create their own light, as opposed to LCDs, which require a separate light source: the backlight. As a result, LED displays have much higher contrast ratios, faster pixel switching times, and operational lives in the tens of thousands of hours, and they are capable of higher brightness and a wider color gamut than LCDs.

Individual LED chips or full-pixel clusters of RGB LED chips are mounted into modules that carry physical framing and linkages in addition to power and signal distribution systems. Multiple modules can be connected to form screens of arbitrarily large sizes and shapes. Modular LED screens are often seen in large spaces, in concert productions, and at public events.

MicroLEDs are very small (1 to 10 μm) crystalline LEDs with the same characteristics as their larger brethren. As MicroLED pixels are quite physically small, the spaces between them are relatively large. This enables a black background between pixels to absorb ambient light and produces high-contrast images and much deeper blacks than LCD panels.

OLED

OLED technology is based on organic, carbon-based, chemical compounds that emit light when a current flows through the device. There are separate organic compounds for red, green, and blue.

OLED devices use less power than LCDs, have much higher contrast ratios, and may be capable of higher brightness and a wider color range than LCDs.

OLEDs are imprinted on a very thin substrate. The active matrix silicon-integrated circuits are imprinted directly under the display, controlling the power to each organic point of light diode (pixel) and performing certain image control functions at a very high speed. OLED's capability to refresh in microseconds rather than milliseconds, as LCDs do, creates highly dynamic motion video. As with microLEDs, the small physical size of OLED pixels facilitates high-contrast images with rich blacks.

OLEDs have a relatively short life span, exhibiting progressive degradation of output, particularly in the shorter wavelengths—seen as blue fading.

Video Wiring Schemes

Video systems have a wide range of wiring topologies or schemes. In a broad overview, they can be categorized as follows:

- **Direct connection**   Sends a signal from device A to device B. For example, a signal is sent from a media server to a display.
- **Looping scheme**   Sends a signal from a single source device to multiple devices. For example, a signal is sent from a media server to display 1, then to display 2, and then to display 3.
- **Distributed scheme**   Sends a signal to multiple devices simultaneously. For example, a signal is sent from a media server to a distribution amplifier and from there to each of three displays.
- **Network distribution scheme**   Sends a digital video stream across a data network to each of the designated receiving devices. For example, a media server sends a video stream to many distributed displays.

Your client's application will indicate which mode or scheme is to be used, and the selection will usually be determined by the AV system designer, but you are required to ensure that all the devices are placed in their designated locations and wired correctly.

Direct Connection

As shown in Figure 10-14, in a direct connection, the video source, such as a live video feed or tuner, connects directly to a video display input such as a monitor, a projector, or a recording device.

These systems are installed simply by connecting the video source to a target device with the appropriate cable.

Looping Scheme

In a looping scheme, the devices that are connected have the ability to pass on or loop the video signal to the next connected device. As shown in Figure 10-15, in a looping scheme, a video signal from a source device is distributed to multiple devices (usually displays) simultaneously.

You might use a looping scheme in a corporate lobby where multiple flat-panel LCDs display a "welcome" video with company information.

Figure 10-14

Direct connection for sending a signal between a single source device and a display

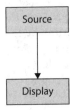

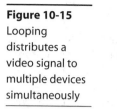

Figure 10-15
Looping
distributes a
video signal to
multiple devices
simultaneously

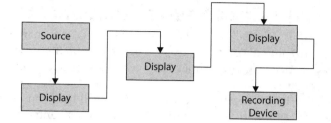

Distributed Scheme

A distributed scheme uses a distribution amplifier (DA) to route the signal to multiple devices simultaneously (see Figure 10-16) while maintaining signal quality. For instance, a signal from a computer video output may be connected to a DA and then to three different displays and recorders. DAs may have as few as 2 or more than 12 outputs.

Network Distribution Scheme

A data network distribution scheme, as shown in Figure 10-17, uses a digital data network such TCP/IP over Ethernet to broadcast a stream of video data packets across a network for individual devices such as displays or recorders to receive. Any number of receiving devices may be simultaneously connected to the network, which may be, wired, fiber, or wireless. For instance, the output of a camera in a lecture theater may be distributed to displays and recorders located in any room with a network connection.

PART IV

Figure 10-16
A distributed
scheme with
distribution
amplifier

Figure 10-17
A network
distribution
scheme provides
feeds to multiple
displays and a
recorder

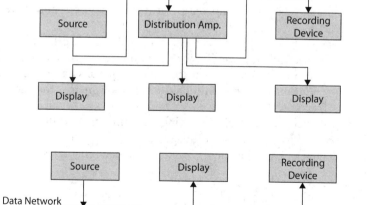

Installing Video Source Components

In this section, you will learn about some of the factors that affect the installation of certain video source components. The order in which you will install these components will depend on your project. Always follow the steps outlined in the manufacturer's literature.

What you have learned in previous chapters will help you in many installation situations, such as mounting components in a rack, on a conference table, or on the wall. For example, when installing a camera into plaster sheeting (sheetrock, drywall, plasterboard, gypsum board, etc.) not located near a wall stud, you will need to use hollow wall anchors (snap toggles, butterfly toggles, gravity toggles, spring toggles, etc.) to provide additional support.

First take a look at some of the video source components that you may have to install. Common types of video sources include cameras, computers, media servers, and media players. Additionally, the AV system may require bring your own device (BYOD) support for the display of content from mobile devices onto the large-format displays in meeting rooms.

Cameras

Video cameras are used to record and stream events in lecture halls and presentations, capture video for IMAG in live events, and show the participants at remote locations in video conference and distance-learning applications. Placement of cameras is usually indicated in the AV design drawings, and installation and wiring are fairly simple and straightforward.

You may need to make some adjustments to the camera's settings if the image produced with default settings yields poor-quality video. You will need to know about changing light levels and color temperatures in the room and their impact on camera performance.

Similar to other AV devices, cameras may be fixed or portable. Issues that require special attention may include cabling for video, image return, talkback, control, and power. Fixed cameras should also be mounted securely to walls or ceilings in a manner that will minimize vibration and unintentional camera movements from nearby speakers, doors, or other objects.

You will need to verify the following adjustments: framing, zoom, focus, white balance, gain, and iris for optimum image quality. You will learn about white balance and how to make camera adjustments in the subsequent chapters on setup and verification.

Media Players

In a simple direct-connection setup, a media player may be connected to a display device with an HDMI cable. In a more complex installation, the media player will need to be connected to a video switcher. Media players are often installed in an AV rack or cabinet at eye level for easy access by the end user. Device control may be via handheld remote or a custom touch-panel system.

Media Servers

As most media servers are designed around a computer-based processing system, they are usually mounted in an AV rack with wired network connections for configuration, data ingress, and operation. The outputs are either direct video streams such as HDMI or DisplayPort or some format of network-streamed video over Ethernet. Mobile media servers for events and short-run productions are usually mounted in dedicated mobile AV racks, often together with DAs, an uninterruptable power supply (UPS), a hot-tracking duplicate unit, and an automated failover system.

BYOD Support Connections

In AV environments with BYOD support, a user can either show content from personal mobile devices on the meeting room display by plugging their device into the inputs available at the conference table or by plugging in a wireless collaboration device. Currently, the most common output for a laptop PC or tablet is via an HDMI, USB Type C, or DisplayPort connection. The meeting room table box shown in Figure 10-18 provides video (VGA), audio (TRS), and HDMI (audio-video combined) connectors.

A variety of adapters and software systems are also available for sharing content from personal devices onto large displays. An increasingly common BYOD strategy is to provide wireless connection to a network interface via Wi-Fi. This approach comes with increased security risks from using open wireless networks. Secure applications with strong authentication protocols are essential.

Table connections should connect with the room system or display device as indicated in the AV system drawings. As always, if you spot a problem with the setup, contact the system designer before you make any changes.

One common challenge to seamless BYOD support is to assure the room's display or switching system network is compatible with a broad range of video output formats, particularly with the tendency for members of the public to shoot video on their personal devices in nonstandard resolutions and often in odd, portrait-mode aspect ratios (9:16 or 1:2).

Figure 10-18
BYOD support at a meeting room table (© 2020 Lew Electric Fittings Company/Kirk Luttrell)

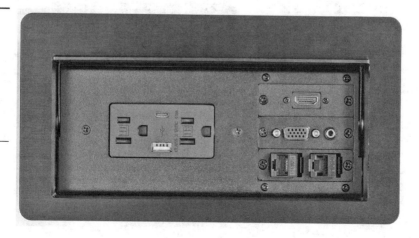

Installing Video-Processing Components

Video-processing components include distribution amplifiers, switchers, scalers, caption generators, and scan converters. In certain applications requiring HDMI cables, where there is a long distance between video source and sink devices, you can troubleshoot by using an HDMI transmitter and receiver.

Distribution Amplifiers

Video distribution amplifiers are used to send the same signal to multiple destinations, as shown in Figure 10-19:

- They distribute a single input to multiple outputs or display devices.
- They distribute audio via embedded audio channels or via separate audio input and outputs.
- They may include gain options for video distribution amplifier channels.

Most distribution amplifiers provide unity gain to the signal, while others can provide gain with adjustable output levels to compensate for cable attenuation.
Signal distribution design objectives typically include the following:

- Efficiently directing the signal to the appropriate destinations
- Monitoring signal strength and how strength diminishes with distance
- Pre-amplifying the signal to compensate for cable attenuation (gain adjustment)

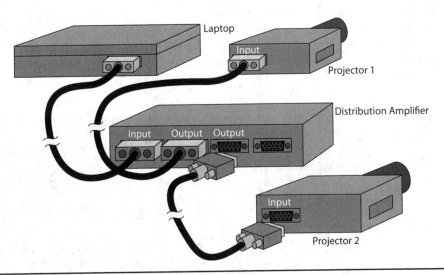

Figure 10-19   A distribution amplifier can send one input to multiple outputs

The purpose of the distribution system is not only to direct the signals to the right destination but also to care for the signals as they are passed from one component to the next. You must consider the video feeds, cabling, and intermediate components and how these elements will affect or degrade the video signal.

Many HDMI distribution amplifiers exactly replicate the input signal on the output ports, but do not handle the negotiation of high-bandwidth digital content protection (HDCP) or extended display identification data (EDID) between the source and the multiple destinations. Without resolving HDCP and EDID negotiations, the resulting failure to display images at the destination may give the appearance of a faulty DA or faulty cables, as the display device will usually work when plugged directly into the source device.

Matrix Switcher

In AV systems designed for sending multiple AV sources to multiple AV destinations, you will probably be required to install a matrix switcher.

Matrix switchers are usually installed at the rack and can be quite complex to install, with additional ports for remote configuration and control, so make sure you read your equipment manuals and install the cables according to the block and wiring diagrams.

Transmitters and Receivers

HDMI cables have significant length limitations. The signal will begin to noticeably degrade beyond the 5-meter (16-feet) mark, and HDMI cables cannot be extended and re-terminated in the field.

If your client wants to plug a laptop into a system and run it to a projector but the cable route between the laptop and projector is longer than 5 meters (16 feet), you can run a 1.8-meter (6-foot) HDMI cable to a transmitter. The transmitter processes the HDMI signal and passes it to a Cat 5e+ UTP network cable or an optical fiber.

Cat 5e+ UTP network cables are lightweight and inexpensive, can carry multiple signal types, and can be field terminated, but are limited in the bandwidth of HDMI signal they can reliably transport.

Optical fibers may be more expensive than UTP network cables such as Cat 5e but can extend the distance between HDMI devices by very much longer distances. Fiber can also handle much higher bandwidths than most twisted-pair network cables and so offers the possibility to extend HDMI 2.1+ at resolutions up to 10K at high frame rates, substantially beyond the limits of even Cat 8 UTP network cable.

The UTP network cable or optical fiber typically runs to a receiver/converter near the projector, where the signal is converted back to HDMI, and the receiver is then connected to the projector via a short HDMI cable.

HDMI extended via video-over-IP transmitters and receivers allows the source device and the display device to be anywhere on the same IP network, provided sufficient bandwidth can be reserved for the signals.

NOTE HDMI transmitters and receivers are typically sold as a pair, and you cannot reliably mix and match transmitters and receivers from multiple manufacturers.

Installing Video Display Components

This section will provide general steps for installing projection screens, projectors, and flat-panel displays. You will use the knowledge you have gained from previous chapters to better understand the installation procedures and concerns discussed here. Your study and work experience will ease the way in your future installations. For example, you will need to calculate the combined weight of the projector and mount you will be installing and know what weight the substructure of the surface you will be mounting the gear to can support to determine what additional measures you should take to secure the component. First take some time to familiarize yourself with the basic principles of determining throw distance and other essentials such as projector offset and keystone correction.

Front-Projection Screens

Before you mount the projectors in the AV system, you will need to set up the projector screen. As noted earlier, with front projection, the image is viewed from the same side of the screen as the projector's location.

A screen's surface coating greatly influences the quality and brightness of a video projector's image. A well-designed projection system incorporates a screen type that reflects projected light over a wide angle to the audience while minimizing the reflection of stray light. Stray light causes a loss of contrast and detail, most noticeably in the dark areas of the image.

Projection screens are passive devices and cannot amplify or create light rays. They can reflect light rays back at wide or narrow angles, thereby providing gain, or brighter images.

Screen gain is the ability of a screen to redirect light rays into a narrower viewing area, making projected images appear brighter to viewers sitting on-axis to the screen. The higher the gain number of a screen, the narrower the viewing angle over which it provides optimal brightness.

A matte white screen is usually made of magnesium carbonate or a similar substance that provides a perfect diffuser for the redistribution of light. In other words, the light energy striking the screen surface is scattered identically in all directions. Figure 10-20 shows the wide dispersion pattern. Matte white is a reference surface, with a screen gain of 1.0 (also known as *unity gain*).

Matte white screens provide good color rendition, and they are well suited to data and graphic applications, while matte gray screens reflect less light but can produce images with higher contrast in dark areas of the image.

Since light reflecting from the screen will diffuse equally in all directions, room lights can be a problem with this type of screen.

Figure 10-20
As light hits a matte white screen, it will evenly disperse, creating a wide viewing area

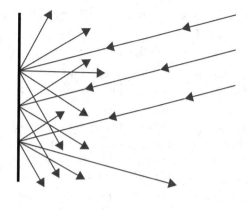

Figure 10-21
Angularly reflective screen surfaces reflect light at the same angle that it strikes the screen

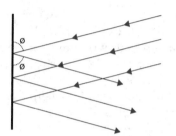

Angularly reflective screen surfaces, as shown in Figure 10-21, reflect light at various angles rather than a uniform 180-degree reflection. Such surfaces perform similarly to a textured mirror. The light is reflected back at the same angle that it strikes the screen, but on the other side of the screen's axis. If a projector is mounted at a height equal to the top center of the screen, the viewing cone's axis would be directed back at the audience in a downward direction. Most screens with a gain greater than five are of this type. This screen type works well for motion video.

Mounting an Electrically Operated Front-Projection Screen

Here are some tips for handling electrically operated front screens:

- Verify that the mounting position is as specified in the AV design and that the mounting points have been approved by the building owner or the project manager and by a structural engineer.

- When the screen arrives from the manufacturer, keep it horizontal while storing and transporting it to avoid wrinkles and tears.

- Avoid turning the screen vertically when transporting it. If it is carried vertically, the screen material can slide down on the roller and wrinkle, tear, or smudge the material.

- When handling screens with an electric roller, keep the slot down, or the bottom bar (called the *batten*) can become jammed.

- Properly handle the screen so that it is mounted as a clean, flat surface, free of dust.

- In a return-air plenum ceiling, local regulations will generally require you to encase the screen in a fire-rated enclosure or have a screen designed for use in a plenum space.

- Before mounting the screen housing, be aware that some motorized screens have a motor inside the housing. Usually the motor is placed to one side of the screen. If this is the case, make sure you are aware of the location of the center of the screen, not just the center of the housing.

- Ensure that you leave access for servicing the motor assembly.

For wall and ceiling installations, use the appropriate fasteners to mount to the material (masonry, plaster board, and so on). Use brackets for flush ceiling or corner mounting.

Electrically operated screens have limits, such as physical switches that turn off the motor. Make sure you follow manufacturer specifications for setting them. The top-limit switch prevents the screen from rising too high. The bottom-limit switch prevents the screen from lowering too far down. Generally, a full wrap to a wrap and a half of screen material should remain on the roller when the screen is at its viewing position. If you allow more to come off the roller, you run the risk of the fabric coming loose.

Depending on the condition of the screen, you may need to clean it before signing off on the installation. Always refer to manufacturer instructions for screen cleaning. For general screen care and cleaning, use a clean, soft-bristled brush to blot the edges of the screen. Work from the bottom of the screen to the top, collecting water droplets with a clean, highly absorbent sponge. Do not use abrasive cleaners or solvents, and never touch the screen with a sharp tool.

Mounting a Manual Wall-Mounted Projection Screen

These instructions give the typical steps required for mounting a manually operated wall screen.

Step 1   Determine the general location within the room for the projection screen from the AV drawings and plans. Verify that the mounting points have been approved by the building owner or the project manager and by a structural engineer. Check that the screen will not cover any of the following: alarm indicators, thermostat, clock, light switches, wall boxes, IR receivers, and power sockets.

Step 2   Ensure that the screen will lie flat, clear of any wall-mounted furniture, marker trays, wall outlets, and so on.

Step 3   Verify that you have the proper screen by checking your paperwork. Clear an area to work, and neatly unpack the screen product.

PART IV

Figure 10-22
Screen center point

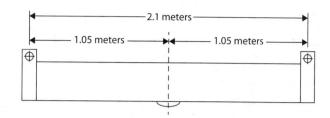

Step 4 Measure the distance between the center of the screen and the suspension points on the screen casing, as shown in Figure 10-22. The halfway point between the two mounting points will be the screen center.

Step 5 Using light pencil marks, transfer this screen center line to the wall using a vertical line so that the screen location will be aesthetically pleasing and balanced within the room. Pay most attention to the ceiling grid, if applicable. However, do not overlook the primary objective: placing the screen in clear lines of sight of the projector and audience. Place this vertical line near the anticipated mounting height.

Step 6 Measure the difference between the mounting hole and the top of the screen surface, as shown in Figure 10-23. Add this measurement to the desired height of the final display surface. The sum will be the height of the screen-mounting points.

Step 7 Measure up the wall the distance discovered in the previous step, using the floor as your reference. Make a horizontal mark at this point that intersects your vertical screen centerline. Extend your centerline if necessary, as shown in Figure 10-24. This mark represents the center of the screen, between the mounting points.

Step 8 Using appropriate mounting hardware, fasten the corresponding screen end to the wall at that point.

Step 9 Elevate the unfastened screen end. Using a level, position the screen case so it is straight. Again, using appropriate mounting hardware, fasten the loose screen end to the wall at that point.

Step 10 Verify that the screen operates as expected.

When you have finished installing the screen, your final task is to verify that the room will be presentable to your clients. Clean up all debris, gather tools, and replace any moved furniture items.

Figure 10-23
Measuring the height of the screen-mounting points

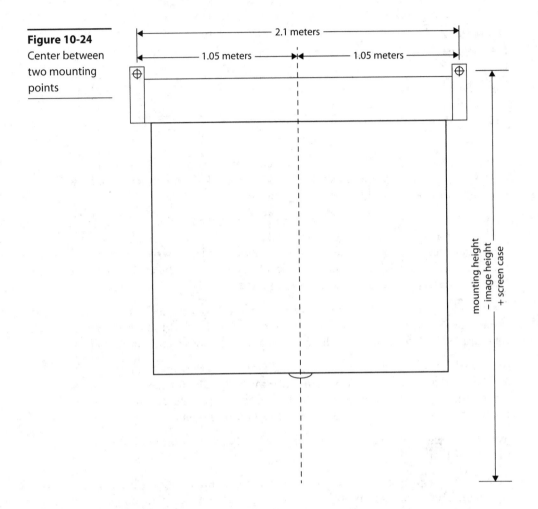

Figure 10-24
Center between
two mounting
points

Rear-Projection Screens

Rear-screen configurations can use a wide range of screen materials, each with different gain and directional properties. Rear-projection screens have either a narrow viewing area with high brightness or a wide viewing area with reduced brightness.

Diffusion screen material is used in some rear-projection screen applications. The screen substructure may be rigid acrylic, glass, or, for portable applications, a vinyl fabric. This material provides a diffused, coated, or frosted surface on which the image is focused. It provides a wide viewing angle both horizontally and vertically but with little or no gain. There may be some hot-spotting because of the transparency of the screen fabric, depending on the vertical placement of the projector with relationship to the audience. The light from the projector is transmitted through the screen with relatively little refraction. The ambient light rejection of this material is moderate and is based on the viewer-side material's reflectivity or sheen.

Optical screen material is used in permanent rear-screen applications. This screen system consists of a series of lenses formed into the screen material. The most common is the two-lens system, in which the following is true:

- The lens that faces the projector is a Fresnel lens. It is a flat glass or acrylic lens in which the curvature of a normal lens surface has been collapsed in such a way that concentric circles are impressed on the lens surface. This lens gathers the light from the projector.

- The second lens, molded to the other side of the piece of transparent screen material or to another sheet, is a vertical lenticular lens (it must be held flat to avoid hot spots). It is a screen surface characterized by silvered or aluminized embossing, designed to reflect maximum light over wide horizontal and narrow vertical angles. This lenticular lens faces the viewing audience and spreads the light horizontally, providing a relatively large horizontal viewing angle.

Optical pattern rear-projection screens capture and concentrate the light to maximize the amount of light passing through to the audience. The image can lose brightness if the viewers or projector are moved off the axis. Therefore, the projector must be at the focal point indicated by the manufacturer.

Rear-Screen Installation

For all rear screens, the surfaces are delicate and can be damaged. Take care to protect them from scratches and abrasions.

Flexible polyvinylchloride (PVC) screens are popular in rental and staging applications but also can be permanently installed. The PVC material must be stretched firmly when installing. The surface can attract dirt easily and should not be placed on the floor during assembly. Flexible rear-projection screens can be easily punctured by sharp objects such as tools and ladders.

Rigid rear-projection screens must be transported vertically. There is a proper side of the screen to face the audience; for coated screens, it is the coated side, and for Fresnel lenticular screens, it is the lenticular side. Note that glass rear-projection screens can break with rough handling.

Ensure wall openings are square cornered and have adequate dimensions to accommodate the screen and frame. Screens must be held in position in the wall by brackets or wooden trim.

After installation is complete, the screen should be protected during further construction operations in the facility.

Rear-Screen Mirror Position

Mirrors are often used to save space in rear projection rooms. A first-surface mirror optically folds the light path from the projector to the screen.

Some first-surface mirrors are made of thin reflective polymer material that is tightly stretched over a frame: a much lighter alternative to glass. Extra caution must be taken not to damage this material because a screwdriver can easily poke a hole into its surface.

It is important to protect the mirror surface during installation. Glass mirrors typically come with a colored plastic coating that protects their surface. It is a good idea to keep this cover on until you are prepared to adjust the projector.

The mirrors used for rear projection must be of sufficient size to reflect the entire image. The position of the mirror in the optical path determines the size of the mirror needed. Because the image size reflected by the mirror will be in the shape of a trapezoid, the mirror should be slightly larger than the longest side of the trapezoid formed.

The AV system designer is responsible for providing detailed plans and elevations of the positions of all of the optical components in a rear-projection installation.

Rear-Screen Mirror Mounting

The optical path can be theoretically folded many times; however, three mirrors is the practical limit, with one- or two-mirror systems being the most common. As you add mirrors, the potential alignment problems are multiplied. Figure 10-25 shows a rear screen set up with two mirrors for a "double-bounce" to reduce the throw distance of the lens. This type of setup is becoming rare because of the availability and convenience offered by short-throw-lens projectors.

A mirror mounting system must be rigid once in place and adjustable during installation. Mounting is done in a way that positioning adjustments can be made. The ability

Figure 10-25

Rear screen with two mirrors for "double bounce" (image courtesy of rp Visual Solutions)

to reposition the mirror in the x-, y-, and z-axes is provided. The mounting system must also have the ability to be locked into position once adjustments are complete.

Glass mirrors are mounted to a solid surface. Avoid wood mounting surfaces because wood can warp over time. Anchor the bottom of the mirror stand to the floor and brace the top.

Finally, check for any secondary reflection problems. Secondary reflection occurs when light rays at certain angles reflect off the backside of the screen. They then strike the mirror and reflect a second time onto the screen. This secondary reflection travels farther than the original image so it is proportionally larger.

Projector Installation

Projectors may be used in a wide variety of applications, and they are often mounted overhead. Since they can be heavy, you will have to make some special considerations as you set them up.

You need to know how the projector will be oriented before you mount it. For a given screen size or display image size, the projector is intended to be placed in a specific position. Precise projector placement is of critical importance to the eventual quality of the displayed image and should be accurately documented in the AV system design.

Projector Position

Calculating the correct projector position can be complex. The formula for projector position depends on the specific equipment. Manufacturer specifications include instructions for positioning the gear.

Projection positioning calculations are usually made in three dimensions, as shown in Figure 10-26. The dimensions represent the following:

- **H** The horizontal offset of the projector to the screen
- **V** The vertical offset of the projector to the screen
- **T** Throw distance

If you do not have access to the manufacturer's specific formula (web search engines are your friends), estimating the third variable, throw distance, can be useful. Estimating

Figure 10-26
There are three dimensions relevant to projector positioning: horizontal offset (H), vertical offset (V), and throw distance (T)

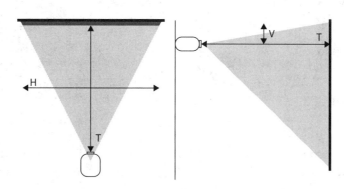

throw distance reveals how far away the projector needs to be from the screen. Your goal is to have an image fill the entire screen surface. Normally, the system designer would provide you with the specific distance. However, you still need to know how to interpret that information.

Throw distance is the distance from a projector to a focusing surface or the screen. The measurement point on the projector varies by manufacturer. Generally, the larger the image to display, the greater the throw distance will be. To reference a manufacturer's throw distance calculation, you need to know the following:

- Projector model
- Lens model and focal length; variable (zoom) or fixed focal length
- Screen's display surface dimensions

As shown in Figure 10-27, throw distance can be simply expressed and estimated as a ratio of projector distance to image width.

Many factors are specific to each room design and projector model that affect the throw distance for a projector and a desired image. The primary resource is the projector manufacturer's manual. The manufacturer will provide a chart or formula to calculate the projection distance or throw distance based on the screen size. The chart will be relevant only to that projector and lens.

You can estimate the throw distance from a projector to a screen using the following formula:

$$\text{Distance} = \text{Screen Width} \times \text{Throw Ratio}$$

where:

- Distance is the distance from the front of the lens to the closest point on the screen
- Screen width is the width of the projected image
- Throw ratio is the ratio of throw distance to image width

Figure 10-27
Projector throw distance can be expressed as a ratio of projector distance to screen width

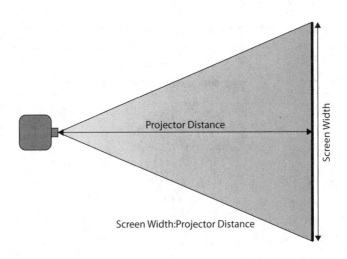

Projector Distance

Screen Width

Screen Width:Projector Distance

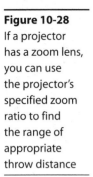

Figure 10-28
If a projector has a zoom lens, you can use the projector's specified zoom ratio to find the range of appropriate throw distance

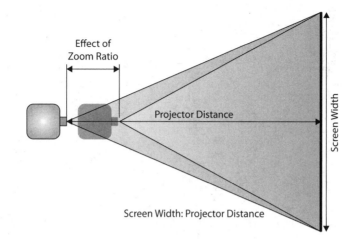

As shown in Figure 10-28, projectors with a zoom lens will affect the throw distance, resulting in a minimum and a maximum throw distance.

Refer to the owner's manual of the projector and lens combination to find an accurate formula for the specific projector. Often, the projector manufacturer will also provide a software solution or online calculator on their website. To use this calculator, simply plug the desired image size into the appropriate lens and projector combination, and the software will give the throw distance result.

Applying Throw Distance

After you have determined the range of the throw distance, you will need to determine from which two points to measure that distance. The installation manual should provide you with this information. As shown in Figure 10-29, the front of the lens to the bottom center of the screen is often used for projectors that have a standard built-in lens. For most projectors that use a range of lenses, the manual may instruct you to measure from a point on the front of the projector body. As with any calculation, there can be a certain margin of error. The manufacturer lists the margin of error as a percentage value, typically 3 to 5 percent. Therefore, you may need to tweak the projector setup a bit.

The location of the projector to the left or to the right of the projected image center is also critical. Typically, the lens of the projector is oriented at the bottom of the screen, equal distance from the left and right, as shown in Figure 10-29.

Best Practice for Applying Throw Distance

When using a zoom lens, it is best to plan to use the center two-thirds of the zoom range. Ignore the extreme positions of the zoom lens. This allows you to avoid any deficiencies in the optical system and room to correct for the margin of error.

Figure 10-29
Projector lens
oriented at the
bottom center of
the screen

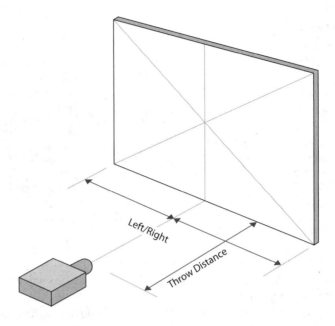

Projector Offset

Along with the minimum and maximum throw distance for a given projector/lens combination, there is a third dimension with which to work, called the *vertical offset*. The vertical offset indicates the orientation of the projector on the vertical axis, as shown in Figure 10-30.

Figure 10-30
Projector
orientation on
the vertical
axis showing
projector's
vertical offset

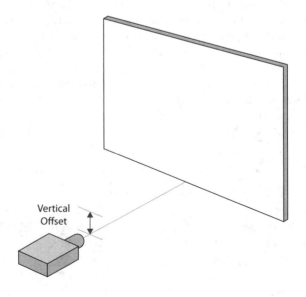

When table mounted, your reference point is typically the bottom-center point of the projected image area. This would be at the bottom of the screen but equal distance from the left and right. The offset value tells you how high or low the projector should be moved in reference to this point on the projection screen.

The projector needs to be located at the correct point in reference to three axes (left/right, distance from screen, and offset), or the image will show keystone distortion.

Best Practice for Projector Offset

A professionally installed projection system will display an image that fills the screen to the borders.

Keystone Correction

If you do not place the projector in the optimal vertical position, you will have a distorted image on the screen, as shown in Figure 10-31. This is called *keystone distortion*.

There will be instances when you need to mount a projector in a way not anticipated by the manufacturer. In this case, some options are available on many projectors that will help correct any keystone distortion that may result. The two main types are electronic keystone correction and optical keystone correction.

Figure 10-31   Keystone distortion caused by the placement of the projector

Electronic keystone correction attempts to square the corners of the displayed image by electronically distorting the image in the opposite direction. Depending upon the critical nature of the displayed image, electronic keystone correction may not be an acceptable solution. If a highly detailed image is displayed, it is possible that critical information could be lost as the projector electronics attempt to correct the image.

Optical keystone correction, also referred to as *lens shift,* is preferred over electronic keystone correction. Some projectors can electronically shift the optics vertically, while a few others have vertical *and* horizontal shifts. The lens shift function is more expensive to manufacture and will be found on high-end projectors. This optical form of keystone correction allows you to move the image to correct for the placement of the projector.

Ceiling-Mounted Projector

Projectors are commonly mounted to the ceiling of a room. This configuration is practical because it locks the projector in place, high out of viewer sightlines. It also prevents the projector from getting bumped in high-traffic areas.

Because a ceiling-mounted projector hangs above the heads of the viewers, security, strength, and safety are critical concerns in the installation of these devices. Other key installation concerns include placement of projectors relative to electrical outlets, sprinkler heads, microphones, loudspeakers, cameras, and lighting fixtures.

For ceiling-mounted projectors, determining throw distance and offset are just the first steps of placement.

Best Practice for Ceiling-Mounted Projectors

Check the structure for vibrations by resting your hand on the beam or bar-joists. Sometimes the HVAC unit is mounted on the roof above, creating vibrations that will cause the image to appear unfocused.

Mounting a Projector to the Ceiling

Follow these general steps to ceiling-mount a video projector. The procedure that follows describes a typical installation in a space with a suspended drop ceiling.

Step 1 Determine the general location within the room for the projector with respect to the projection screen from the AV drawings and plans. Verify that the mounting point has been approved by the building owner or the project manager and by a structural engineer.

Use the project documentation and/or the information supplied by the projector manufacturer to establish precisely where the projector should be located. This information can be from a software program, a chart, a formula in the manual, or even the manufacturer's website.

Step 2   Familiarize yourself with the manufactured mount, how it connects to the projector, and the mounting structure you will create. Identify the center point of the manufactured mount, preparing to transfer that point to the ceiling.

Read and follow all manufacturers' mounting precautions and instructions.

Step 3   Analyze the feasibility of mounting in the position identified.

- Carefully remove ceiling tiles for clear visibility.
- Verify that there is adequate ceiling space at the mounting location.
- Decide on the proper mounting method by observing the structure available to which the mount may be fastened. (This example uses a ceiling plate and pipe connecting the manufactured mount to the building structure.)
- Using the required 5:1 safety margin, verify that the integrity of the structure you anticipate will support the projector and mount.
- Verify that you have run all necessary signal and control cables to the projector location and that the cable types conform to local regulations.
- Check for necessary power and that the projector's input power lead conforms with local regulations.
- Consider serviceability of the installation; how would a service technician gain access to the device for periodic maintenance?

Step 4   Take inventory of the tools and materials you need to accomplish mounting at the determined location. Be sure that you have all critical elements that are required before you start.

Step 5   Clear ample space to do the installation. Construct personnel barriers and remove furniture that either would be damaged or in the way. Cover other furniture or equipment to protect it from dirt, dropped material, and so on.

Step 6   Make clear, erasable pencil marks indicating the center point location of the mounting assembly identified in step 2. Do not confuse this with the projection throw distance calculation reference point in step 1.

Step 7   As an industry best practice, construct an adjustable mounting support structure centered at the location indicated in the previous step.

Step 8   Fasten a manufactured ceiling plate/pipe flange to the adjustable mounting support, centered on your marks.

Step 9   Reinstall the ceiling tile below the mount assembly.

Step 10   Using a plumb bob that is suspended from the center of the pipe flange on the mount assembly, locate the spot on the ceiling tile and mark it.

Step 11   Cut a circular hole in the ceiling tile for the pipe to pass through, as shown in Figure 10-32. Keep the tile clean! It is best to remove the tile from the ceiling first.

Figure 10-32
An installer uses
a hole saw to
cut a circular hole
in a ceiling tile

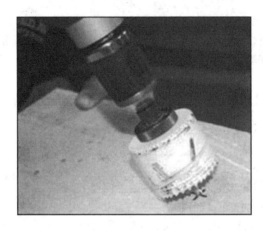

Step 12   Select a pipe length that is appropriate to your specific situation. You can use a threaded rod for adjustment, but keep it as short as possible. Try to use steel-channel strut (Unistrut, Kindorf, EzyStrut, or something similar) to give structural strength yet some flexibility to get the projector into the proper position. Insert the pipe through the ceiling tile, threading it into the pipe flange. Tighten securely. Use a setscrew in the pipe flange to maintain the threaded position.

A decorative ring may be applied to the pipe at this time. (Some can be installed after the mounting is complete.)

Consideration should be given to the final appearance of the pipe. Raw black iron pipe detracts from the appearance of a finished room; it should be cleaned and painted.

Step 13   Align pipe threads with the manufactured projector mount. Thread the mount onto the pipe until it is hand tight.

Step 14   Locate the front of the mount and rotate in the opposite direction until the front of the mount is parallel to the screen surface.

Step 15   Tighten the setscrew onto the threads of the pipe to prevent turning.

Step 16   As shown in Figure 10-33, join the projector adapter bracket to the projector.

Figure 10-33
The projector
adapter bracket
joined to the
projector

CAUTION Do not use screws or bolts that protrude too far into the projector because this will damage the product.

Step 17   Place the projector with the adapter into the secured mount assembly.

Step 18   Connect the signal and control cabling as required.

Step 19   Connect the power cable. Apply power to the projector.

Step 20   Supply a reliable test pattern and select the proper input.

Step 21   Use the zoom function to size the image for the screen and the focus function to focus it properly.

Step 22   Make fine adjustments to the ceiling mount assembly to align the image onto the screen surface and eliminate any image keystone distortion.

It is important to make sure the projector is mechanically configured before any electronic adjustments are made.

Step 23   Check all inputs to the projector, making electronic adjustments as necessary.

Step 24   Verify the accuracy of any documented operational instructions that will be provided to the end user.

When you have finished mounting the projector, complete the process by leaving the room in a professional state. Clean up thoroughly, check the ceiling area for tools, and remove all construction debris. In addition, remove all packing material from the premises, retain all installation and operational instructions, and replace all equipment and furniture that was moved. In addition to executing a safe, sturdy, ceiling-mounted projector installation, leaving the worksite in a clean state is the hallmark of a professional installer.

Flat-Panel Displays

The following are general steps required for mounting a typical flat-panel display to a wall.

Step 1   Determine the general location within the room for the flat-panel display from the AV drawings and plans. Verify that the mounting points have been approved by the building owner or the project manager and by a structural engineer. Check that the panel will not cover any of the following: alarm indicators, thermostat, clock, light switches, and IR receivers.

Step 2   Analyze the feasibility of mounting at the location, including the following:

- Locating preplanned construction blocking and cable pathways
- Locating designated electrical outlets
- Verifying there will be no restriction of air flow to the mounted panel

In the case of no preplanning, determine the structural integrity of the proposed location. When in doubt, always consult the project manager and a professional engineer. Large flat-panel displays can be difficult to handle and could cause serious injury. Verify proper visibility of the image. Ensure that a pathway can be created for signal cabling. Determine electrical power practicality.

Step 3   Familiarize yourself with the mounting hardware. The typical mount system will be made in two separate pieces: the wall bracket and the panel bracket. These couple together to hold the flat-panel display on the wall. (Use only premanufactured mount assemblies. Do not attempt to create custom mounting devices in the field.) Follow the manufacturers' instructions and warnings.

Step 4   Secure the area around the construction zone for safety.

Step 5   Carefully lay the flat-panel display face-down on a flat, clean, shipping blanket.

Step 6   Connect the mounting bracket to the flat-panel display as directed by the mount installation manual.

Step 7   Determine the desired center point of the flat-panel display when the installation is completed. Mark this center reference point on the wall.

Step 8   If mounting to a plaster board wall, use a stud finder to locate the wood or metal studs behind the mounting wall. Using a pencil, make light marks to indicate the center of the studs. The wall bracket must be positioned to connect to at least two studs.

Step 9   Hold the wall bracket up to the wall and locate the center reference point with the center of the bracket. Using a level and with the bracket centered, mark a minimum of six mounting points where the bracket will be connected to the studs.

Step 10   Drill pilot holes and secure the wall bracket to the wall with appropriately sized mounting hardware.

Step 11   Carefully raise the flat-panel display to the wall bracket and connect the two brackets. Some mounts will have security screws that will also be installed after connection.

Step 12   Connect all signal and control cables and electrical power.

Step 13   Supply a reliable test pattern and select the proper input.

Step 14   Adjust the display based on the viewing environment.

Step 15   Clean up. After the installation has been completed, remove all construction debris and all packing material from the premises. Retain all installation and operational instructions for system documentation. Replace all equipment and furniture that was moved. Pack up all tools and remove personnel barriers.

The installation of flat-panel displays in a video wall configuration will require careful attention to additional details such as the mounting structure and where the mounting point for each display is located, as well as the placement and wiring of outboard components, including controller devices that are typically mounted in an AV rack. Flat-panel displays for touch-interactive applications also have special considerations based on size, placement, and data return feeds.

Chapter Review

In this chapter, you studied some of the basic concepts and terms used in video capture, processing, and signal wiring and routing to display devices. You learned about video system components and how to install some of them, including ceiling-mounted projectors and rear-projection screens.

Upon completion of this chapter, you should be able to do the following:

- Mount a projector to a ceiling
- Install manual and electric projection screens for front-projection applications
- Install a rear-projection screen and mirrors
- Install a direct view or flat-panel display as specified in the AV design

You have learned about the installation of video components. Equally important are the setup and verification processes discussed in Part V.

Review Questions

The following questions are based on the content covered in this chapter and are intended to help reinforce the knowledge you have assimilated. These questions are similar to the questions presented on the CTS-I exam. See Appendix E for more information on how to access the free online sample questions.

1. What does the first number in an aspect ratio represent?
   A. Depth
   B. Height
   C. Width
   D. Length

2. A 1080p HDTV signal with a frame rate of 30Hz will reproduce images at _____.
   A. 30 lines interlaced
   B. 30 frames per minute
   C. 30 lines per scan
   D. 30 frames per second

3. The number of times per second a display will draw the image sent to it is the _____.

    A. frame rate

    B. refresh rate

    C. signal-to-noise ratio

    D. aspect ratio

4. _____ draws the video image on the display in a sequence of lines from top to bottom.

    A. Progressive scanning

    B. Interlaced scanning

    C. Refresh rate

    D. Frame rate

5. Which component would you use to distribute video from multiple sources to many displays or in different zones?

    A. Distribution amplifier

    B. Scan converter

    C. Matrix switcher

    D. Scaler

6. The picture shows a(n) _____ connector.

Image: Dan Totilca / Getty Images

    A. HDMI

    B. DisplayPort

    C. SDI

    D. DVI-I

7. Which device is used for sending the same video signal to multiple destinations?

   A. Distribution amplifier

   B. Media player

   C. Matrix switcher

   D. Scan converter

8. What are key considerations when determining projector placement? (Choose all that apply.)

   A. Throw distance

   B. Type of lens

   C. Offset value

   D. Standard deviation

9. Which is the preferred method for the correction of keystone distortion?

   A. Optical

   B. Electronic

   C. Digital

   D. Analog

10. What should you be aware of when installing an electric front-projection screen? (Choose all that apply.)

   A. Keep it horizontal when stored or transported.

   B. Some motorized screens have a motor inside the housing.

   C. In a return-air plenum ceiling, local codes apply.

   D. Use the manufacturer's specification for setting physical switches.

Answers

1. C. The first number in an aspect ratio represents the width of the display.

2. D. A 1080p HDTV signal with a frame rate of 30Hz will reproduce images at 30 frames per second.

3. B. Refresh rate is the number of times per second a display will draw the image sent to it.

4. A. Progressive scanning draws the video image on the display in a sequence of lines from top to bottom.

5. C. A matrix switcher is used to distribute video from multiple sources to many displays.

6. B. The picture shows a DisplayPort connector.

7. **A.** A distribution amplifier is used for sending the same video signal to multiple destinations.

8. **A, B, C.** Throw distance, fixed or zoom lens, and the offset value are key considerations when determining projector placement.

9. **A.** Optical keystone correction or lens shift is the preferred method for correcting keystone distortion.

10. **A, B, C, D.** You need to be aware of all the points listed.

PART V

System Setup, Verification, and Closeout

CHAPTER 11

Introduction to System Verification

In this chapter, you will learn about
- Standards compliance
- Audiovisual performance verification standard
- The system verification process
- Verifying wiring, power, and earthing/grounding
- Troubleshooting methods

As the audiovisual (AV) industry establishes standards and best practices on and off the job site, owners, consultants, and integrators will need to conform to the standards and follow the guidelines for the proper design, fabrication, installation, and integration of AV systems. In this chapter, you will learn about the system performance verification standards and resources provided by several organizations. You will also learn how to troubleshoot common problems. These resources and skills will help ensure the systems you work on are complete, without defects, well documented, and can readily be maintained over the life of the system.

Subsequent chapters in Part V will discuss the basics of control systems and networking technologies for AV systems, setting gain and equalization (EQ) in audio systems and creating an extended display identification data (EDID) strategy for video.

For starters, you should familiarize yourself with various AV-related standards. This is an excellent time to verify that the system you installed is compliant with all relevant standards. You will also need some specific test and measurement tools for calibration and verification of data, control, and audio and video.

Your knowledge of the basic principles and theories involved in electronics, audio, video, and networking technologies will be well utilized during the final stretch of AV system integration. Once you have studied the theory and conducted verification of the system on the job site, you will be able to set up and verify new AV systems with confidence.

Duty Check

This chapter relates directly to the following tasks on the CTS-I Exam Content Outline:

- Duty C, Task 2: Wire the Audiovisual Equipment Rack
- Duty C, Task 8: Test the Audiovisual System
- Duty C, Task 9: Calibrate the Audiovisual System
- Duty E, Task 3: Address Needed Field Modifications
- Duty E, Task 4: Repair Audiovisual Systems

These tasks comprise 20 percent of the exam (about 20 questions). This chapter may also relate to other tasks.

Standards Compliance

Upon completion of AV equipment installation, you are ready to turn the system on for tuning, optimizing, and performance verification. Throughout the project and especially during the setup and verification process, you will need to reference and comply with established standards and performance specifications.

In the AV industry, technical standards help create consistency and interoperability. Performance verification standards assure that the AV system is functioning as expected, and management standards increase efficiencies.

Performance Verification Standard

As AV systems have increased in complexity, the possibility of misconfiguration, improper installation, and failure to conform to project requirements has increased. To assist with the process of verification of a system's correct performance, AVIXA has developed the standard ANSI/InfoComm 10:2013, Audiovisual Systems Performance Verification.

The standard provides a framework and supporting processes for determining elements of an AV system that need to be verified, the timing of that verification within the project delivery cycle, a process for determining verification metrics, and reporting procedures. It identifies AV system performance evaluation requirements under these functional categories:

- Audio performance
- Video performance
- Audio/video performance

- Cable management, termination, and labeling
- Control performance
- Electrical
- Information technology
- Operations and support
- Physical environment
- Physical installation
- Serviceability
- Wireless
- System and record documentation

One of the many planning tools provided in the standard is a reference list of approximately 150 verification items and a description of the verification process for each item. This list serves as a template from which you can develop a performance verification list for your project. Some of the relevant verification items for installation are included in the chapters that follow.

A formal systems performance verification (commissioning) process is a critical element of an AV system installation for several reasons:

- It provides a comprehensive step-by-step approach to testing all the elements of what may be a complex system.
- It provides a means for the AV company to objectively demonstrate that the installed system meets the system performance specifications that were defined in the system proposal and installation contract.
- It identifies any problems or issues that should be addressed and corrected prior to handing the system over to the client.

Systems performance verification confirms that the installation tasks are complete and the system is ready for operation.

At the time of publication, the ANSI/AVIXA 10:2013 standard was undergoing a major revision.

System Verification Process

The performance verification process involves many phases, starting with pre-integration and all the way through to final acceptance by the client, as shown in Figure 11-1. The actual testing of system performance starts with testing and verifying system components and subsystems in the workshop and continues through to handing over the controllers and final documentation to your client.

PART V

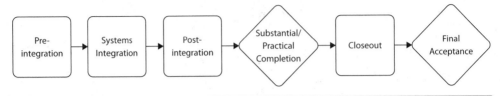

Figure 11-1 Performance verification phases and milestones

The verification process in each of the phases is as follows:

- **Pre-integration verification**   Refers to items that take place prior to systems integration. These items will generally verify existing conditions, such as the presence of device enclosures, or items such as backing/blocking/framing that require coordination among the trades for AV system installation.

- **Systems integration verification**   Refers to items that take place while the audiovisual systems are being integrated or built, including offsite and onsite work. These items will generally verify proper operation or configuration so that the system can function (for example, equipment mounted level, phantom power, termination stress, AV rack thermal gradient performance).

- **Post-integration verification**   Refers to items that take place after the audiovisual system integration has been completed. These items will generally verify system performance against verification metrics, as defined in the project documentation (for example, image contrast ratio, audio and video recording, and control system automated functions).

- **Substantial/practical completion verification**   Indicates conditional acceptance of the project has been issued by the owner or owner's representative, acknowledging that the project or a designated portion is substantially or practically complete and ready for use by the owner; however, some requirements and/or deliverables defined in the project documentation may not be complete. This milestone occurs at the end of the post-integration verification phase.

- **Closeout verification**   Refers to items involved with closing out the project. These items will generally be related to documenting the as-built/as-is status of the systems and transfer of system software, among other items, such as control system test reporting, as-built drawings complete, and warranties.

- **Final acceptance verification**   Indicates that acceptance of the project has been issued by the owner or owner's representative, acknowledging that the project is 100 percent complete; all required deliverables, services, verification lists, testing, performance metrics, and sign-offs have been received; and all requirements defined in the project documentation that occur at the completion of the closeout verification phase have been satisfied and completed. No further project activity will take place after this milestone is verified.

Certain items may need verification multiple times during a project. The reference verification items provided in the ANSI/AVIXA 10:2013 Standard for Audiovisual Systems Performance Verification define the verification phases at which those items should be tested. Where additional items are added on a project-specific basis, they will also need to be allocated a verification phase.

Regional Regulations

You need to know how codes, regulations, and safety procedures apply to your job site. To do that, you need to identify in the location of your project the *authority having jurisdiction* (AHJ), or the *regional regulatory authorities* as they are known in some places. These organizations typically monitor compliance with codes and laws. Standards and best practices are typically established by organizations consisting of representation from various sectors of the industry.

Codes, regulations, and laws are mandated methods, practices, and collections of standards that are enforceable by law. You can be legally punished for not following them. Generally, an inspector will be present to verify that the work is being done according to the law. Each jurisdiction may have its own set of regulations.

If you encounter conflicting regulations, follow the most restrictive regulatory code for the region in which you are working. In other words, if you find the interpretations of applicable regulatory codes and standards are in conflict, follow the requirements of the more stringent code or standard.

Standards are documents that provide requirements, specifications, guidelines, or characteristics that can be used consistently to ensure that materials, products, processes, and services are fit for their purpose. They are often prepared by a standards organization or group and published with an established procedure.

Best practice is the best choice of the available methods for accomplishing a specific task in an industry. Best practices are recognized as the optimum way to do a certain task. They can be a generally accepted industry practice or unique to a specific company.

Resources for Regional Codes

Many different codes and standards apply to various regions in the world. Every country has its own resources for standards. If you need to access standards by your region or industry, always remember that web search engines are your friends. You might begin by researching on the following websites:

- **The International Organization for Standardization**   ISO (www.iso.org) publishes standards to ensure that products and services are safe, reliable, and of good quality. For businesses, they are strategic tools that reduce costs by minimizing waste and errors and increasing productivity. The ISO has technical committees working on standards for a large number of industries.

- **International Electrotechnical Commission**   IEC (www.iec.ch) publishes consensus-based international standards and manages conformity assessment systems for electric and electronic products, systems, and services, collectively known as electrotechnology.

PART V

Performance Verification checklist	Impedance meter	SPL meter
Computer device with interfaces/adapters for Wi-Fi, fiber and wired LAN, serial data, audio, and video	Copper and fiber-optic cable certification testers	Multimeter
Signal analyzer and/or signal analysis software	Network cable testers	Continuity tester
Test signal generators and/or signal generation software	Measurement microphone	Headphones

Table 11-1 Test and Measurement Equipment for AV System Verification

- **Institute of Electrical and Electronics Engineers Standards Association**
  IEEE SA (https://standards.ieee.org) is the standards group within the IEEE, a worldwide professional association for electrical and electronic engineering, telecommunications, computer engineering, and allied disciplines.
- **StandardsPortal** StandardsPortal (www.standardsportal.org) is owned and maintained by the American National Standards Institute (ANSI) and provides standards information on the United States, the People's Republic of China, the Republic of India, the Republic of Korea, and the Federative Republic of Brazil.

Verification Tools

In your pre-installation preparations, you will have developed a verification checklist based on the templates in the ANSI/AVIXA 10 Performance Verification Standard; revisit it to make sure you have the tools for calibration and verification of audio and video signals listed in Table 11-1, including general items such as crimpers and tape that should be in your AV tool bag (see Table 5-1 in Chapter 5). In addition, you will use software-based signal generators and analyzers for some of the systems you have installed. Note that Table 11-1 does not provide an exhaustive list; you will probably have more tools that are specific to your work on various projects.

You may prefer to use tools, interfaces, or software from certain companies. You may trust only certain manufacturers. Ultimately, you will need to select the tools that you are the most comfortable with to get the job done correctly.

Circuit Theory

Installation, verification, troubleshooting, and maintaining AV equipment requires an understanding of the fundamentals of electronic circuit theory. From the work you undertook to obtain your CTS certification, together with the material revised in the section on electrical basics in Chapter 2, and the section on connecting loudspeakers in Chapter 9, you should by now have a basic understanding of the following electronic concepts:

- Voltage
- Current

- Power
- Resistance
- Impedance
- Ohm's law and the power equation
- Transformers
- Series and parallel circuits
- Alternating current (AC) and direct current (DC)

If you are uncertain about how well you understand these ideas, it is advisable to revise your CTS notes or Exam Guide and to check back over Chapters 2 and 9 before proceeding with the more detailed sections that follow.

AC and DC Currents

There are two types of electrical current: direct current and alternating current, as shown in Figure 11-2.

Direct Current

Direct current flows in one direction. Batteries and USB chargers are examples of DC devices; they have both a positive electrode and a negative electrode. In a DC circuit, the negative electrode releases electrons into the circuit, where they travel to the positive electrode, which attracts electrons.

In a DC circuit:

- The current always flows in a single direction (negative to positive).
- The current may not be continuous and may vary in level.
- Most electronic equipment runs on a DC power supply, which comes from batteries, an external power supply device, or a mains-powered DC power supply section within the equipment.
- Most digital data signaling is pulsed DC (which has many of the characteristics of AC).

Figure 11-2

The difference in voltage over time in DC (left) and AC (right)

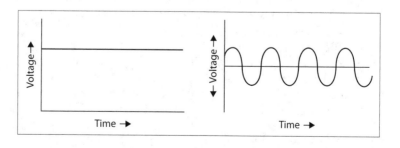

Alternating Current

Alternating current refers to an electrical current that periodically alternates (reverses) its polarity and its direction of flow. The frequency of the reversals is measured in hertz (cycles per second). Alternating currents found in the AV world include all analog audio and video signals, the feeds to and from RF transmitters and receivers, and the mains supply that provides the power to most AV equipment.

The voltage of the AC mains power from general-purpose outlets is typically 120V at a frequency of 60Hz in North America, and 220V at 50Hz in Europe. In other places, mains voltages may be 100V, 120V, 220V, or 230V at frequencies of either 50Hz or 60Hz. Although definitions vary between regulatory jurisdictions, low voltage (LV) is usually considered as ranging between approximately 50V and 1000V AC. Extra low voltage (ELV) generally includes anything under 50V AC. Each jurisdiction has rules or codes on which type of workers are allowed to install or work with differing voltage levels. You must check with your AHJ to see who is able to do what kind of work on your project.

The values quoted for AC currents, power, and voltages are generally not the peak positive or negative values, but *root mean square* (RMS) values. RMS is an algorithm that calculates an average effective value, taking into account both positive and negative peak values.

Circuits: Impedance and Resistance

The amount of current that flows in a circuit is limited by the opposition to current flow in the circuit. In a continuous-DC circuit, resistance is the only form of opposition to current flow. In AC circuits and pulsed-DC circuits, impedance is also a factor in the opposition to current flow.

Resistance

The resistance (R) of a conductor is a property of the material it is made from, its length, and its cross-sectional area. For a given material, resistance decreases as conductor size increases. Conductor size is measured in cross-sectional area (or sometimes by wire gauge).

One way to reduce circuit resistance is to use a cable with a larger cross-sectional area. Resistance increases with cable length; therefore, a longer cable run will result in a lower voltage at the end of the cable (voltage drop). When resistance is too high, it will degrade the signal or even prevent an effective current from flowing.

Impedance

Impedance (Z) is the total opposition to current flow in a circuit. It includes the resistance of the materials in the circuit, plus their *reactance* (X). Reactance is the opposition to current flow in the circuit caused by reactions to the changes in the voltage and current, and is directly related to the frequency of the changes in the current. In a continuous-DC circuit, where there are no changes, the frequency is zero, and hence the reactance is zero, so the impedance is equal to the resistance.

Figure 11-3

A series circuit with an AC signal source, capacitor, and resistor

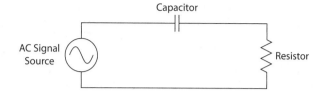

Reactance has two components: *capacitive reactance* (X_C) due to the capacitance in the circuit and *inductive reactance* (X_L) due to the inductance in the circuit.

Although audio signals are AC, for loudspeaker calculations it is simpler (and a reasonable approximation) to treat impedance as resistance when making loudspeaker impedance calculations.

Capacitance

Capacitance occurs when two electrically charged conductors are separated by a non-conductive material. The electrostatic field set up between the conductors stores some charge. The amount of charge stored is governed by the area of the conductors, the distance between them, and the dielectric strength of the material separating them. An electronic component that uses this principle to store charge is called a *capacitor*. The construction of a capacitor usually employs two large-area conducting electrodes, known as plates, to maximize the amount of charge that can be stored.

A charged capacitor discharges when a conducting path is present between the plates. Figure 11-3 shows a capacitor in series with a load, shown as a resistor.

The capacitor affects the flow of current through this circuit. You can see this impact on the graph in Figure 11-4. The current flowing in the circuit is represented by the y-axis, and frequency of the AC is shown on the x-axis. As frequency increases in this circuit, so does current flow. Capacitors have less effect at high frequencies and more effect at lower frequencies.

The capacitance of an electric circuit acts to oppose changes of voltage in the circuit by storing some of the energy. This effect is called capacitive reactance (X_C). The higher the frequency of the signal in a circuit, the lower the capacitive reactance losses.

Significant capacitance between the wires in a cable can attenuate and distort the signal carried in the cable. More cable creates more capacitance, resulting in a greater loss of high frequencies. One means of reducing the capacitance in a cable is to keep the wires well separated, such as in the coaxial cables used for SDI and RF applications.

Figure 11-4

The impact of a capacitor on an AC circuit. As frequency of the AC increases, the current will also increase.

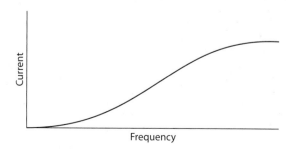

Figure 11-5
A series circuit
with an AC
signal source,
an inductor,
and a resistor

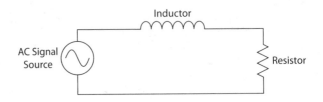

Inductance

When current flows through a conductor, it produces a magnetic field around the conductor. Conversely, when a conductor is placed in a changing magnetic field, a current is *induced* to flow in the conductor. However, the induced current flows in the opposite direction to the current that produced it, and therefore acts to reduce to the original current flow. This opposition to current flow is *inductive reactance* (X_L). The higher the frequency of the signal in a circuit, the higher the opposing current from inductive reactance.

An electronic component that uses the principle of inductance to create a magnetic field is called an inductor. As the magnetic effect around a conductor is quite small, most practical inductors are constructed from a long length of conductive material formed into a coil to concentrate the magnetic field. Many inductors also incorporate an iron (or other ferrous material) core at the center of the coil to further concentrate and intensify the induced magnetic field.

As the induction effect occurs in a changing magnetic field, induction can be produced by an alternating or pulsed direct current in a stationary conductor, or by the relative physical movement of a conductor and a magnetic field. Electric motors, alternators, and generators are devices where a conductor and a magnetic field are in relative motion. Figure 11-5 shows a simple circuit with an AC signal source, an inductor, and a resistance.

Figure 11-6 shows the relationship of the current and frequency in an AC circuit. In the graph, the current flowing in the circuit is represented by the y-axis, and frequency of the AC is on the x-axis. As the frequency of the AC in this circuit increases, the current flow decreases.

Figure 11-6
The impact of
an inductor on an
AC circuit. As the
frequency of the
AC increases,
the current flow
will decrease.

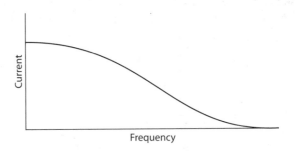

The changing magnetic fields around a conductor carrying an alternating (or pulsed direct) current induce currents with similar waveforms in nearby conductors, even if there is no direct electrical connection. The substantial induced currents from mains power cables can result in unwanted mains-frequency signals (and their many harmonics) being injected into signal cables carrying data, control, video, and audio. This manifests as hums and/or buzzes in analog signals or as random corruption in digital signals.

Induced currents from adjacent signals in multicore cables can cause signal distortion and other forms of interference, the phenomenon known as *crosstalk*. Methods of reducing induced crosstalk in multicore cables include using grounded shields around signal pairs to carry the induced current away to ground, as found in audio multicores and Category 7+ twisted-pair data cables. Another crosstalk reduction approach is to physically separate the wire pairs by including spacer material between the signal pairs, as found in Category 6A+ twisted-pair cables.

Induced currents are the operating principle for the transformer, where the current flowing through a primary coil produces a magnetic field that induces a current in a secondary coil, without any direct electrical connection. Transformers are used to electrically isolate one part of a circuit from another or to change voltages between circuits. In the section on constant-voltage loudspeakers in Chapter 9 you will find more detailed material about transformers and about their application in drive circuits for loudspeaker systems.

Verifying Wiring and Power

During setup and verification of AV systems you require a basic understanding of voltage and current, electric circuits, power and grounding, and electric safety. As an AV technician, you will encounter circuit breakers, electrical outlets, and grounding/earthing systems. Dealing with these systems requires great responsibility. Although you will work only on low-voltage wiring, knowledge of high-voltage best practices and potential problems could help you avoid unsafe conditions.

System Wiring

Wiring issues are the source of many problems in an AV system, so it is important for an AV technician to verify the signal path from source to destination during setup. This step confirms that all the connections have been completed and that the wiring methods and cable pathways are correct. The lead installer will compare the technical drawings against what they discover in the cableways, racks, patch fields, software settings, and elsewhere. During the inspection process, look for the following:

- All cables are of the specified types and have been installed as indicated on the drawings.
- Connectors have been terminated appropriately.
- Cables have been connected to appropriate terminals on equipment as indicated on the drawings.

- Cables have been organized in pathways in the walls and ceiling.
- Cables have adequate service loops where required.
- Cables are dressed correctly for neatness, signal type, and maintenance access.
- Signal separation has been maintained according to design documentation.
- All cables have been labeled with a consistent, and preferably standardized, labeling scheme.
- Cable bend radius has been maintained.

As you work through this process, any errors or discrepancies should be corrected or noted for immediate repair. In addition, if permanent changes were made during installation, document them and submit them to the designer for updating.

Powering the System On/Off

Once you have installed all the equipment and verified that it is wired correctly, it is time to turn the system on.

It is possible to damage equipment, in particular loudspeakers, by improperly powering on and powering off a system. When the power is turned on or off from some audio electronics, a transient pulse may be generated at the output, which, if reproduced by the amplifier, may damage the loudspeakers. Therefore, the following sequence should always be used:

- **When powering a system on**   First, turn on all other electronics; then, turn on power amplifiers.
- **When powering a system off**   First, turn off all power amplifiers; then, after waiting 15 seconds, turn off all other electronics.

When large numbers of analog power amplifiers, high-output projectors, dimmers, or luminaires are switched on at one time, the inrush current can be very large. This may cause circuit protection devices to open, cause voltage sags on the line, or momentarily overload auxiliary generators, which could affect other equipment. If this is a concern, these high-current devices must be switched on in a delayed sequence. Some power distribution systems include a configurable power sequencer for this purpose. Programming the power-up sequence may be part of commissioning the project.

Electrical Safety

Whether the AV project is a fixed installation or temporary for a live event, the electrician or electrical contractor is one of the most important allies of the AV contractor. The electrical contractor is in charge of installing the electrical power from where it enters the building to the outlets you plug into or to directly wired equipment, such as electric projection screens, winch systems, lighting dimmers, or equipment racks.

As every building's power system has capacity limits, the operational power requirements for the AV system will have been discussed and specified between the AV system designer, the project electrical engineer, the project manager, the owners' representative, the electrical contractor, and the power supply provider.

One way to ensure safety is to follow electrical safety practices, including the following:

- Maintain applicable regulatory standards for electrical supply systems.
- Follow proper procedures for grounding/earthing all equipment, including racks.
- Verify that there is compliant current protection on all branch circuits.
- Verify that there are compliant safety switching devices, such as ground-fault circuit interrupters (GFCIs), core balance relays (CBRs), earth leakage circuit breakers (ELCBs), or residual current devices (RCDs), on all branch circuits.
- Know where the power supply cutoff is located. This may be labeled *main breaker* or *main switch* in a local distribution panel.
- Verify that there is adequate lighting and safe access to electrical system service rooms.
- Ensure that all equipment safety earthing/grounding is properly connected, and remove any "earth lifting" or "ground lifting" devices, adapters, or equipment modifications.
- Where enclosed raceways are used for power distribution, verify that the number of enclosed cables does not exceed regulatory requirements.
- Verify that sufficient power has been provided in accordance with the electrical system design, and confirm proper wiring and safety earthing/grounding with the electrical contractor.

Electrical Terminology

Electrical technology, just like AV, has its own jargon and particular meanings for words that you may think you already understand. When it comes to communicating with the electrical engineers and electricians involved in a project, it is important to confirm that you have established the precise meanings of the technical terms being used to describe the project, particularly if you are engaged on a project outside of your usual geographic or linguistic region.

Verifying Power and Earthing/Grounding

To be sure that electrical power is properly supplied to the AV system and that important issues such as earthing/grounding have been correctly addressed, it is useful to understand some of the key principles and best practices in this area.

Power Distribution Basics

Mains power distribution requires two conductors:

- **Supply**   A line/live/active/phase/hot conductor that connects to the power network grid or a local power source such as a generator set or a standby power source. All switches, disconnection, and current protection devices are inserted in this supply conductor. In most large installations there are three sets of live/phase conductors to distribute the load more evenly across the power generation network.

- **Return**   A neutral/return/ground conductor that provides a current return path from connected devices to the power source via a connection to the ground (planet Earth). Regardless of which supply conductor they are connected to, all devices are connected to the same neutral or return path.

Safety Earth/Ground

The safety earth/ground connection in a power distribution system serves the single purpose of providing an uninterrupted low-resistance path to ground/earth from any conductive parts of electrical systems and devices that can be accessed by an end user. The function of the earth/ground is to carry away to the ground (planet Earth) any voltage that may arise on conductive parts of the installation that may otherwise cause harm to the end user. Figure 11-7 shows the earth/ground connection point for an installation.

The safety earth/ground path must not include any switching or current protection devices. In a plugged electrical supply connection, the safety earth/ground pin must connect first on insertion and disconnect last on removal. Figure 11-8 shows a simplified power supply and earthing/grounding system.

Components of an electrical supply system that must have a safety earth/ground connection include:

- Electrical equipment enclosures
- Distribution boards/panelboards and switchboards
- Junction boxes and backboxes
- Receptacles and audiovisual plates

Figure 11-7
An earth spike/
grounding
electrode outside
a building

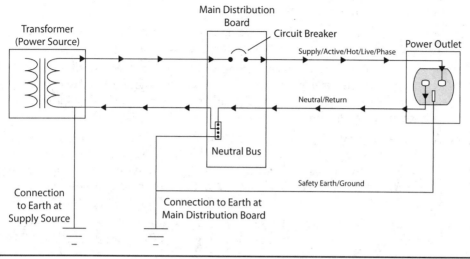

Figure 11-8 The connections for a power distribution and safety system

- Equipment racks
- Luminaires
- Raceways, ducts, conduit, and cable trays
- Dimmer racks
- Electrical load patch systems
- Any other metallic or conductive housing that has the potential to become energized in a fault situation

Best Practice for Power Distribution

Wherever practicable, all circuits for the AV system should originate from a single distribution board. This minimizes the likelihood of any differences in earth/ground potential between the separate circuits used for the AV system.

Safety Reminders

Here are some steps to follow when verifying that the power distribution and safety systems have been correctly installed:

- Make sure all equipment is securely connected to the safety earth/ground.
- Determine what power loss recovery is in place, verified, and documented. If power is interrupted, in which state will the system return?

PART V

- Always maintain proper power sequencing.
  - Power amplifiers should be turned on last and turned off first to avoid any "thumps" or "pops," which may cause loudspeaker damage.
  - Equipment with large inrush currents, such as luminaires, dimmer systems, atmospheric effects devices, amplifier stacks, and high-output projectors, should be powered up sequentially, rather than simultaneously, to avoid tripping current protection devices.
- Always check all AC power outlets for reversed supply lines and a continuous earth/ground connection. This can be done using an inexpensive three-indicator tester. These testers can reveal a live/neutral swap, live/ground swap, or open earth/ground wiring error. They will *not*, however, reveal a neutral/ground transposition error, which should only be diagnosed with the aid of a qualified electrician.
- More sophisticated testers may test for voltage drop under load, equipment grounding, conductor impedance, neutral/ground shorts, and GFCI/AFCI/RCD/ELCB/CBR operation.

Troubleshooting Methods

There are going to be times when the AV system you installed will not work correctly. When this happens, you will need to diagnose where the issues are coming from.

Troubleshooting is a process for investigating, determining, and resolving problems. A reliable troubleshooting approach follows a systematic, logical process intended to clearly define the nature of the problem and narrow down the potential issues until the actual cause of the problem is identified. There are three steps involved in troubleshooting.

Step 1: Symptom Recognition and Elaboration

To clearly identify the problem, you must be able to differentiate normal system performance from abnormal performance. Understanding what the system should do when operating properly requires knowledge about the characteristics and design of the system. This information should be found in the system documentation prepared during the initial verification process. To decide whether the system is malfunctioning, you need to be able to recognize if it is truly a system problem or if it is the result of user error.

Asking questions helps to define before and after points of demarcation to better develop a picture of what the system did, or did not do, prior to the problem. Here are some questions you can use to help you define the problem.

Symptom recognition:

- What is the exact manifestation of the problem?
- What does it normally do?
- When did it last work correctly?
- When did it stop working correctly?

- Did any other equipment or anything related to this system change between the time it last worked correctly and when it failed?

- Who was using it when it stopped working correctly?

- What were they doing when it stopped working correctly?

- What happened when it stopped working correctly?

- If there are error messages, error indicators, or error screens, what exactly do they display?

- Did it stop working correctly immediately, or did this gradually happen over time?

- What does it look or sound like when it is functioning normally?

- Can the problem be reproduced?

- Does documentation exist for the system?

The next step is to elaborate on the symptoms by asking the following questions.

Symptom elaboration:

- Is it a failure of an entire system or a single device within the system?

- Is the system or device turned on?

- Is the switch on the microphone in the ON position?

- Is the master fader or main output controller actually open?

- Is it plugged in and switched on at the power outlet?

- Does it have a fuse or reset button?

- Does the electrical outlet have power?

- Can you reset the system or cycle the power?

These questions will help you gather important information about the symptoms quickly and efficiently.

Step 2: List and Localize the Faulty Function

The best way to streamline the process and narrow the investigation is to continue getting answers to questions. Here are some questions you can use to help you identify and localize faulty functions.

List probable faulty functions:

1. First identify the major functions and subsystems present within the system. What types of failures can occur within these functions and subsystems that may result in the identified symptoms?

2. Make a list of where the problem could be occurring based on the system configuration.

3. Use this information to create a list of potential sources of the system fault.

4. Finally, create a list of the possible faulty components or functions that can be related to the symptoms. These will become the subjects of the testing process during the next step in the troubleshooting process.

Localize the faulty function:

1. Test the items from your list of potentially faulty functions to rule out properly functioning areas and identify the faulty areas. Tests can be simple, such as checking to see whether cables are connected, or more complicated, like using a test device to check for proper output characteristics.

2. Select the specific sequence of items to test based on a number of factors. You can select your test sequence based on any of the following:

   - Select a function or device that will eliminate other possible faulty functions.

   - Select a test that is easy to accomplish.

   - Select a specific item to test based on prior experience with the specific system or component.

3. If you are unable to create a detailed list of all the probable faulty functions, you can begin to analyze the system in your head. Given the symptoms, what would most likely cause the failure?

4. If the cause is not obvious, begin tracing the signal path from an accessible midpoint in the system by substituting known-working system elements for similar elements in the faulty system. For example, cross-plug microphone inputs on a mixing console, or cross-plug inputs on a matrix switcher, and see if the faulty channel moves or remains. You will immediately be able to exonerate half the system, knowing that problem is before or after the cross-plugging point. You can continue with substitutions at midpoints in the section of the system that has now been identified as containing the fault.

5. You may find it helpful to communicate sympathetically with the end users, especially during time-sensitive presentations or live events. Explain to them what you have been doing to solve the problem and what you are going to do next. Ask the end user if they want to implement a backup plan or have you continue to troubleshoot the problem.

6. If you cannot resolve the problem within a reasonable amount of time, consider setting up a temporary, substitute system to accommodate immediate needs. Contact any colleagues to help you fix the problem to accomplish the task as efficiently as possible.

Step 3: Analysis

An analysis of the information you now have will help you further localize the faulty function. Keep on the task by performing the following steps:

1. While you are tracing the signal, check the connectors to make sure they are plugged in all the way. The majority of problems in electronic systems are the result of faulty connections.

2. Continue to test to isolate the problem to the specific faulty component configuration point.

3. While checking, look for exposed wire and damaged insulation. Look for bent or broken pins on the connectors. If there is damage to any of the components, replace them.

Failure analysis:

1. After you have determined the problem, it is time to correct it and work out how to prevent it from recurring. Is the problem the result of any of the following?

   - Inadequate training?

   - Inadequate preventive maintenance?

   - Improper component selection?

   - Disruption in signal flow?

2. If you cannot tell what is creating the problem, carefully observe the equipment. Do you see or hear anything out of the ordinary such as clicks, flashing indicators, humming, or a buzz? If you hear unusual sounds from a particular component, try to replace it first.

3. If the problem has not been solved and you have implemented a backup plan, follow up with your client and apologize for the inconvenience. Explain how you plan to prevent this problem from occurring in the future. Tell the client only what he or she needs to know to understand what the problem is. Too much technical detail may confuse the client. Assure the client that the problem will be solved as soon as possible. Ask when they plan to use the equipment again and ensure it is fixed before that date.

4. In a situation where the service request was left unresolved, think about what you might do differently to speed up the process next time.

5. Ask yourself several questions:

   - What can you learn from the situation?

   - What can you do to prevent it from happening again?

   - When can you show the customers how the equipment works?

   - Can you think of ways you could be better prepared for this situation?

   - Do you know how the system is wired?

- Are there schematics for the system so you trace the signal path?
- Do you have a crash/emergency maintenance kit?
- Are there more items you should include in the kit next time?
- Have you created a log of past problems that you or others have encountered?
- Is this a recurring problem?
- What were some solutions you or others used in the past?

6. Finally, contact your suppliers and colleagues in the industry to see whether they have faced similar problems and how they responded. Take time to learn about the equipment.

In AV installation, troubleshooting starts before the gear leaves the warehouse by determining which critical systems should be redundant. On site, the technical crew has tools, troubleshooting devices, and spare equipment specific to the systems involved. Troubleshooting test points should be established in each system for quick location of a signal or equipment failure.

In live events production, there is often little time to effect a solution when a technical problem occurs. This is why critical systems are often designed with redundant components, live synchronized show-tracking, and automated cut-over systems. When touring, a local AV supplier should be secured in advance as backup for equipment and services.

If you turn on an AV system and nothing happens, make sure the power for each component is turned on and that there are no blown fuses or tripped over-current devices. Check all of the settings and controls on each piece of gear. Look for master switches or faders in the wrong position and muted channels. Next, check all of the connections and look for missing or unplugged cables. Trace the signal from the input to the output.

Always remember it is better to be safe than sorry. Treat every circuit or power receptacle as live. When in doubt, talk to the electrical contractor about any electrical concerns you have. Taking a moment to ask the right questions could be a lifesaver.

Best Practice for Troubleshooting
The following are some basic troubleshooting best practices:

- Assume nothing. Anything that *can* go wrong *will* go wrong.
- Change only one thing at a time.
- Test after each change.
- Use signal generators to provide a known reference to measure against.
- Use signal analyzers to obtain quantifiable information.
- Document your procedures and findings.

Chapter Review

In this chapter, you learned about industry standards and the importance of compliance with standards, codes and laws, and best practices on AV system installation projects, from the pre-installation phase through setup and verification to final handoff to the client. You studied the general structure of the ANSI/AVIXA Performance Verification Standard and should refer to this document for further study. Additionally, resources for searching regional regulatory codes were presented for projects you will be working on in the future.

Since many problems encountered during setup and verification stem from electrical issues, you studied some of the basics of electrical wiring, power distribution, and grounding.

Upon completion of this section, you should be able to do the following:

- Identify the tools an installer will use when verifying an AV system
- Verify that an AV installation is compliant with all applicable standards
- Verify that each AV component is wired correctly and receives power
- Explain the process for troubleshooting a problem

You have learned the process for system verification, which will help you verify the audio and video equipment you have installed.

Review Questions

The following questions are based on the content covered in this chapter and are intended to help reinforce the knowledge you have assimilated. These questions are similar to the questions presented on the CTS-I exam. See Appendix E for more information on how to access the free online sample questions.

Note that some of the questions in the CTS-I exam are based on content you would have studied in preparation for the CTS exam, and with at least two years of work experience, you should be well versed in that content by now.

1. What are the benefits of using the ANSI/AVIXA 10:2013 Standard for Audiovisual Systems Performance Verification? (Choose all that apply.)
   A. Reducing project risk
   B. Aligning outcome and performance expectations
   C. Providing a verifiable outcome
   D. Creating reporting that completes the project documentation

2. During which phases of an AV project should performance verification be conducted? (Choose all that apply.)
   A. System design
   B. Pre-installation
   C. Completion
   D. System closeout

3. How can you access regional codes for safety procedures?

    A. Check with the client

    B. Check with local residents

    C. Check websites of standards organizations

    D. Check websites of manufacturers

4. When powering a system on, which device from the following list should you turn on last?

    A. Pre-amplifiers

    B. Wireless mic receivers

    C. Blu-ray players

    D. Power amplifiers

5. When powering a system off, which device from the following list should you turn off first?

    A. Pre-amplifiers

    B. Wireless mic receivers

    C. Blu-ray players

    D. Power amplifiers

6. What is the point at which mains power enters a building?

    A. The feeder

    B. The main distribution board

    C. The electrical entrance

    D. The main line

7. In an electrical power distribution installation, the safety earth/ground connects to___. (Choose all that apply.)

    A. the ground (planet Earth)

    B. raceways, ducts, conduit, and cable trays

    C. exposed conductive parts of the installation

    D. equipment racks

8. If a fuse blows and you suspect there is a problem with high-voltage (AC mains) wiring, who should you contact immediately?

    A. Video engineer

    B. Licensed electrician

    C. Manufacturer's rep

    D. Design consultant

9. What causes the majority of problems in electronic systems?

   A. The end user

   B. An impedance mismatch between devices

   C. Faulty connections

   D. Poorly written system documentation

10. What is the industry best practice when troubleshooting? (Choose all that apply.)

   A. Change only one thing at a time

   B. Change the faulty device right away

   C. Change the signal analyzer

   D. Change the signal generator

Answers

1. **A, B, C, D.** Using the ANSI/AVIXA 10:2013 Standard for Audiovisual Systems Performance Verification will help reduce project risk, align outcome and performance expectations, provide a verifiable outcome, and create reporting that completes project documentation.

2. **A, B, C, D.** Performance verification should be conducted during all phases of an AV system installation project.

3. **C.** Check the websites of standards organizations such as the ISO, IEC, ANSI, IEEE, and NEC for regional codes for safety procedures.

4. **D.** Power amplifiers should be turned on last when powering on an AV system.

5. **D.** Power amplifiers should be turned off first when powering off an AV system.

6. **B.** The main distribution board is the point at which power enters the building.

7. **A, B, C, D.** In an electrical power distribution installation, the safety earth/ground provides an uninterrupted low-resistance path to ground (planet Earth) from all conductive parts of the electrical system and devices that can possibly be accessed by an end user.

8. **B.** Contact the electrical contractor or licensed electrician if you suspect there is a problem with the electrical wiring.

9. **C.** The majority of problems in electronic systems are the result of faulty connections.

10. **A.** Best practice when troubleshooting is to change only one thing at a time.

PART V

Networks

In this chapter, you will learn about
- Network basics
- Common network topologies
- Network cables and connectors
- IP address schemes
- Data link protocols for AV systems

The audiovisual (AV) and information technology (IT) industries have converged. As an AV professional, you need to know the basics of how computer networks are set up and how Internet Protocol (IP) networking works. Configuring network-attached devices is a collaborative effort. Often, the integrator's role is to communicate system needs at meetings and through documentation and to verify that the system performs as expected rather than to directly assign IP addresses or alter network component settings. You need to be able to identify an IP address, test for network connectivity, and perform basic network troubleshooting. Most importantly, you need to have the knowledge and vocabulary to discuss the network requirements of the AV system you are installing with the IT stakeholders and other vendors on the project.

You have learned about installing audio and video technologies, and as you prepare to set them up for optimal performance, you will need to know how they work on your client's computer network. Many AV components need IT network support to function, as well as for remote monitoring, control, and troubleshooting. Many projectors, for example, use data networks to report on light source condition, filter status, and other important functions. In large screen and digital signage applications, computers and networked media players store and forward content to digital displays. Conferencing systems are, by their very nature, networked audio, video, and messaging systems. Training, huddle, and collaboration spaces use networks to communicate and share resources. Streaming video can be distributed over network infrastructure in real time to remote audiences around the world or to desktops and tablets around a campus.

This chapter describes the different types of networks, devices, and connections used for the transmission of audiovisual data. You will learn about the layers of a network that affect networked AV systems and about the control systems that provide the critical link between AV systems and the user. Since information technology is a vast field of study and one that is continually evolving, this chapter will focus your study on topics related to networked AV systems.

When setting up and verifying networked AV systems, you will need to collaborate with IT professionals. Being able to explain the network requirements of the AV systems you will be installing, as well as being attentive to the concerns of IT personnel, is an acquired skill. To communicate effectively, you will need to speak their language and have an understanding of their responsibilities within your client's organization. Let's first take a look at the role and responsibilities of an IT network manager.

Role of IT Network Managers

IT network managers often have multiple responsibilities. Different parts of organizations have different expectations network managers must address on a daily basis. Consider how busy they may be while serving as the main point of contact on all IT-related services and troubleshooting for a corporate office or academic campus. They must communicate effectively not only with internal clients such as the human resources, finance, operations, and accounts departments within their own company but also with external clients such as partners and customers. They may have to manage a staff of IT professionals and be able to escalate issues in a timely fashion for the infrastructure of their company's business to run smoothly.

Network managers often also oversee the development and maintenance of the overall IT strategic plan. They must therefore approve and monitor major projects, the IT budgets, and priorities, as well as coordinate these strategies between departments and users within the organization. Out of these needs, a network manager seeks to meet three major goals:

- Efficiency of the network
- Security of the network
- Maximum uptime or use on the network

Consider how installing AV devices on the network may complicate or influence these goals. As you dive deeper into how networks are put together and how they work, keep in mind the significant role of a network manager. The more you know about their job, the easier it will be to earn their trust and help you complete your tasks on the AV project.

Duty Check

This chapter relates directly to the following task on the CTS-I Exam Content Outline:

- Duty C, Task 6: Configure Network Properties of Equipment

This task comprises 4 percent of the exam (about four questions). This chapter may also relate to other tasks.

Network Device Inventory

Your principal task in configuring attached AV components is to communicate effectively with those responsible for managing the network. The Network Device Inventory document is an inventory of networked devices in an AV system. It is a means of "starting a conversation" with IT management at the earliest possible stages of the networked audiovisual system design process. It is usually developed in a spreadsheet by the AV designer in collaboration with the client's IT network manager; you will need to refer to this document during setup and verification.

AVIXA recommends including the following information in the document:

- **Device narrative** Record what devices need to be connected to the network and why. What information do these devices need to send and receive? What other devices do they need to "talk" to?

- **Interdevice communication** Record the transport protocols each device will use, including the Transmission Control Protocol (TCP) and User Datagram Protocol (UDP) port numbers. Is quality of service (QoS) supported? Will the device's traffic be multicast or unicast? What codec will be used? How much bandwidth will it consume?

- **Device properties** As actual components are selected, record their relevant manufacturer specifications, including the name of the manufacturer, model number, Media Access Control (MAC) address, and software and firmware version numbers.

- **Routing and addressing** As actual components are selected, record the information needed to address each device and route data to and from the address. This section includes the device's physical location, as well as all applicable IP addressing, Domain Name System (DNS), and port and protocol information.

- **Conferencing addressing** If applicable, record key data for all conferencing devices, including the gatekeeper address, system name, and E.164 address.

- **Passwords** If requested by the customer, provide any username and password information for each device.

Each device's properties and requirements will be determined throughout the needs assessment, system design, product selection, and installation processes. Recording all this information in one document allows discussions between system designers and network managers or internal and external service providers to be conducted quickly and with clarity.

Ports and Protocols

All network-ready devices will have specifications such as those detailed in Table 12-1. These specifications will be documented at different stages of AV system design and integration. For example, the decision to use multicast or unicast for streaming audio or video would be made during the design phase and be based on the operational requirements, the number of connected devices, available bandwidth, and overall capabilities of the network.

The MAC address, on the other hand, will usually be documented during the installation process, after the specific piece of equipment is onsite and its location in the network

PART V

Device	Rationale			
Manufacturer	Model No.	Software version	Firmware version	MAC address

Routing and Addressing

Location:	☐ Static ☐ Dynamic ☐ DNS	IP Address	☐ Wired ☐ Wireless	Host:	Ports:
Gateway IP:		Subnet Mask		Server(s):	Protocols:

Inter-device Communication

TCP/UDP	QoS Tag	Multicast Unicast	Codec	Bandwidth Incoming:	Bandwidth Outgoing:

If Applicable: SNMP (Simple Network Management Protocol)

SNMP Server	SNMP Community Name	Admin Username	Admin Password

If Applicable: Conferencing Addressing

Gatekeeper Address	System Name	E.164 Address

If Requested: Username and Password

Device Username	Device Password

Notes

Table 12-1   The Ports and Protocols Document Contains Device Properties

has been decided. You will get it from the physical device itself, and it may be one of the last specifications to be documented.

The AV designer in collaboration with the network manager creates the ports and protocol spreadsheet, which is used by the installer and network manager to set up the networked AV system.

Network Components

The first step to understanding an AV system's network requirements is mastering the terminology associated with IT networks. This section will explore basic terms, devices, and functions.

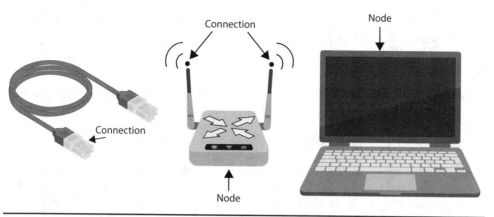

Figure 12-1 Nodes are active devices that send and receive data. Connections are the physical means by which data is transmitted.

Nodes and Connections

In the IT world, *network* refers to a computer network or data network. All networks have two main parts: nodes and connections, as shown in Figure 12-1.

Nodes are the devices used for sharing data. A node could be any active device on a network that sends and receives data. In the early days of networking, a node was basically a computer. Today, a node could be a computer, a mobile phone, a media server, an IoT device, a mixing console, a video projector, or any other content-carrying device. On IP networks, a node is a device that has an IP address. Active network components such as routers, wireless access points, and switches are also nodes.

Connections are the means by which data travels from one node to another. The base-level connection could be any physical signal transmission medium such as radio frequency (RF) waves, copper cable, or fiber-optic cable. Passive devices such as patch panels also fall into the category of physical connections or links.

Network Interface Controllers and MAC Addresses

Every device that connects to the network must have a Network Interface Controller (NIC) with an associated MAC address. A NIC is an interface device that allows you to connect a device to a network. Many devices have multiple NICs: one for each of the networks (wired, fiber, or wireless) it can connect to.

A MAC address is the actual hardware address, or number of the NIC device. Every device has a globally unique MAC address that identifies its connection on the network. MAC addresses use a 48-bit number, expressed as six groups of two hexadecimal numbers, separated by a hyphen or colon. For example, *bc:5f:f4:56:a7:3d* is the MAC address of the Ethernet port connecting the author's computer to its local network. The first part of the address (*bc:5f:f4*) indicates the interface controller manufacturer (in this case, ASRock, Inc., in Taiwan), and the second part of the address is a serial number for the product or circuit component. In devices with multiple NICs, each NIC will have its own associated MAC address.

Network Devices

As shown in Figure 12-2, devices that are used in data networks include switches, routers, gateways, wireless access points, and servers. Blended devices combine the functionality of several devices.

Take a look at each device's functionality:

- **Blended devices**   It is important to note that while the functionality of gateways, routers, and switches are discussed separately here, they do not have to be separate physical devices. Much like an audio digital signal processor (DSP) combines the separate functions of a mixer, equalizer, and compressor/limiter, networking devices often include several functions in a single box. For example, routers often include the functionality of a switch and a Wi-Fi controller. On a small network, there will probably not be a separate gateway, router, Wi-Fi controller, and switch. In fact, there may be only one router, which acts as a gateway to the Internet, a wireless access point for local Wi-Fi connections, and a switch directing traffic among the devices on the network.

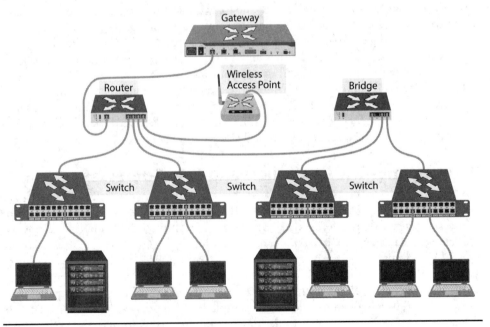

Figure 12-2   Devices used in data networks

- **Switch**   A network switch provides a direct physical connection between multiple devices through storing and forwarding data. As each device is connected, the switch collects and stores its MAC address. This allows devices connected to the same switch to communicate directly with each other via Ethernet, using only MAC addresses to locate each other. A switch acts as a multi-port bridge between the network segments connected to its ports. There are two types of switches: unmanaged and managed.

  - Unmanaged switches have no configuration options. You just plug in a device, and it connects.

  - Managed switches allow the network technician to perform functions such as adjust port speeds, set up virtual local area networks (VLANs), configure QoS, and monitor data traffic. Managed switches are commonly used in corporate and campus environments and are required for some networked AV applications.

- **Router**   A router forwards data between devices that are not directly physically connected. They mark the border between a local area network and a wide area network. After data leaves a local area network (that is, travels beyond the switch), routers direct it until it reaches its final destination. When data arrives at a router, it examines the packet's logical address, the IP address, to determine its final destination and/or next stop.

- **Gateway**   A gateway connects a private network to outside networks. Similar to a router, a gateway examines a data packet's IP address to determine its next destination. A gateway can also forward data between devices that are not directly physically connected. All data that travels to the Internet must pass through a gateway. Routers below the gateway will forward packets destined for any device that cannot be found on the private network to the gateway. When traffic arrives from outside the private network, the gateway forwards it to the appropriate router below. Gateways also translate data from one protocol to another. For example, when data leaves a private network to travel across the Internet, a gateway may translate it from a baseband to a broadband protocol.

- **Bridge**   A bridge forwards data between multiple networks or network segments, creating a single aggregated network. The bridge reads the destination on an incoming packet, stores the data, then forwards it to whichever port connects to the destination device. In a local area network (LAN), the devices are identified only by their MAC address. Most network switches are multi-port bridges.

- **Wireless access point (WAP)**   A WAP is a common blended device, combining a Wi-Fi controller and an Ethernet bridge. It forwards packets between connected Wi-Fi devices and a LAN, based on the MAC address of the connected devices. In an access point, the wireless devices are in the same subnet as the Ethernet port.

- **Server** Almost any network of significant size uses servers to share resources among connected nodes. A server is not really a piece of hardware. Rather, it is the role a piece of hardware can play. If a piece of computer hardware does nothing but provide services, that computer is called a *server*. The term *server* can also refer to a program or service that runs on a computer alongside other programs. In a client-server network architecture, a server provides services to dependent nodes. It does resource-intensive or remotely hosted tasks that those nodes cannot perform for themselves.

Servers on larger networks are usually hosted on many separate computers or virtual machines. An organization may host several servers on a single computer or may have a dedicated computer for each service. A computer whose only task is to perform a particular service is known as a *thin server*. A thin server is a server that offers only one service. Typically, a thin server resides on a dedicated computer, virtual machine, or container, configured with only the functionality required to perform the service. This increases the resources available for the server's dedicated task. It also minimizes the risk that a malicious intruder could exploit one of the operating system's unused features. It is a good idea to turn off any features that will not be used to ensure that they are not exploited.

The OSI Model

The Open Systems Interconnection (OSI) Model defines seven layers of networking functions and provides a common reference for describing how data is transported across a network. Figure 12-3 shows the layers in the OSI Model.

The OSI Model can be used to describe the functions of any networking hardware or software, regardless of equipment, vendor, or application. However, not all networking technologies or AV devices fit into the strict categories of the OSI Model. Many operate at several different layers. For example, the TCP/IP model, the basis for most AV data

Figure 12-3

Layers in the OSI Model

7. Application

6. Presentation

5. Session

4. Transport

3. Network

2. Data Link

1. Physical

OSI Model		TCP/IP Model
7	Application layer	Application layer
6	Presentation layer	
5	Session layer	
4	Transport layer	Transport layer
3	Network layer	Internet layer
2	Data Link layer	Network Access layer
1	Physical layer	

Table 12-2   The seven layers of the OSI model compared to the TCP/IP model

communications, compacts all the functions into just four layers, as shown in Table 12-2. Still, the OSI Model provides a useful shorthand for discussing networking software and devices. You will often hear networking professionals and manufacturers talk about the "layer" at which a technology operates or the layer at which a problem is occurring.

Layers of the OSI Model

Knowing the OSI layer (or layers) at which a technology operates can be useful in several ways. A layer in the OSI Model can reveal what a technology does and when those events occur in the data transfer process. For instance, Application layer error checking occurs at the host and may be aware of the kinds of errors that really matter to the software application. Transport layer error checking has no awareness of the application; it just looks for any missing packets.

The OSI Model provides a road map for troubleshooting data transfer errors. It describes the signal flow of networked data. Just as you would use a signal flow diagram to troubleshoot a display system in a conference room, checking at each point in the path, you can troubleshoot a network by observing the data transfer process one layer at a time.

The OSI Model can also indicate which service providers are responsible for each stage of data transfer. Layers of the OSI Model often represent a service provider handoff. For instance, an AV technology manager may be responsible for layer 1 and 2 devices and layer 5 to 7 software, while the network manager controls layer 3 and 4 technology.

The Layers

The OSI Model uses a stack of seven layers to communicate or transmit a file from one computer to the next. Essentially, layers 1 to 3 get data from point A to point B, layers 5 to 7 define what the data does when it gets there, and layer 4 does a little bit of both.

- **Layer 1 – Physical layer**   The Physical layer can be copper, fiber, or even a wireless link; the devices need to be plugged into the network. The Physical layer sends and receives electrons, light, or electromagnetic flux.

- **Layer 2 – Data Link layer**   This layer is the interface to the Physical layer; it uses frames of information to talk back and forth. The addressing scheme uses MAC addresses. One MAC address talks to another MAC address. Switches use layer 2 to send and receive frames.

- **Layer 3 – Network layer**   This layer addresses data packets and routes them to addresses on the network. Packets "ride" inside layer 2 frames. Layer 3 adds IP addressing. Routers can send, receive, and route IP addresses. Note that many routers include an integrated switch.

- **Layer 4 – Transport layer**   This layer works with ports to identify the destination of the data within the host. These virtual ports can be found at the end of an IP address such as 192.168.1.35:80. The :80 is the port number. Port 80 is generally associated with Hypertext Transfer Protocol (HTTP), used by the World Wide Web. Routers can route by ports.

- **Layer 5 – Session layer**   This layer controls (starts, stops, monitors, keeps track of) layer 4 and layer 3. For example, Real-Time Streaming Protocol (RTSP) keeps track of the UDP layer 3 traffic. It keeps track of the data as it makes it to the far end address. UDP cannot do it on its own; it requires a session protocol to do the work.

- **Layer 6 – Presentation layer**   This layer transcodes or translates data between the Session layer and the Application layer. It is a kind of go-between. Data encryption and decryption are done at this layer. An application cannot talk to data directly; it needs an interpreter, and this is it.

- **Layer 7 – Application layer**   This layer defines human interaction with data on the network. Using the streaming example, the Application layer is occupied by the Windows Media Player. The application has the controls to manipulate the output of audio and video. It takes *all* the layers for a player application to retrieve the stream and play it on a desktop. No layer is skipped or bypassed. When troubleshooting, all layers need to be considered.

Table 12-3 shows the seven layers of the OSI Model, the type of signal each carries, the type of protocol involved in transport with each layer, and the device or function engaged in each layer.

Layer	Name	Carries	Sent to/From	Device/Function
L7	Application	Data	Application protocols	Programs (HTTP, FTP, e-mail)
L6	Presentation	Data	Data translation protocols	Encryption
L5	Session	Data	Session protocols	Session
L4	Transport	Segments	Ports	Firewalls
L3	Network	IP packets	IP addresses	Switches, routers
L2	Data Link	Ethernet frames	MAC addresses	Switches
L1	Physical	Pulses	NIC, wire, Wi-Fi controller	Copper/fiber/wireless

Table 12-3   OSI Model Layers with Type of Signal, Protocol, and Device or Function Involved in Transport

Data Transmission and OSI

Layers 1 to 3, known as the media layers, define hardware-oriented functions such as routing, switching, and cable specifications. These are the areas that most concern AV professionals.

Layers 4 to 7, the host layers, define the software that implements network services. Each layer contains a broad set of protocols and standards that regulate a certain portion of the data transfer process. A data transfer on any given network likely uses several different protocols at each layer to communicate. Layer 4, the Transport layer, is also important to AV professionals because it is where the transition between gear and software occurs. This layer tells the media layers which applications are sending the data. It also divides and monitors host-layer data for transport. Data is sent across a network by applications. That means when a computer sends a message, that message starts out at layer 7, the Application layer, and moves down through the layers until it leaves the sending device on layer 1, the Physical layer. The data travels to the receiving device on layer 1 and then moves up through the layers until the receiving device at layer 7, the Application layer, can interpret it.

Network Connections

Different physical technologies are used to transport data over long distances on wide area networks, including optical fiber, T-1 lines, coaxial cable, cellular networks, satellite, direct wireless, and digital subscriber line (DSL). The network connections that AV professionals have to deal with directly, though, are primarily those used within local area networks, that is, the physical connections within a system or building. For LAN connections, the three most common physical transmission methods are as follows:

- Over copper, as voltage
- Over fiber, as light
- Over air, as radio frequencies

Of course, it is not as simple as it sounds. Within each transmission medium, there are several options. One copper-wire network cable may have very different capabilities, limitations, and internal design than another. The same is true for fiber and wireless networks. As an AV professional, you should know which medium is best for your client's application and why.

Copper Twisted-Pair Network Cables

The most common transmission medium for LANs is four-pair,100-ohm, copper twisted-pair cable, which is available in several categories based on the signal bandwidth and crosstalk. These categories, often abbreviated as Cat, are Ethernet standards under the EIA/TIA 568 standard. To label a cable as Cat cable, a manufacturer must produce a cable that meets the minimum specified standards for the category. If a cable is installed and terminated correctly, it should be expected to perform as listed in Table 12-4.

Cable Category	Description	Bandwidth	Data Speed
Cat 5e	Currently defined in TIA/EIA-568-B	Up to 100MHz	Typically both 100Mbps and 1Gbps
Cat 6	Currently defined in TIA/EIA-568-B	Up to 250MHz	Typically both 100Mbps and 1Gbps
Cat 6a	Currently defined in TIA/EIA-568-B.2-10	Up to 500MHz	Typically 10Gbps
Cat 7	Informal ISO/IEC 11801 Class F cabling	Up to 600MHz	Typically 10Gbps
Cat 7a	Informal ISO/IEC 11801 amendment 1 Class F cabling	Up to 1GHz	Suitable for 40Gbps
Cat 8	Defined in ANSI/TIA 568-C.2-1	Up to 2GHz	Suitable for 40Gbps at distances up to 30–36m (100–120ft)

Table 12-4 Category Cable Designation with Corresponding Bandwidth and Speed

More information on the construction and performance of these cables can be found in the section in Chapter 6 on twisted-pair network cables.

RJ-45 (8P8C) Connectors

The connector used for Cat cabling is the 8P8C, known widely as RJ-45 (see the picture of the RJ-45/8P8C connector in Table 6-4 of Chapter 6). For some lower-speed Ethernet connections, only four of the eight wires in the cables are actually used for data transmission. Gigabit Ethernet or faster, as well as certain special devices, use the other four conductors for other purposes. The specific pin-out configuration depends on the connection standard you are using.

There are two wiring formats within the IEEE 802 standard: T568A and T568B (see Figure 12-4). Both formats are acceptable to use; ask the IT manager which format is used within the facility.

If a cable is terminated with a T568A on one end and T568B on the other, the cable is known as a *crossover* cable. A crossover cable allows two devices, such as two computers, to connect and share information at up to 100Mbps, without the use of a switch or router that normally does the crossover electronically.

Crossover cables are not used as network infrastructure cables. However, they may be used to connect a network device console port to a computer at up to 100Mbps so configuration changes can be made. As a result, it is common to find a crossover cable in the data closet, usually well identified as being nonstandard. Never make assumptions about a cable; always test it before using it. Most modern Ethernet devices will connect automatically, whether or not a crossover cable is used.

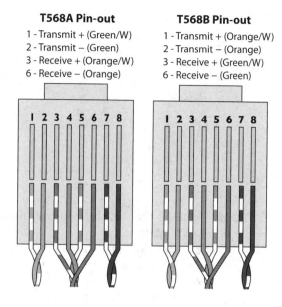

Figure 12-4

The two wiring termination formats within IEEE 802 standard

T568A Pin-out

1 - Transmit + (Green/W)
2 - Transmit – (Green)
3 - Receive + (Orange/W)
6 - Receive – (Orange)

T568B Pin-out

1 - Transmit + (Orange/W)
2 - Transmit – (Orange)
3 - Receive + (Green/W)
6 - Receive – (Green)

1 2 3 4 5 6 7 8

1 2 3 4 5 6 7 8

Fiber-Optic Cable

As noted in Chapter 6, fiber is a transparent optical medium used for transmitting modulated light from point A to point B. Fiber-optic cable offers high bandwidth and maintains total electrical isolation. This allows fiber to maintain signal integrity in electromagnetically noisy environments, experiencing little signal degradation over long distances. It does not burn, and it withstands aging and corrosion.

Fiber-optic connections are also popular for security reasons. It is much more difficult to covertly capture data from a fiber-optic line than from copper wires or radio transmissions. You have to physically intercept the path of the light.

You may want to review the advantages and limitations of using fiber-optic cable, including single-mode and multimode, and the connectors used in AV in Chapter 6.

Wi-Fi Connection

In addition to wired connections, you are likely to encounter the need for wireless connection from clients who want to control devices, connect to portable devices, or send content wirelessly over the network. Figure 12-5 shows a typical wireless access point.

The wireless connection known as Wi-Fi is defined by the IEEE 802.11 standard. This standard is a set of MAC and Physical layer specifications for implementing wireless LANs. It has been amended several times to provide the basis for the latest wireless network products. The speed of a Wi-Fi connection depends on the RF signal strength and the revision of 802.11 with which you connect. As signal strength weakens, the speed of the connection slows. The number of users accessing the wireless devices also affects connection speed. The most commonly used versions of Wi-Fi today are 802.11a, 802.11g,

PART V

Figure 12-5
A typical wireless access point (Image: thongseedary / Getty Images)

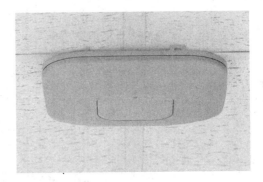

802.11n, and 802.11ac. Table 12-5 shows the frequency band, typical throughput, and maximum throughput for each standard.

Wi-Fi is extremely popular, both among users and manufacturers. It is difficult to find laptop and desktop computers today that do not have factory-integrated wireless technology. Computer manufacturers assume users would rather connect wirelessly than via a network cable. This might be true for the home user. For the enterprise network, however, Wi-Fi can create more problems than it solves. Before choosing Wi-Fi as a physical connection medium, you should carefully weigh its advantages and disadvantages.

Revision	Release Date	Frequency Band	Typical Throughput	Maximum Throughput
802.11a Wi-Fi 2	October 1999	5GHz	27Mbps	54Mbps
802.11b Wi-Fi 1	October 1999	2.4GHz	~5Mbps	11Mbps
802.11g Wi-Fi 3	June 2003	2.4GHz	~22Mbps	54Mbps
802.11n Wi-Fi 4	September 2009	5GHz and/or 2.4GHz	~144Mbps	600Mbps
802.11ac Wi-Fi 5	December 2013	5GHz	~433Mbps	3.5Gbps
802.11ad	December 2012	60GHz	~1.5Gbps	6.7Gbps
802.11ax Wi-Fi 6	December 2018	5GHz and/or 2.4GHz	~4.8Gbps	11Gbps
802.11ay	November 2019	60GHz	~6Gbps	20Gbps
802.11be Wi-Fi 7	2021 (estimated)	5/6GHz		40Gbps

Table 12-5 802.11 Wi-Fi Standard Revisions

Wi-Fi Advantages

Wi-Fi allows users great flexibility. It is a lot cheaper than cable. It is also the default customer expectation. Users expect to be able to walk into a meeting room or presentation space, connect to the Internet, and launch a presentation from their laptops, tablets, or mobile phones. You could try to address this requirement by providing a connection for every device imaginable. Or, you could let users connect to the presentation system using an ad hoc Wi-Fi network. Wi-Fi just seems easier for several reasons. The following are the advantages of Wi-Fi:

- Wireless hotspots are everywhere. If the user's device is Wi-Fi capable, they can nearly always find a way to connect to the network.

- The convenience of Wi-Fi encourages mobility and therefore productivity. Workers can reasonably expect to walk to a park, sit in a cafeteria, or ride in an airplane and still conduct business. Their work could feasibly be accomplished from anywhere.

- Wi-Fi requires little infrastructure. Installing a small-scale wireless connection is almost as simple as pulling an access point out of a box and plugging it into the wall. Most wireless access points are sold with a functional default installation for small networks.

- Wireless networks are reasonably scalable. If you need to add more client nodes to a wireless network, all you have to do is add more access points and the supporting infrastructure.

- Wireless networks are cheaper to install than copper or fiber networks, which require the installation of cables, connectors, wall jacks, switches, and patch panels for each connection point. With Wi-Fi, all you need is an access point.

Wi-Fi Disadvantages

Although Wi-Fi offers numerous benefits, the disadvantages can be potentially catastrophic. For some applications, you simply cannot use Wi-Fi. Here are some of the disadvantages:

- Wi-Fi has a limited range. Restrictions on the range of Wi-Fi devices are established by the 802.11 equipment standards and the International Telecommunications Union Radiocommunication Sector (ITU-R).

- Wi-Fi devices are susceptible to radio frequency interference, intermodulation, and jamming, particularly in the 2.4GHz band.

- Equipment selection and placement can be difficult. Proper placement, antenna selection, and signal strength are key. Building construction materials affect RF propagation. Some construction materials will dampen, or even block, RF signals. Others act as reflectors, enhancing signal quality inside the space.

- Wi-Fi devices can be expensive. Although a Wi-Fi network may seem cheap at first, the purchase of additional access points, repeaters, or highly directional antennas to expand the network's range will increase costs rapidly. The cost and complexity of building an extended wireless network can grow quickly.

- Even as Wi-Fi speeds increase, the fastest Wi-Fi is no match for the dependability and throughput of wire or fiber networks. If you need to stream live, high-quality UHD video, Wi-Fi is often not a dependable option.

- Wi-Fi networks have finite limits on the numbers of connections that can be supported by each access point. If public access Wi-Fi is to be provided, resource planning and access-management policies are required to ensure that satisfactory services can be delivered.

- Wi-Fi networks are not secure. They are far more susceptible to malicious attacks than wired networks because it is easy for devices to connect with each other. For this reason, they are often severely restricted or completely prohibited in certain business, financial, and government/military facilities.

If your installation requires wireless connectivity, you may simply need to coordinate access and bandwidth requirements with network management. Be prepared for potential objections. You may even find yourself in the position of explaining why Wi-Fi may not be the best option.

Ethernet

When data is transmitted within an enterprise, it is typically sent via Ethernet across a local area network. Here, you will explore typical LAN topologies and the capabilities and limitations of Ethernet data transmission.

Local Area Networks

Data sent across a network must be sent to a destination address. That address will be either physical or logical. Local area networks use physical addresses to communicate. The physical address is the MAC address; the unique address that is hard-coded into each node and never changes. LANs require devices to be directly physically connected by wire, fiber, or wireless, which effectively limits their geographical size.

Stated as simply as possible, data travels across a LAN in this manner:

- Data sent across a LAN is addressed to the MAC address of one of the devices on the LAN.

- A switch or bridge receives the packet and examines the MAC address to which it is addressed.

- The switch/bridge forwards the packet to the appropriate device.

LANs are usually privately owned and operated. They are fast and high capacity.

Most real-time AV network protocols, including AVB, SDVoE, SMPTE ST 2110, Dante, and RAVENNA, are designed for LAN speeds and capacity.

Topology

One of the most important characteristics of any given network is its topology. Network topology is a determining factor in how far data must travel to reach its destination. A network's topology will show which nodes and connections data must pass through to get from its source to its destination and how many stops there will be along the way. It will also reveal which network devices and connections will have to carry the most data. This helps network engineers calculate which parts of their network need the most capacity. Both of these factors are crucial in determining whether and how to send AV signals over a network.

All networks, local and wide area alike, have a layout or *topology*.

Physical topology maps the physical placement of each network device and the physical path (see Figure 12-6). Where will the devices and cables actually be located? Physical topology is constrained by the actual space the network equipment occupies.

Logical topology maps the flow of data within a network (see Figure 12-7). Which network segments and devices must data pass through to get from its source to its destination. Logical topology is not defined or constrained by physical topology. Two networks with the same physical topologies could have completely different logical topologies, and vice versa.

The topology can reveal the location of a network's potential weak spots. When you look at a local area network topology, you should always look for single points of failure. A single point of failure is any one device or component whose failure will cause the entire system to fail. A single point of failure could be any device that a number of other devices depend on. For example, it could be a network switch that many other devices are connected to, or it could be an audio DSP that handles inputs and outputs from many other devices. Whenever possible, you want to create network redundancy at every potential single point of failure, so if one device or signal path fails, another device or signal path is available and waiting to take its place. As a best practice, no more than 20 devices should be affected by any one single point of failure.

Figure 12-6 Physical topology maps the physical placement of each device

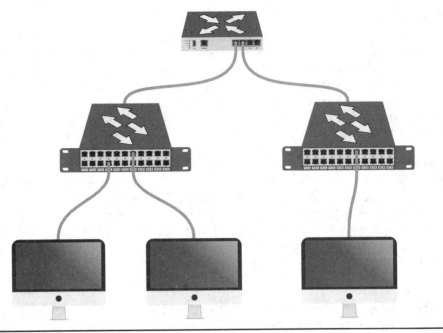

Figure 12-7   Logical topology is not constrained by the physical location of the device

LAN Topologies

Devices can be connected on a local area network in several arrangements, and the layout of connected devices is identified as the network's topology. The basic types are bus, ring, star, extended star (sometimes referred to as *tree topology*), and mesh. Figure 12-8 shows some common local area network topologies you may encounter.

- **Star topology**   All nodes connect to a central point; the central point could be a router or switch. Star networks are hierarchical. Each node has access to the other nodes on the star through the central point. If any node fails, information still flows. The central device is a single point of failure; if it fails, communication stops.

- **Extended star topology**   When there is more than one level of star hierarchy, it is known as an extended star topology. If any device fails, access to the devices below it is cut off, but the rest of the network continues to work. The central device remains a single point of failure; if it fails, communication stops.

- **Meshed topology**   Each node connects to every other node. Meshed topologies provide fast communication and excellent redundancy, ensuring that the failure of no one device can bring down the whole network. Providing physical connections between every device is really expensive, though. Fully meshed networks are rare.

- **Partially meshed topology**   Each node connects to several other nodes, but not all. Partially meshed topologies provide good redundancy, ensuring that several devices must fail at the same time (see Murphy's Law) before communications cease.

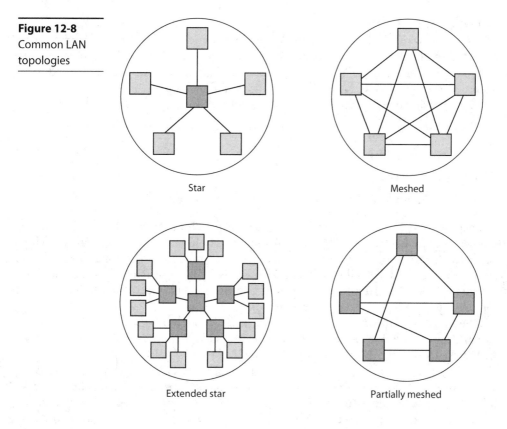

Figure 12-8
Common LAN
topologies

Star

Meshed

Extended star

Partially meshed

What Is Ethernet?

Ethernet is a standard for how data is sent across LANs from one physically connected device to another.

Ethernet has become the de facto standard for LANs. It is defined in the IEEE 802.3 suite of standards, which define a data frame format, network design requirements, and physical transport requirements for Ethernet networks.

When IP data is sent across a LAN, it is encapsulated inside an Ethernet frame, as shown in Figure 12-9. The Ethernet frame is generated by the device's NIC. The frame has a header and footer enveloping the packet, making sure it gets to its destination and arrives intact.

| Preamble | SFD | Receiver Mac Address (Destination) | Sender Mac Address (Source) | VLAN TAG | Type Field | Ethernet Data 42-1500 Bytes IP Packet 192.168.72.220 [- - IP Payload - -] 192.168.1.25 | PAD | CRC Checksum |

Figure 12-9 Ethernet frame with VLAN tag encapsulating the IP data packet

Ethernet Speed

The speed at which a device sends data over Ethernet depends on the capability of its NIC and the bandwidth of the connecting cable. Not every device can handle sending and receiving data at a rate of 1Gbps or more. Bear in mind that the overall speed of an Ethernet connection is no faster than the slowest link in its path.

The following types of Ethernet are in use today:

- 100Mbps Ethernet
- 1Gbps Ethernet
- 10Gbps Ethernet
- 40/100Gbps Ethernet

Each of these types of Ethernet has its own capabilities, intended applications, and physical requirements. You will likely encounter only legacy systems that are 100Mbps Ethernet. The common 1Gbps Ethernet switches are generally compatible with the legacy 10Mbps and 100Mbps formats with either T568A or T568B pin-outs and are referred to as 10/100/1000Mbps devices. Gigabit Ethernet (1Gbps and 10Gbps) is used for delivery of audiovisual data over a network.

The higher-bandwidth 40/100Gbps Ethernet, although originally only intended for use in the network backbone to connect devices such as routers and switches, is now frequently used to transport uncompressed UHD (4K/8K) video.

All of these Ethernet technologies are interoperable. A frame can originate on a 100Mbps Ethernet connection and later travel over a 1Gbps or 10Gbps Ethernet backbone. As a general rule, you need a faster technology to aggregate multiple slower links. If your end nodes communicate with their switches by using 1Gbps Ethernet, the next layer of the network hierarchy may need to be 10Gbps Ethernet, and so forth. As devices become capable of faster network speeds, even faster backbones are required.

Ethernet Types and Their Meanings

In the field, you will hear the type of Ethernet referred to by common abbreviations, such as 1000Base-T. To decipher the meaning, it helps to know the following:

- The number indicates the nominal transmission speed in megabits or gigabits per second. For example, 100Base-T transmits at 100Mbps, 1000Base-T transmits at 1,000Mbps or 1Gbps, and 10GBase-T transmits at 10Gbps.

- The word in the middle indicates whether the connection is baseband or broadband. Almost all Ethernet connections are baseband, so you will usually see *Base* with Ethernet speeds.

- The characters at the end give you information about the Physical layer technology used. This varies from standard to standard, but in general, *T* stands for twisted-pair cable, *S* stands for short wavelength (multimode) fiber, *L* stands for long wavelength (single-mode) fiber, and most other character combinations refer to other types of fiber media.

There are many IEEE standards for Gigabit Ethernet, and different characters are used to indicate the type of cable. For example, the IEEE 802.3z standard includes 1000BASE-SX for transmission over multimode fiber and uses 1000BASE-LX for transmission over single-mode fiber.

Isolating LAN Devices

For security or other reasons, your client may not want all the devices on the LAN to be able to communicate directly with each other via Ethernet all the time. Some servers and devices host confidential information that other devices should not be able to access.

You may need to ensure that certain devices have access only to one another and that outside devices cannot easily talk to them. You may also want to isolate certain groups of devices to limit the traffic they receive.

All networked devices send out occasional administrative messages, called *broadcast* messages, to every device on the LAN. Sometimes you want to limit the amount of broadcast traffic a group of devices receives. The need to limit traffic is particularly relevant to networked audio and video. For some protocols, broadcast traffic from other devices on the LAN can cause serious problems to the quality of the audio or video. You need a way to isolate certain AV devices in a cone of silence, shutting out traffic from outside the AV system. One way to handle the issue is to isolate AV devices on a separate LAN by using dedicated switches behind an isolated router port. Figure 12-10 shows a simple networked AV system with a dedicated switch with a direct connection to the network. This simple solution may sometimes be a viable option.

Virtual Local Area Networks

A VLAN is a way to isolate certain devices from others connected to the same network. The devices in a VLAN can be configured on a managed switch, which forwards the data from a device in a VLAN by adding the VLAN identifier to the Ethernet frame, as shown in Figure 12-9.

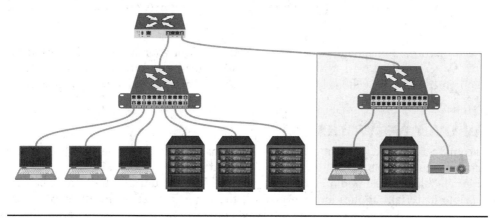

Figure 12-10   Isolation of AV devices (right) on the network with a dedicated switch

Devices in a VLAN can be connected to different switches in completely different physical locations. As long as those switches are in the same LAN (that is, they do not have to go through a router to contact each other), the devices can be placed in a VLAN together. Because VLANs are virtual, you can even place a device on more than one VLAN on the same physical LAN. Devices in a VLAN can send each other Ethernet traffic directly. They also receive each other's broadcast traffic. They do not receive any broadcast traffic from the other devices on the LAN, and they cannot communicate with devices outside the VLAN via Ethernet. Traffic in and out of the VLAN has to go through a router.

Applications of VLANs   Communication among devices on a VLAN is switched rather than routed. This makes communication among VLANs very efficient. Because devices in a VLAN may be connected to different switches, a VLAN allows you to directly connect devices in different physical locations. For example, in a campus-wide deployment of digital signage, a centrally located media server could send content to display screens at various locations. In a simple signage application with only store-and-forward content to display screens, the devices probably belong on a VLAN. Isolation of the signage devices will enable the transport of video content without issues, and you are unlikely to need to access data on other network devices.

Some locations have a single high-bandwidth LAN covering an entire campus with a separate VLAN for each application group, such as accounting, ticketing, IT, VoIP, operations, streaming video, AV, paging, outside broadcast, audio recording, lighting, stage machinery, system control, BIM, HVAC, communications, hearing assistance, production management, CCTV, security, etc.

Requesting VLANs   If you determine that a set of devices should be isolated on a VLAN for the good of either AV signal distribution or the devices on the network, you will need to provide the network manager with the following information:

- Why a VLAN is necessary and what devices will be in it
- Whether any routing between the VLAN and other network locations is required or permitted

This information should all be documented as part of your system device inventory. The AV team has to work with the IT network manager and team to identify VLAN requirements, but IT will likely choose the VLAN name and configure the network.

AV over Networks

Streaming audio and video in real time requires bandwidth allocation to deliver all the data. If insufficient bandwidth is available, the data gets lost or delayed, resulting in unintelligible, clipped audio and blocky video. Some amount of latency may be permissible in certain applications. In general, the more interactive the AV application, the less latency is acceptable in the displayed content. For example, in a lecture that is being streamed live to a remote location, several seconds of latency would be tolerable.

However, in a two-way live conversation, even a couple of seconds delay would be unacceptable. Similarly, a networked audio system in a stadium cannot tolerate much latency because the announcers' commentaries should not lag behind the action on the field.

Many protocols are designed to deliver real-time audio and/or video over LANs. Some are proprietary, which means a private company developed the protocol and owns the rights to its use and then licenses the technology to manufacturers to use in their products. This enables the licensed manufacturers to produce interoperable products. Other protocols are either free, open-standards or proprietary protocols that have been released into the public domain for general use (which is not exactly the same thing). Open protocols may be developed, improved, extended, and released by anyone who has the inclination. Proprietary protocols, whether strictly licensed or available for public use, are maintained, updated, and extended by the original developer and released at that developer's discretion, which reduces the possibility of slightly incompatible variants circulating in the wild.

AV over Ethernet

Early on in the development of networked audio protocols, a number of proprietary protocols were created that squeezed digital audio over the Physical layer (layer 1) and the Data Link layer (layer 2) of the 10Mbps, and later, the100Mbps networks then coming into use. The most widely adopted of these were the CobraNet protocol from Peak Audio and Ethersound from Digigram, both of which could handle up to 64 channels of 20- to 24-bit audio over dedicated audio networks. These layer 2 protocols may still occasionally be found in existing installations.

AVB

In 2011 the IEEE released 802.1-AVB into the public domain. This Audio Video Bridging (AVB) suite of Ethernet (layer 2) standards enables the transport of low-latency AV on 100Mbps and faster Ethernet networks. Unlike its predecessors on layer 2, AVB can coexist on an Ethernet network with other traffic. To achieve this feat, the AVB standards include prioritization and traffic-shaping functions to ensure that AV data is not unduly delayed by other traffic. As standard Ethernet switches and bridges do not usually include the required quality of service prioritization capabilities (see "Quality of Service," later in this chapter), only AVB-qualified bridges and switches can be used in AVB networks. Over Gigabit Ethernet, AVB can carry uncompressed video and up to 200 channels of 48kHz, 24-bit audio, plus embedded control and monitoring, in real time. AVB has also added ASE67 interoperability to its audio protocols. As a level 2 protocol, AVB data cannot be routed outside its LAN.

AV over IP Networks

As network bandwidth became much less expensive, the early proprietary Data Link layer protocols were replaced by a wide range of IP-based Network layer (layer 3) protocol suites that can be routed across multiple networks, including the Internet, if sufficient bandwidth can be made available. Some IP-based protocols are widely licensed to many manufacturers and have significant ecosystems that provide a huge range of solutions for control and system integration, while some others are tightly limited to a particular manufacturer's range of products. System designers have a wide range of AV over IP network architectures to choose from, so you are likely to meet several of the networks mentioned here. Each protocol and its accompanying technologies use a different approach to installation, configuration, and maintenance and will therefore require research, reading, training, and field experience before you can implement it with confidence.

AES67 Interoperability

Most of the early networked AV protocols were incompatible, leaving isolated islands of one or another technology scattered throughout the AV world. To alleviate this chaos, in 2013 the Audio Engineering Society (AES) released AES67, the AES Standard for Audio Applications of Networks – High-Performance Streaming Audio-Over-IP Interoperability, which has now been incorporated into the major audio-over-IP protocols from all developers. This allows substantially different protocol suites to include gateways and bridges for the seamless movement of audio between varying networks and AV architectures.

Each AES67 link carries up to 120 channels of 16- or 24-bit audio at sample rates of 44.1, 48, or 96kHz.

Dante

Dante is a proprietary IP-based, Network layer (layer 3) bi-directional protocol developed by Audinate and used in thousands of products from hundreds of manufacturers. Dante is a combination of hardware, control software, and the transport protocol itself.

Dante requires a switched Ethernet network of 100Mbps or better, with at least a 1Gbps Ethernet backbone. Dante does not require dedicated bandwidth, as it uses the QoS DiffServ VoIP category to prioritize AV and control data over other traffic. Over a 100Mbps Ethernet network, a single Dante connection can carry 96 channels of 24-bit, 48kHz audio or 48 channels of 24-bit, 96kHz audio. Over Gigabit Ethernet, a single Dante connection can carry 1,024 channels of 24-bit, 48kHz audio or 32-bit video with eight channels of 16-, 24-, or 32-bit audio at sample rates of 44.1, 48, 88.2, or 96kHz.

SMPTE ST2110

Developed by the Society of Motion Picture and Television Engineers (SMPTE), SMPTE ST 2110 Professional Media Over Managed IP Networks is a suite of industry standards aimed toward a single, common IP-based delivery mechanism for the professional media industries. Its target applications include film, Internet streaming and broadcast production and post production, live event production, museums, digital video distribution, and theme parks. It is intended by SMPTE to replace its traditional SDI transport protocols for high-quality video and audio distribution over dedicated coax and fiber networks. SMPTE ST2110 is a publicly available suite of standards that can be incorporated into products by any manufacturer.

The suite of protocols specified include a transport mechanism for uncompressed video and audio streams in any resolution and format, control signals and metadata, traffic shaping and delivery timing for the data, and frame and clock synchronization across the network. It also allows the synchronized splitting, combining, embedding, and de-embedding of clock, control, and audio from the video stream. The audio components of the suite are AES67 compliant.

SMPTE ST2110 is a publicly-available suite of standards that can be incorporated into products by any manufacturer.

RAVENNA

Realtime Audio Video Enhanced Next generation Network Architecture (RAVENNA) is an IP-based audio protocol suite developed by ALC NetworX, who continue to contribute to its development, even though it is now an open technology. Initially developed for the broadcast market, RAVENNA is based around a range of existing IP standards for clock synchronization, QoS, and interchange via AES67. It includes direct compatibility with Q-SYS, Dante, and the audio sections (30 and 31) of SMPTE ST2110 protocols.

Each RAVENNA device may have two independent network outputs, which can be connected to independent physical networks, providing for high levels of redundancy. In-built functionality provides seamless receiver failover between independent streams.

RAVENNA is implemented by many manufacturers in a broad range of products across the professional audio spectrum. It is used in production facilities, live events, television outside broadcasts, recording facilities, and interstudio links across WANs.

RAVENNA requires Gigabit Ethernet (1Gbps, 10Gbps, 40Gbps) to handle up to 768 channels of uncompressed audio data at bit depths up to 24 bits and at sample rates up to 384kHz.

Crestron DM NVX

AV control and automation system supplier Crestron includes the proprietary DM NVX video over IP technology as part of its all-encompassing AV ecosystem. The 1Gbps Ethernet-hosted protocol suite includes low-loss, low-latency video compression for a range of video formats, together with Dante/AES67-compliant audio, USB 2.0 signaling, and the Crestron family of control protocols.

DM NVX AV over IP is a component of Crestron's integrated environment and is usually installed, configured, and maintained by Crestron-trained and certified technology specialists.

AMX SVSI

AV control and automation system supplier AMX includes the proprietary SVSI video over IP technology as part of its all-encompassing AV ecosystem. The 1Gbps Ethernet-hosted protocol suite includes low-loss, low-latency video compression for a range of video formats, together with AES67-compliant audio, EDID and HDCP management, complex matrix switching, and the AMX family of control protocols.

SVSI AV over IP is a component of AMX's integrated environment and is usually installed, configured, and maintained by AMX-trained and certified technology specialists.

Extron NAV

AV control and automation system supplier Extron includes the proprietary NAV Pro video over IP technology as part of its all-encompassing AV ecosystem. The 1Gbps or 10Gbps Ethernet-hosted protocol suite includes its PURE3 codec for low-loss, low-latency video compression for a range of video formats, together with an AES67-compliant audio, USB 2.0, RS-232 and IR control, EDID and HDCP management, complex matrix switching, and the Extron family of control protocols.

NAV Pro's AV over IP is a component of Extron's integrated environment and is usually installed, configured, and maintained by Extron-trained and certified technology specialists.

BlueRiver and SDVoE

The BlueRiver FPGA-based AV over IP transmitter and receiver chipset was originally developed by Aptovision for the transport of uncompressed 4K HDMI 2.0a video, 32 channels of HDMI audio, USB 2.0, RS-232, and infrared signaling via IP over Gigabit Ethernet networks with a 10Gbps backbone.

Semtech took BlueRiver and developed it into a more energy-efficient ASIC chipset, together with a set of application programming interfaces (APIs) that operate in the Session, Presentation, and Application layers of the network. This more advanced AVP version of BlueRiver technology has been incorporated into products from many different original equipment manufacturers (OEMs) and has become the core technology for the SDVoE (Software Defined Video over Ethernet) alliance. BlueRiver is targeted at a broad range of networked AV applications, including live events, e-sports, medical imaging, high-end residential, and command and control.

SDVoE

SDVoE is a combination of 10Gbps Ethernet and BlueRiver technology to produce a standard method for the distribution of video (and its audio) over Ethernet networks. An SDVoE network includes a control server providing device discovery, video scaling and cropping, matrix switching, signal routing, and EDID processing. All SDVoE sources and display devices are equipped with BlueRiver chipsets to handle signal encoding and decoding. At the time of publication, the SDVoE Alliance included dozens of high-profile AV technology companies.

Q-SYS

Audio systems company QSC has extended its original proprietary Q-SYS networked audio platform to become a fully fledged, IP-based network audio, video, and control ecosystem. The 1Gbps Ethernet-hosted protocol suite includes low-loss, low-latency video compression for a range of video formats, together with Dante-based AES67-compliant audio, USB integration, AVB compatibility, acoustic echo cancellation, a VoIP/SIP conferencing interface, complex matrix switching, and a complete suite of control protocols.

Q-SYS is at the core of QSC's integrated environment and is usually installed, configured, and maintained by QSC-trained and -certified technology specialists.

Wide Area Networks

When traffic needs to be transported from one LAN to another, it travels over a wide area network (WAN) like the Internet. The difference between LANs and WANs can seem hazy. Some "local" area networks belonging to universities or large enterprises are vast and are often known as campus area networks (CANs). Larger networks that cover a geographic area such as a suburb or a city are identified as metropolitan area networks (MANs). The true distinction between a LAN and WAN lies in how the data is addressed and transported: WAN data travels in IP packets rather than Ethernet frames. A wide area network is a network that connects one or more LANs, as shown in Figure 12-11.

The nodes on a WAN are routers. If a LAN is connected to outside networks via a WAN, a router sits at the top of its network hierarchy. Any data that needs to travel to a device outside the LAN gets forwarded to the router. The router strips the packet of identifying LAN information, like the MAC address of the originating device, before forwarding the packet to its intended IP address.

A WAN can be any size. It may connect two LANs within the same building, or it may span the entire globe, like the world's largest WAN, the Internet. Unlike LAN connections, long-distance WAN connections are rarely privately owned. Usually, WAN connections are leased from Internet service providers (ISPs). The speed of a WAN connection is often directly related to its cost.

Figure 12-11

A simple partially meshed WAN

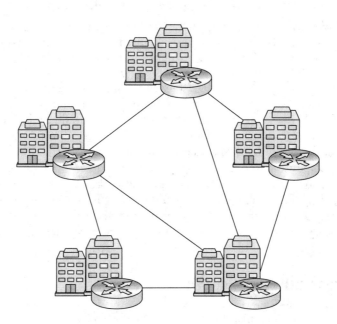

PART V

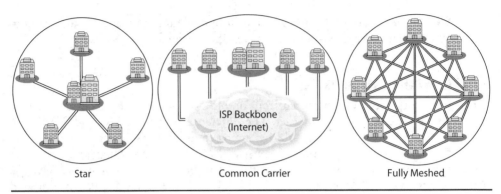

Figure 12-12   Three common WAN topologies

WAN Topologies

Wide area network topologies can be placed into three common categories, as shown in Figure 12-12:

- **Star topology**   In a star WAN, each LAN connects to a central location. For example, several branch offices may connect to a corporate headquarters. Like a star LAN, a star WAN may have several nested layers of hierarchy, with several hubs connecting to several spokes.

- **Common-carrier topology**   In a common-carrier WAN, each LAN site connects to an ISP backbone. The backbone itself is likely to be part of a meshed WAN.

- **Meshed WAN topology**   In a fully-meshed WAN, every LAN connects to every other LAN. This provides excellent redundancy. Most meshed WANs are only partially meshed, although this also provides a degree of redundancy.

One of the major advantages of networking is the ability to share resources. When examining a WAN topology, try to identify the most effective place to locate shared resources. With a star topology, shared resources should be located at the most-central hub of the nested stars. In a common-carrier topology, the enterprise may choose to lease space at the ISP and host shared resources there, or it may choose to pay for a lot of bandwidth to and from one of its sites and locate resources there. In a meshed WAN topology, resource location is generally accomplished by building a network map, including the data throughput of all connections and locating resources "in the middle."

Shared AV resources, such as streaming servers or multipoint control units (MCUs), will ideally be stored in the same physical location as shared IT resources. In some cases, however, you may not have access to IT server spaces for security reasons.

Private and Public WANs

A network could be private or public, but the access or connection to a WAN is usually owned and provided by an ISP. An ISP may lease secure, dedicated communication connections to individual organizations. Some companies and organizations that need to ensure secure connections may own the access and connections to their long-distance communication networks; these are often referred to as *enterprise networks*.

Virtual Private Networks

A virtual private network (VPN) uses the Internet to create a virtual (software) tunnel between points on two or more LANs by adding a layer of secure encryption to the data packets sent between the endpoints. A VPN is used to create a virtual wide area network for managed communication and for remote monitoring, troubleshooting, and control. VPNs are typically controlled and configured by the enterprise network administrator. Each host or user requires the proper software, access rights, and password to log into the client network.

Virtual private networks provide support for services that are particularly necessary for AV applications, such as QoS, low-latency streaming, managed routing, and multicast transmission. VPN endpoints may be built into routers, network servers, or security appliances such as firewalls. Organizations using VPNs on a large scale may require dedicated VPN devices.

Using a VPN increases the required bandwidth because an encryption and tunneling wrapper must be added to each data packet. This additional overhead may not be significant in its bandwidth requirement, but it can increase the Ethernet frame size to the point where packets must be fragmented before they can be sent across the network. Packet fragmentation can be disastrous for the quality of streamed video or videoconferences. Always be sure the frame size is set low enough to account for VPN overhead. If you are having difficulties with streaming content over a VPN, this may be the cause, and the problem should be brought to the attention of the network manager.

Internet Protocol

Internet Protocol is the communications protocol for relaying data across an IP network. It establishes how data packets are delivered from the source host to the destination host, based solely on the IP addresses in the packet headers. It defines the rules for addressing, packaging, fragmenting, and routing data sent across an IP network.

This protocol is on the Network layer (layer 3) of the OSI Model and addresses these crucial functions that make wide area networking possible:

- **Addressing**   Rules for how each host is identified, what the addresses look like, and who is allowed to use which addresses
- **Packaging**   What information must be included with each data packet
- **Fragmenting**   How big each packet can be and how overly large packets will be divided
- **Routing**   What path packets will take from their source to their destination

Internet Protocol is basically the postal service of the Internet. A postal service sets rules for how to package and address mail. If you do not include the right information in the address, your package will not be delivered. If you do not package it correctly, it may get damaged in transit. Just like a postal service, IP assumes responsibility for making sure your data arrives at its destination, although, just like with a postal service, some messages do get "lost in the mail." Addressing, packaging, fragmenting, and routing all help ensure that as many messages as possible reach their intended destinations.

IP Addressing

An IP address is the logical address that allows devices to locate each other anywhere in the world, no matter where they are physically located. The numerical address defines the exact device and its exact location on the network. Even if AV devices such as videoconferencing systems are isolated on a LAN, they are assigned an IP address.

An IP address has three components:

- **Network identifier bits**   These bits identify the network. They help the IP packet find its destination LAN. The network bits are always the first digits or prefix in an IP address.

- **Host identifier bits**   These bits identify a specific network node. They help the IP packet find its actual destination device. The host bits are always the last bits or least-significant bits in a network address.

- **Subnet mask**   These bits tell you which bits in the IP address are the network bits and which bits are the host bits. The subnet mask also reveals the size of the network. The subnet mask is a separate address that must be included with the IP address.

IPv4 Addressing

An IP address can look very different depending on which version of Internet Protocol addressing is used to create them. Two versions of the protocol are currently in use: version 4 (IPv4) and version 6 (IPv6). Originally defined in the IETF standard RFC 760 in 1980, Internet Protocol version 4 addressing is slowly and cautiously being phased out in favor of IPv6. IPv4 remains the most prevalent addressing scheme by far, and with many ancient IPv4-only devices liberally scattered throughout the Internet, rumors of its death have been greatly exaggerated. During the transition to IPv6, you will need to know the IPv4 addressing scheme and structure.

An IPv4 address consists of four 8-bit groups, known as *octets*, or bytes. These four octets are usually expressed as decimal numbers (from 0 to 255) separated by dots: a format known as *dotted-decimal* or *dotted-quad* notation. Hence, an IP address looks like this:

192.168.1.25

An IPv4 address given to a device on the network has a network prefix (number on the left) and a network host number on the right. Note that each decimal number actually represents eight bits. That same address, written in binary, looks like this:

11000000 10101000 00000001 00011001

The entire range of IPv4 addresses includes every possibility, from all 0 bits to all 1s. In dotted-decimal notation, that range is expressed as follows:

0.0.0.0 to 255.255.255.255

In total, there are almost 4.3 billion possible IPv4 addresses. Although a few addresses are reserved for specific purposes, there are still a lot of possible addresses. However, 4.3 billion addresses will not be enough for the ever-growing Internet of things (IoT) and other networks of the future. That is why IPv4 is being phased out, although the widespread adoption of network address translation (NAT) has slowed the adoption process.

IPv4 Subnet Masks

Looking at an IPv4 address on its own, you cannot know which groups of numbers are the network ID and which ones are the host ID numbers. To interpret any IPv4 address, you need a separate 32-bit number called a *subnet mask*.

A subnet mask is a binary number whose bits correspond to IP addresses on a network. As shown in Figure 12-13, the structure of an IPv4 subnet mask looks a lot like an IPv4 address. It consists of four octets, expressed in dot-decimal notation. Bits equal to 1 in a subnet mask indicate that the corresponding bits in the IP address identify the network. Bits equal to 0 in a subnet mask indicate that the corresponding bits in the IP address identify the host.

For example, subnet mask 255.255.255.0 shown in Figure 12-13 would indicate that the first three octets of any corresponding IP addresses are the network address and the last octet is the host address.

Also notice the difference in the binary IPv4 network ID and IPv4 host ID in Figure 12-13. The IP address assigns the numbers 1 and 0 in combination, but in the subnet mask, the first part of the subnet is all 1s and the second part (right) is all 0s. All the devices on the same network have the same network identifier bits in their IP addresses. Only the host bits will differ.

In IPv4, there are two ways to express a subnet mask:

- You can write it out as its own full dotted-decimal number.

- You can attach it to the end of an IP address using the shorthand format: Classless Inter-Domain Routing (CIDR) notation. CIDR notation is simply a slash, followed by a number that indicates how many of the address bits are network bits, with the remaining bits being host bits.

PART V

Figure 12-13
IPv4 address and subnet mask

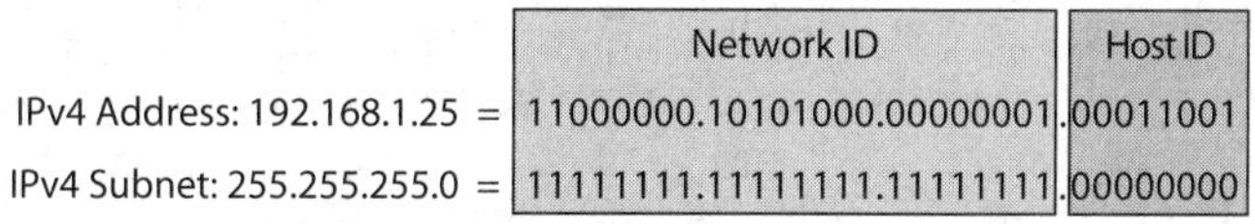

	Network ID	Host ID
IPv4 Address: 192.168.1.25 =	11000000.10101000.00000001.	00011001
IPv4 Subnet: 255.255.255.0 =	11111111.11111111.11111111.	00000000

Here is an example of the two ways the same subnet mask can be written:

- Dot-decimal number: 255.255.192.0
- CIDR notation: /18

In binary, both are equal to 11111111 11111111 11000000 00000000.

Networks are both physical and logical. Two devices may be physically attached to the same switch, but that does not mean they are logically on the same network. They must also have the same network identifier bits and subnet mask. When you are working with network management to obtain IP addresses for your network-connected devices, make sure everyone knows which devices need to be in the same subnet.

Types of IP Addresses

The Internet Assigned Numbers Authority (IANA) is in charge of issuing IP addresses or reserving them for specific purposes. The IANA maintains three categories of addresses: local, private or reserved, and global.

Table 12-6 shows the more important reserved IP addresses, their ranges, sizes, and purposes.

Network Address	Address Range	Number of Addresses	Application	Purpose
0.0.0.0/8	0.0.0.1–0.255.255.255	16,777,216	Software	Used as a placeholder to represent an unknown IP address.
10.0.0.0/8	10.0.0.0–10.255.255.255	16,777,216	Private network	Private (unrouted) network address range for very large networks.
127.0.0.0/8	127.0.0.0–127.255.255.255	16,777,216	Local host	Used as a loopback address for the current host.
169.254.0.0/16	169.254.0.0–169.254.255.255	65,536	Private network	Addresses used for local (unrouted) Automatic Private IP Addressing (APIPA). Used in the absence of a static IP addressing or DHCP address assignment.
172.16.0.0/12	172.16.0.0–172.31.255.255	1,048,576	Private network	Private (unrouted) network address range for large to medium networks.
192.168.0.0/16	192.168.0.0–192.168.255.255	65,536	Private network	Private (unrouted) network address range for medium to small networks.
224.0.0.0/4	224.0.0.0–239.255.255.255	268,435,456	Internet	Used for multicast communications.
240.0.0.0/4	240.0.0.1–255.255.255.255	268,435,456	Internet	Reserved for experimental purposes/future use.
255.255.255.255/32	255.255.255.255	1	Local network	Broadcast address used to talk to every host on a local network.

Table 12-6 Examples of Some IANA-Reserved IP Addresses

Local Addresses

Not all devices need to access the Internet directly. Many devices need to communicate with other devices only on their local area network. The Internet Assigned Numbers Authority reserves three IPv4 address ranges and one IPv6 address range for local networking. The addresses in these ranges are private and are not routed outside the network by any routers. Devices with private IP addresses cannot access the Internet or communicate with devices on other networks directly.

IPv4 Private Address Ranges

Private addresses are not routed on the public Internet. As can be seen in Table 12-6, there are three IPv4 private address ranges: 10.0.0.0/8, 172.16.0.0/12, and 192.168.0.0/16. The range your client will use depends on the size of their network.

The principal advantage of private network addresses is that they are reusable. Global addresses have to be unique so no two hosts can use the same IP address to access the Internet. Otherwise, when data is sent to or from an address, there would be no way to know which host was the intended recipient. Since private addresses are not exposed to the Internet, different organizations can use the same private address range. Devices on different networks can have the same private IP address because those devices will never attempt to communicate with each other. As long as no devices on the same network have the same IP address, there is no confusion.

Subnetting Local Networks

Subnetting is the act of logically dividing a network into smaller networks. Each smaller network is called a *subnet*. The networks to which you are connecting AV devices may be divided into several subnets. You will need to work with network management to make sure that any devices that need to communicate via Ethernet are in the same subnet.

Subnets are created when the subnet mask of an IP address is extended. If the subnet mask is extended by one bit, you end up with two subnets that are each approximately half the size of the original network. Similarly, if the subnet mask is extended by two bits, you end up with four similarly sized subnets, each one-quarter the size of the original network. For devices to communicate directly by Ethernet or to belong to the same VLAN, they must be in the same subnet.

The main reason to create subnets is to increase network efficiency. A subnet has fewer addresses than a full class network, so address resolution is faster. Fewer devices also mean less broadcast traffic, as devices on a subnet receive broadcast messages only from other devices on the same subnet, not from the entire undivided network.

Global Addresses

Most network devices will need to connect to the Internet at some point. To access the Internet, a device needs a global IP address, one that any Internet-connected device can locate.

Global addresses go by many names in the networking community, including *globally routable addresses, public addresses,* or *publicly routed addresses*. Any IP address that is not in one of the local or reserved address ranges can be a global address. If an IP address is not a local or reserved address, it is a global address.

Network Address Translation (NAT)

The obvious disadvantage of private IP addresses is that they can communicate only with devices on the same network because their addresses cannot be routed to the Internet. Initially, this made the networking community reluctant to use them. However, that problem has been solved through the use of NAT.

NAT is a method of altering IP address information in IP packet headers as the packet traverses a routing device. NAT is used to allow devices with private, unregistered IP addresses to access the Internet through a device with a single registered IP address. NAT conserves address space, which is a concern in IPv4 implementations (though it's also used for IPv6 networks). NAT hides the original source of the data. From outside the network, all data appears to originate from the NAT server. Any data that arrives at the NAT server without being requested by a client has nowhere to go; it has the address of the building but not the apartment number. Using NAT, all unrequested data is blocked by the firewall or router, and a malicious intruder can't trace the data's path beyond the edge of the network.

Broadcast Addresses

Broadcast addresses are used as the destination IP address when one node wants to send data to all network devices. Broadcast messages are one-way; there is no mechanism for the other nodes to reply. An IPv4 broadcast address is any IPv4 address with all 1s in the host bits. When data is sent to that address, it goes to every device with the same network bits.

For example, if your network address is 192.168.0.0 and your subnet mask is 255.255.0.0, your broadcast address is 192.168.255.255. Any data sent to that address will go to every device in the address range from 192.168.0.1 to 192.168.255.254. If instead the subnet mask was 255.255.255.0, your broadcast address is 192.168.0.255, and data sent to that address will go to every device in the address range 192.168.0.1 to 192.168.0.254.

Loopback Addresses

Data addressed to a loopback address is returned to the sending device. The loopback address is also known as the *localhost,* or simply *home.*

The loopback address is used for diagnostics and testing. It allows a technician to verify that the device is receiving local network data. Essentially, it allows you to ping yourself.

Any IP address in the range 127.0.0.0 to 127.255.255.255 can be used for loopback, but most network devices automatically use 127.0.0.1.

IPv6 Addresses

IPv6 addresses look very different from IPv4 addresses. An IPv6 address consists of 16 bytes (128 bits), four times as long as an IPv4 address. Because IPv6 addresses are so long, they are usually written in eight lowercase, four-character hexadecimal groups (hextets), separated by a colon, as shown in Figure 12-14.

Figure 12-14

An IPv6 address is written in eight 4-character hexadecimal groups

To compress the representation of an IPv6 address, the leading zeros may be omitted from each hexadecimal group. For example:

> The address 2006:0fe8:85a3:0000:0002:8a2e:0a77:c082
> may be written as 2006:fe8:85a3:0:2:8a2e:a77:c082

The representation may be further compressed by replacing a single sequence of consecutive zero-value hexadecimal groups by a double colon (::) symbol. For example:

> The address 2006:0fe8:0000:0000:0000:8a2e:0000:00c8
> may be written as 2006:fe8::8a2e:0:c8

Note that to avoid any ambiguity, only one run of consecutive zero-value hexadecimal groups may be replaced by a :: symbol.

Since each hexadecimal character represents 4 bits, each group represents 16 bits. You can interpret an IPv6 address as follows:

- The first three hexadecimal groups are the network identifier bits. (48 bits ~ 281 trillion networks).
- The next hexadecimal group identifies the subnet within the network (16 bits = 65,536 subnets).
- The last four hexadecimal groups are the host identifier bits within the network (64 bits ~ 18,446,744 trillion hosts).

The host identifier portion of an IPv6 address is long enough to include a 48-bit MAC address, so IPv6 can actually use a device's MAC address as part of the host identifier. Some IPv6 implementations even do this automatically. Because a MAC address uniquely identifies a device, using the MAC address in the host identifier should ensure that no two devices ever have the same IPv6 address.

IPv6 Subnet Masks

An IPv6 subnet mask can be written out in eight full hexadecimal words, but the first three words of the netmask will always be all 1s and the last four will always be all 0s. As a result, many implementations of IPv6 allow you to enter the subnet mask as a single four-character hexadecimal word. The subnet mask could be written as simply c000, as shown in Figure 12-15. You can also express an IPv6 subnet mask using CIDR notation. The CIDR suffix for an IPv6 subnet mask will almost always be between /48 and /64.

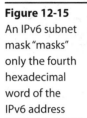

Figure 12-15
An IPv6 subnet mask "masks" only the fourth hexadecimal word of the IPv6 address

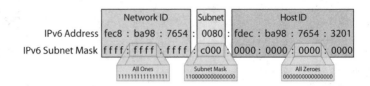

Address Assignment

Every device connected to the network must have an IP address. That address may be obtained automatically or configured manually. Either way, you will need to work with network management to make sure there are enough addresses available for your devices. You will also need to make sure your devices have the right kind of address (local or global). Finally, you need to make sure that any device that needs a permanent address gets one.

Static and Dynamic IP Addresses

IP addresses are either static or dynamic. Here are some of the characteristics of each.

A static IP address is assigned manually to a device, and the address will not change. It is necessary to have a static address for certain devices such as those used in video-conferencing systems or IP-controlled AV equipment so that they can be easily found on the network. As shown in Table 12-7, the network manager has to document and manually keep track of which devices have static addresses and what addresses they have been assigned.

If any two devices have the same addresses, most likely neither will work because of the conflict or perhaps either device may work intermittently. As a result, keeping track of static addresses can be a pain point for network managers.

A dynamic IP address is a temporary IP address that is automatically assigned to a device when it connects to the network. Dynamic addresses can change and therefore are not practical for applications that require locating a device through its IP address. If you need a control system to be able to locate a device by its IP address, you should assign the device a static address.

Manufacturer	Model #	Software Version	Firmware Version	MAC Address	IP Address	Subnet Mask	Gateway WAN IP Address
ProjectTech	4000ZT	8.0	11.4.5	78:ab:0f:23:32:89:0c:7a	192.168.38.4	255.255.255.224	202.38.192.1

Table 12-7   Log of Static IP Addresses and Related Data

Dynamic Host Configuration Protocol

Dynamic Host Configuration Protocol (DHCP) is an IP addressing scheme that allows network administrators to automate address assignment. When a device connects to the network and the device has the "obtain IP address automatically" option activated, the network DHCP service will read the MAC address of the device and assign it an IP address. The pool of available IP addresses is based on the subnet size and the number of addresses that already have been allocated.

A DHCP server will allow a device to hold the IP address for only so long; the amount of time is called the *lease time*, which is set by the network administrator. After the lease time has expired, the lease will usually be renewed automatically if the device is still connected to the network; otherwise, another device connecting to the network can reuse that same address. There is no guarantee that a device will be allocated the same address when it next connects to the network.

A single DHCP server can assign addresses to devices on multiple subnets. The server keeps track of the following:

- The range of available network addresses
- Which addresses are available for DHCP assignment
- Which addresses are currently in use by which devices
- The MAC addresses of the connected devices
- The remaining lease time on each allocated address

DHCP Advantages

DHCP is simple to manage. It takes care of making sure no two devices get the same address, relieving potential conflicts. It allows for more people to connect to the network, as the pool of addresses is continually updated and allocated. For example, a conference center may have thousands of visitors using their Wi-Fi in the course of a week. With DHCP, you only need enough host addresses to cover the devices that are using the network at any given time.

A disadvantage of DHCP is that you never know what your IP address will be from connection to connection. If you need to reach a certain device by IP address, you must have a high level of confidence that the device will be there all the time, and DHCP may not give you that confidence.

Reserve DHCP

Reserve DHCP is a hybrid approach to DHCP that reserves a block of addresses for static addresses and dynamic addresses. The pool of addresses for DHCP is reduced by the number of addresses reserved for static devices. To make this happen, an IT manager will need the MAC address of each device that must be statically set. The static (manually assigned) IP address and MAC address are entered into a table.

PART V

When a device configured for automatic addressing connects to the network and reveals its MAC address, the DHCP server will see that an IP address is reserved for the device and will assign it. The IP address cannot be given to any other device or MAC address. This eliminates the possibility of devices being manually assigned static addresses that have also been allocated to the DHCP pool.

Note that if a device with a reserved IP address is replaced, the MAC address of the replacement device needs to be reported and reconfigured on the DHCP server so that the new device can assume the old IP address.

Best Practice for Reserve DHCP

If you have a DHCP server, it is best to use reserve DHCP rather than manually assigning an IP address to each AV device.

Automatic Private IP Addressing

A device on a DHCP-enabled network can fail to get an address from a DHCP server under the following conditions:

- The DHCP server is down.
- DHCP is not configured properly on the client.
- The DHCP server has exhausted the IP address pool.
- The DHCP server is improperly configured.

In these circumstances the device will not be able to access other network segments or the Internet, but it may still be able to communicate with other devices on its own network segment using Automatic Private IP Addressing (APIPA), if it has been enabled.

Intended for use with small networks with fewer than 25 clients, APIPA enables Plug and Play networking by assigning unique IP addresses to computers on private local area networks. APIPA uses a reserved range of IP addresses (169.254.0.1 to 169.254.255.254) and an algorithm to guarantee that each address used is unique to a single computer on the private network.

It works seamlessly with the DHCP service, yielding to the DHCP service when DHCP is deployed on a network. A DHCP server can be added to the network without requiring any APIPA-based configuration. APIPA regularly checks for the presence of a DHCP server, and upon detecting one replaces the private networking addresses with the IP addresses dynamically assigned by the DHCP server. A device with an APIPA address may not be able to communicate outside the local network, but it can communicate with other devices (if any) on the same subnet.

When a device has an IP address but cannot connect to the Internet, network trouble-shooters often look for link-local addresses such as APIPA addresses.

Domain Name System

Devices on a network must have unique identifiers. At the Data Link layer (layer 2), devices are uniquely identified by their MAC addresses. At the Network layer (layer 3), devices are uniquely identified by their IP addresses. Naming services allow people to identify network resources by a name instead of a number. From the human perspective, names are a lot easier to memorize than numeric or alphanumeric addresses. For example, it is quite easy to remember *store.avixa.org*, but not so easy to remember 101.53.188.4.

DNS is the most widely used system for name-to-address resolution. It is a hierarchical, distributed database that maps names to data such as IP addresses. The web addresses you type into a browser are not actually addresses; they are DNS names. Every system that connects to the Internet must support DNS resolution.

The goal of DNS is to translate, or resolve, a name into a specific IP address. DNS relies on universal resolvability to work: every name in a DNS must be unique so that information sent to a domain name arrives only at its intended destination.

A DNS uses domain name servers to resolve names to addresses. The server contains a database of names and associated IP addresses. These servers are arranged in a hierarchy. Each server knows the names of the resources beneath them in the hierarchy and the name of the server directly above them in the hierarchy. No one device has to keep track of all the names and IP addresses on the Internet. That information is distributed across all the DNS servers on the network.

As shown in Figure 12-16, a domain name or alias has three main parts:

- A computer name or alias
- The domain itself
- The top-level domain (TLD)

A domain may be further divided into subdomains. This system helps prevent any two devices from being assigned the same name. Many World Wide Web servers are assigned the subdomain "www," many e-mail servers are assigned the subdomain "mail," and many e-commerce servers are assigned the subdomain "shop." Any number of computers may be called www, mail, or shop, provided they belong to different domains.

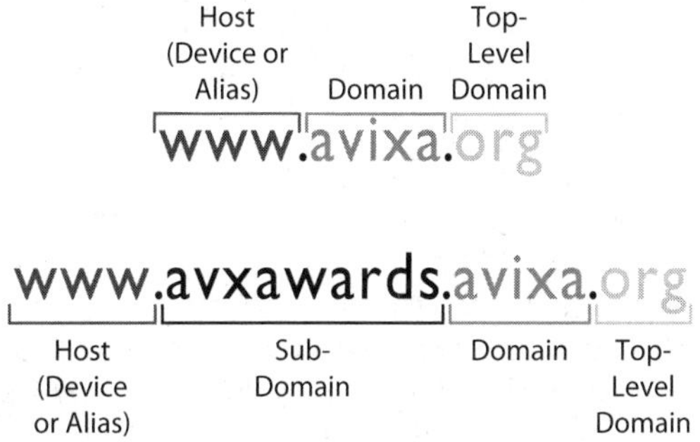

Figure 12-16

Three main parts of a domain name

Internal Organizational DNS

You do not need your own DNS server to resolve the names of web addresses on the Internet. You can do that through your Internet service provider's DNS server. However, many organizations use DNS internally to manage the names and addresses of devices on their private networks. In this case, you will need your own DNS server. Usually, you will have a master DNS server and at least one secondary DNS server that runs a copy of the database stored on the master. This provides a backup in case the master ever fails. If an organization is really dispersed, you may want to locate a secondary DNS server at each physical site. This keeps DNS traffic off the wide area network.

Using a service called dynamic DNS (DDNS), DNS can work hand in hand with DHCP. The service links and synchronizes the DHCP and DNS servers. Whenever a device's address changes, DDNS automatically updates the DNS server with the new address. When DHCP servers and DNS are working together, you may never need to know the IP address of a device, only its name. This makes managing the network simpler, as the IP address does not need to be static. The entire addressing scheme could change without affecting the communication between devices.

Internal DNS Adoption

The advantages of using internal DNS address assignment for a networked system seem obvious:

- Devices can be identified by easily remembered names.
- Dynamic DHCP makes control and remote monitoring systems much easier to maintain.
- DNS includes load balancing functionality.

Many AV devices still do not natively or fully support DNS address assignment. Connecting a device without native DNS support to a network with a DNS addressing scheme may require the network manager to perform many, and ongoing, manual DNS server updates. If your device supports DHCP but not DNS, you are usually better off reserving a pool of DHCP addresses for your devices by MAC address rather than manually configuring the DNS servers.

As IPv6 is adopted alongside and eventually replaces IPv4, DNS should become more commonplace on AV devices. Since IPv6 addresses are long and susceptible to errors when entered manually, the transition to IPv6 will likely encourage the widespread adoption of DNS.

Transport Protocols

Layer 3, the Network layer, handles assigning IP addresses to network devices and identifying paths from one network to another. The actual end-to-end transportation of data, however, is handled by layer 4, the Transport layer. Transport layer protocols fragment IP packets into smaller chunks that fall within the maximum transmission unit (MTU) size of the network connection. This process is known as *segmentation*.

The transport protocol is responsible for segmenting data for transmission and reassembling it at its destination.

A transport protocol may be connection-oriented or connectionless. Connection-oriented transport protocols are bi-directional. The source device waits for acknowledgment from the destination before sending data. It checks to see whether data has arrived before sending more. Connection-oriented transport includes error checking and flow control.

Connectionless communication is one-way. The source device sends. The destination device may or may not receive. Connectionless protocols are less reliable than connection-oriented protocols, but they are also faster because there are no pauses for replies. Many media-oriented applications, including practically all real-time protocols, use connectionless transportation protocols.

In IP networks, the commonly used connection-oriented transport protocol is TCP and the commonly used connectionless protocol is UDP.

TCP Transport

TCP transport uses two-way communication to provide guaranteed delivery of information to a remote host. It is connection oriented, meaning it creates and verifies a connection with the remote host before sending it any data. It is reliable because it tracks each packet and ensures that it arrives intact. TCP is the most common transport protocol for sending data across the Internet.

TCP data transfer involves the following steps:

- TCP communication starts with a *handshake* that establishes that the remote host is there and negotiates the terms of the connection, including the sliding window size (how many packets can be sent at once before verification is required).

- The origin device sends one window at a time to the destination device.

- The destination device acknowledges receipt of each window, prompting the origin device to send the next one. The sliding window cannot move past a packet that has not been received and acknowledged. If any packets are damaged or lost in transmission, they will be resent before any new packets are sent.

Because TCP is reliable and connection oriented, it is used for most Internet services, including HTTP, File Transfer Protocol (FTP), and Simple Mail Transfer Protocol (SMTP).

UDP Transport

UDP is a connectionless, unacknowledged protocol. It begins sending data without attempting to verify the origin device's connection to the destination device and continues sending data packets without waiting for any acknowledgment of receipt.

In UDP data transfer, the following happens:

- The origin computer does not attempt a "handshake" with the destination computer. It simply starts sending information.

- Packets are not tracked, and their delivery is not guaranteed. There is no sliding window.

UDP lacks TCP's inherent reliability. That does not mean all data transmitted using UDP is unreliable. Systems using UDP may manage reliability at a higher level of the OSI Model, such as the Application layer.

UDP is used for streaming audio and video. When packets are lost in transport, UDP transport just skips over missing bits, inserting a split second of silence or a repeated image instead of coming to a full stop and waiting for the packets to be resent.

UDP may also be used to exchange very small pieces of information. In some cases, such as retrieving a DNS name, a TCP "handshake" takes more bits than the actual exchange of data. In such instances, it is more efficient to use the "connectionless" UDP transport.

TCP vs. UDP

TCP transport is used when the guaranteed delivery and accuracy or quality of the data being sent is most important, for example, when sending AV control signals. UDP transport is used when speed and continuity are most important, for example, during any real-time AV communications. However, many enterprises have policies against UDP transport because of security issues. UDP streams can be used in malicious attacks such as denial-of-service attacks, which swamp network equipment with useless requests, or self-replicating Trojan horse viruses. If you recommend the use of UDP transport for streaming media, be prepared to defend its necessity.

Ports

After data arrives at a device, how does the device know what to do with the data? The Transport layer protocol, either TCP or UDP, will include a port number. Transport layer ports are not physical ports: they are logical ports, telling the network data what application it should "connect" to on the network device.

Essentially, the port number indicates to the server what you want the data to do. The Internet Assigned Numbers Authority permanently assigns many port numbers to standardized, well-known services. Every IP network has a Services file that contains a list of permanently assigned ports and their associated services:

- System ports, 0 to 1023, are assigned to standard protocols. These are also known as *well-known* ports.
- User ports, 1024 to 49151, are assigned by IANA upon request from an application developer.
- Dynamic ports, 49152 to 65535, cannot be assigned or reserved. Applications may use any dynamic port that is available on the local host. However, the application cannot assume that port will always be available. Dynamic ports are also known as *ephemeral* ports.

Although you can often identify the originating application by the data packet's port number, this is not always possible. Sometimes, a service with a permanently assigned port has to open one or more dynamic ports to run several instances of the service on the same host. Many applications will choose a port from the dynamic port range at random.

Protocol	TCP Port(s)	UDP Ports
HTTP	80	
Secure HTTP (HTTPS)	443	
File Transfer Protocol (FTP)	20 (data), 21 (control)	
Secure Shell (SSH)	22	22
Network Time Protocol (NTP)	123	
Simple Network Management Protocol (SNMP)	161	161
Domain Name System (DNS)	53	53

Table 12-8   Well-Known Ports Commonly Used in AV

Table 12-8 shows some well-known ports that are commonly used by AV traffic.

Ports are one of the most important points of coordination between AV and IT. You will need to consult the manufacturer's specifications of each piece of networked gear to discover what ports it uses. You and the network manager may also want to test the gear to ensure that port documentation is complete. This information will need to be documented in the Network Device Inventory. The network manager will use this information to make sure the right ports are opened on the right devices. That way, your AV traffic will not be blocked by a firewall or router because it uses an unrecognized port.

A port number may be specified in a URL by appending it to the domain name after a colon. For example, *http://www.domain.com:8080* would direct the web browser to connect to port 8080 on the domain.com web server.

The Host Layers

In the OSI Model, layers 5 to 7 are host layers (layers 1 to 3 are media layers and layer 4 is the transport layer). The media layers are where most of the AV–IT coordination needs to take place. IP and Ethernet networks do not care what kinds of applications they are carrying, as long as those applications are sending out data in the right format. Most network troubleshooting takes place at the lower levels as well. Still, you need to be familiar with the terminology and functions of the host layer.

The Session Layer

Layer 5, the Session layer, manages sustained connections between devices. TCP includes some Session layer functionality in that it verifies that the receiving device is listening and negotiates how much data it can send before transmitting packets. True Session layer protocols negotiate even more parameters. A Session layer protocol formally begins and ends sustained communication among devices. It regulates which devices transmit and which receive. It also regulates what kind of data each device can send and receive and at what bandwidth. Session layer protocols are important in streaming media and conferencing applications. They negotiate to make sure each device sends and/or receives the best quality of which it is capable. In conferencing applications, they also manage which device talks and which devices listen at any given moment.

The Presentation Layer

Layer 6, the Presentation layer, is responsible for making data look the same to the lower-level protocols. The Presentation layer is also responsible for encoding and compressing data to reduce its required bandwidth. Codecs are a Presentation layer technology. The Presentation layer is also sometimes responsible for encrypting and decrypting data for security purposes, although this can also take place at the Application layer.

Codecs

The term *codec* is short for enCOder/DECoder. A codec is an electronic device or a software process that encodes or decodes a data stream for transmission and reception over a communications medium.

Codecs may be one-way or two-way, encrypted or unencrypted, symmetrical or asymmetrical, and compressed or uncompressed. They may also include analog to digital or digital to analog conversion. The decision as to what codecs to use for a client's streaming services will be determined by a number of factors, including the following:

- IT policies. The software that users currently have or may be allowed to have will determine what codecs are available on the end-user playback device, which in turn determines what you can encode to.
- Licensing fees associated with the codec.
- Resolution and frame rate of the source material.
- The available processing power of the encoding and decoding devices.
- Desired resolution and frame rate of the stream.
- Latency introduced by the codec processes.
- Bandwidth required for the desired quality. You may not be able to find specification for bandwidth; therefore, some network testing will be necessary.

The Application Layer

The Session layer hides the differences between data. Layer 7, the Application layer, unpacks them. The Application layer is responsible for presenting data to the right software in a way that the software can understand. It turns the data it receives from the Presentation layer into e-mails, web pages, FTP files, databases, media streams, and so on, depending on the port number identified by the Transport layer. This is the layer that turns network data into data the user can actually interact with. There are as many Application layer protocols as there are software applications; there are too many to count, and more every day.

Bandwidth

Bandwidth is a critical networking concern that spans multiple layers. In fact, as an AV professional, the network's bandwidth is one of the attributes you care most about. If the network does not have sufficient bandwidth, the AV signal quality will plummet or the signal simply will not arrive at its destination. In analog signals, system bandwidth is measured in Hertz (Hz); however, in most complex digital encoding systems, a single signal cycle may encode more than one bit of data, so *data throughput* is measured in bits per second (bps).

The term bandwidth refers to the following:

- The capacity of the network connections, such as "This switch has a bandwidth of 100Gbps."

- The throughput requirements of the data or devices, such as "This videoconference system requires 4Mbps of bandwidth per endpoint."

When using an IT network to transport AV data, your concern should be about bandwidth availability, not about bandwidth capacity. You need to make sure the network has enough free unused bandwidth to handle AV signals.

In reality, no more than 50 percent of a network's ready capacity should be allocated for routine use. Your client's AV devices may be on a separate LAN, in which case you do not have to worry about other data traffic crowding the client's network. This allows you to comfortably plan to use a substantial percentage of the available bandwidth. Work with your client's network engineer or IT manager to find out how much bandwidth you can actually plan to use for AV.

Quality of Service

If you run into bandwidth limitations, there are strategies you can use to make sure your AV traffic gets through. *Quality of service* is a term used to refer to any method of managing data traffic to preserve system usefulness and provide the best possible user experience. Typically, QoS refers to some combination of bandwidth allocation and data prioritization.

Many different network components have built-in QoS features. For example, videoconferencing codecs sometimes have built-in QoS features that allow various devices on the call to negotiate the bandwidth of the call.

Network managers may also use software to set QoS rules for particular users or domain names. During the network design stage or with AV device installation, you should be concerned with QoS policies that need to be configured directly on network switches and routers. You will need to work with network management to make sure they are aware of any networked device QoS requirements and that those settings have been configured on the relevant network devices.

DiffServ

The underlying strategy of network-based QoS is to prioritize time-sensitive traffic over other traffic. One way to accomplish this is to assign each type of traffic on the network to a particular QoS Differentiated Service (DiffServ) class. Each class is handled differently by the managed network switches and routers, which is why it is called differentiated service. Some classes are designed as *low loss* to preserve data without losing any packets. Some classes are designed as *low latency* to transport data as quickly as possible.

Some classes prioritize data arriving in the exact order in which it was sent, that is, *low jitter*. The lowest-priority class is *best effort,* where data in this class will arrive when and how it arrives, with no guarantees of integrity or timeliness.

Each application your customer will use is assigned a DiffServ class on the network routers and switches. When traffic enters the network, these devices automatically detect which application it comes from and tags it with a DiffServ class. The DiffServ class then defines how the network devices prioritize the traffic. The *signaling service* class is for traffic that controls applications or user endpoints. For example, signals that set up and terminate a connection between conference call endpoints would belong in this class.

The *telephony class* is intended for Voice over IP (VoIP) traffic, but it can be used for any traffic, such as streaming video and audio, that transmits at a constant rate and requires very low latency. The *real-time interactive* class is for interactive applications that transmit at a variable rate and require low jitter and loss and very low delay. Examples include interactive gaming and some types of videoconferencing.

The *multimedia conferencing service* class is for conferencing solutions that can dynamically reduce their transmission rates if they detect congestion. If a conferencing class cannot detect and adapt to network congestion, the real-time interactive class should be used instead.

The *broadcast video service* class is for inelastic, noninteractive media, that is, media streams that cannot change their transmission rate based on network congestion. This class is used for live events, AV streaming, and broadcast video.

The *multimedia streaming service* class is for noninteractive streaming media that can detect network congestion and/or packet loss and respond by reducing its transmission rate. This class is used for video-on-demand (VOD) services, video that is stored before it is sent and buffered when it is received to compensate for any variation in transmission rate.

The *low-latency data* class is for applications where data arrives in big, short-lived bursts. The *high-throughput data* class is for longer, high-volume traffic flows. It is used by applications that store data and then forward it, like FTP service and e-mail.

The *low-priority data* class is used for any applications that can tolerate long interruptions.

The *standard* class provides best-effort delivery. Any applications that are not specifically assigned to another class will fall into the standard class.

Security Technologies

Networks make resources easily accessible worldwide. You no longer have to be physically present at a device to use it, configure it, or troubleshoot it. In theory, you can do so from anywhere. The problem is, unless your client's network is properly secured, anyone can gain access to it. If the wrong people gain access to your client's network resources, they can do a lot of damage.

Security is often cited as the number-one concern regarding attaching AV devices to an enterprise network. What are IT professionals so worried about? In this section, you will learn about the common security risks network professionals must face. You will also learn what can be done about those risks.

Network Access Control

User authentication and authorization are key aspects of security on an enterprise network. All the network's known users, including administrators, have a user profile. This profile identifies the user's e-mail account, access privileges, group memberships, and other relevant information. When users log onto the network, they have to prove their identity to gain access to their user profiles. This is usually accomplished with a username and password combination (something you know), although more secure networks often have additional multifactor requirements, such as physical access cards, digital token devices, code generators, or biometric scans (something you have).

The guiding philosophy of user access is *least privilege*. That is, users should have the least level of privileges they can get by with and still do their jobs. From time to time, administrators will scan the network and systems to verify user access. If a user needs additional access, they will go through a formal approval process. Then the administrator will escalate the privileges. The administrator will still enforce least privilege.

Permissions to enter a network and what rights you have once there are governed by a group of technologies and policies known collectively as Network Access Control (NAC). NAC is based on an idea that is simple to understand but challenging to implement: when you log onto a network, who you are should determine what you can do.

In a robust NAC environment, "who you are" is determined by more than your username. Identification and authentication are part of NAC, and the right to access certain VLANs, files, or programs may be directly associated with the user's login. However, NAC may look at other factors to determine what rights a user should have. For instance, NAC may examine the endpoint you are using and limit your rights if it is not sufficiently secure. For example, is the antivirus software on the computer up to date? If not, you may be denied access to sensitive areas of the network. NAC may also examine what type of connection you are using. Are you connected via Wi-Fi, remote VPN, or a cable connected to an onsite wall port? You may have access to more parts of the network via a physical, onsite connection than a remote or wireless one. NAC may examine some or all of these factors when deciding what rights to grant a user.

Access Control List

After a user is granted access to a network in general, their specific rights within the network may be governed by an access control list (ACL). The ACL is typically configured on the network router or on the device being accessed. It controls what is permitted to travel through the router based on type of traffic, source, and/or destination. The ACL may also contain network privileges regarding who can access what parts of the network. If your client's AV system will require special access rights, be sure that network security personnel create an ACL for the system and add the appropriate end users.

Firewall

A firewall is any technology—hardware or software—that protects a network or device by preventing intrusion by unauthorized users and/or regulating traffic permitted to enter or exit the network. Firewalls may control access across any network boundaries,

PART V

including between an enterprise network and the Internet, between LANs within an enterprise, and between a host and its local network.

An enterprise network is usually protected by a dedicated hardware firewall appliance, while LANS are often protected by firewall applications on the gateway router, and individual hosts are usually protected by the firewall service built into their operating system.

A firewall is really a set of policies implemented across a range of devices. In essence, a firewall policy within an enterprise can be distilled to one of two approaches:

- All network traffic will be allowed unless it is specifically forbidden; the default is "allow."

- All network traffic will be forbidden unless it is specifically allowed; the default is "deny."

The former emphasizes ease of use but forces the network administrators to try to predict how the network may be attacked. The latter is more secure but makes new systems more difficult to configure. In either case, the responsibility of the AV designer is the same: document system ports and protocols and coordinate those needs with the network manager.

 CAUTION A firewall cannot protect users from traffic they invite onto the network. For example, it will not stop a virus or malware that they download. This is why user awareness training will always be the front line of network defense.

Types of Firewalls

The number of firewalls an enterprise network will need is a critical network design decision. There will be firewalls protecting gateways to the Internet. Firewalls may also be deployed within an organization's private network to protect certain areas from internal intruders. Firewalls use several different strategies or a combination of strategies to protect the network, including the following:

- Packet filtering rules determine whether a data packet will be allowed to pass through a firewall. Rules are configured by the network administrator and implemented based on the protocol header of each packet.

- Packet inspection tracks the state of ports and protocols in network connections and determines whether the data in each packet is part of a permitted connection initiated by devices behind the firewall. Deep packet inspection may work at the Application layer (layer 7) to verify that the data in the packets matches with the packet headers and is in the permitted format for the intended application.

- Port address translation (PAT) is a method of network address translation whereby devices with private, unregistered IP addresses can access the Internet through a device with a registered IP address. Unregistered clients send datagrams to a NAT server with a globally routable address (typically a firewall). The NAT server forwards the data to its destination and relays responses to the original client.

PAT is also known as *one-to-many NAT, Network and Port Translation* (NAPT), or *IP masquerading*.

By enabling multiple devices to access the Internet without globally routable addresses, PAT conserves address space, which is a concern in IPv4 implementations. Even though the number of available global addresses in IPv6 is effectively unlimited, PAT is used with IPv6 networks because it hides the original source of the data. From outside the network, all data appears to originate from the NAT server. Any data that arrives at the NAT server without a client's request has nowhere else to go (analogous to having the address of a building without an apartment number, resulting in entry blocked by the doorman). PAT blocks all unrequested data with a firewall. A malicious intruder cannot trace the data's path beyond the edge of the network.

Port forwarding combines PAT and packet filtering. The firewall inspects the packet based on packet filtering rules. It is also configured to translate certain ports to private addresses on the network.

By combining packet filtering and PAT, the network administrator can allow incoming, unrequested traffic under controlled conditions. For example, the computer with a specific IP address may be allowed to send Telnet commands over port 23 to AV devices, but port 23 packets from any other address will be rejected. The firewall detects the IP address and the port and translates that port to an address, automatically forwarding the Telnet command to the designated device.

Network Verification and Troubleshooting

When troubleshooting an AV system on an IT network, both AV and IT professionals have to make an effort to understand how both AV and IT systems work separately and in conjunction. It is easy to blame the network for a problem if you do not really understand how or why the problem occurs. On the other hand, no AV system works perfectly all the time, and AV systems that operate over networks are no exception. If you learn some basic network troubleshooting skills, you will be far better equipped to identify the issues caused by the network and those caused by the AV gear. If the problem does lie within the network, you will be able to bring your IT counterparts a lot more information, enabling a collaborative effort to solve the problem much faster.

Troubleshooting is a process involving investigating, determining, and settling problems. It should never involve finger pointing or assumptions. The troubleshooting process should never begin with an accusation, such as "The video conference system is not working. There must be something wrong with the firewall." When you begin an investigation by saying to yourself, "I know what the problem is," you have already made your first mistake. Instead, follow a set, systematic procedure to isolate and repair problems.

If you service AV systems, you probably have your own step-by-step processes for troubleshooting different kinds of products and systems. The same is true for any IT service technician. You may each have elaborate flow charts or action plans to deal with particular problems. These established practices should not stop you from putting your heads together when a problem overlaps the AV and IT worlds. All troubleshooting methods share three basic stages: symptom recognition and elaboration, listing and localizing the faulty function(s), and analysis. You may want to review the general troubleshooting methodologies detailed in Chapter 11.

Network Troubleshooting Tools

Network troubleshooting may be necessary at any layer of the OSI Model. Network hardware and cabling must work if upper-layer protocols are going to function. Most errors observed at the Physical layer are from damaged wiring, defective transceivers, or network interface cards or switch ports going bad. In most cases, a protocol analyzer can detect the presence of physical errors. However, they might not be able to determine which device is failing.

Whether you are verifying or troubleshooting the network infrastructure, there are a few extra tools you should consider having handy. These may be physical devices or software applications that help you get the job done. This section covers some of the most commonly used tools and their functions.

Network Cable Tester

Network cable testers are electronic tools used to verify Ethernet Cat cable installation (see Figure 12-17). The cable tester requires two components: a sending unit and a receiving unit. Each unit is connected to one end of a network cable to assess its connectivity. Depending on the quality of the tester, it can test for simple connectivity, broken wires, crossed wires, and/or mismatched pairs. The tester can also identify attenuation; that is, it can tell you whether the cable length is too long.

Cable Tracer

Cable tracers are used when proper physical topology documentation is not available. A tone cable tracer can help you trace a twisted-pair cable run through a client's site. As shown in Figure 12-18, a tone tracer has two components: a tone generator and a probe.

Figure 12-17
Network cable tester

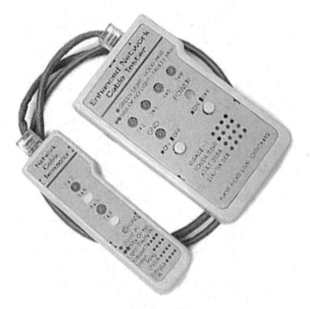

Figure 12-18
Tone tracer

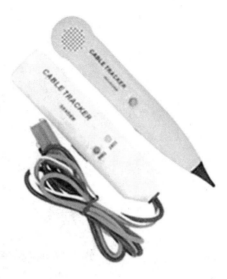

The tone generator is attached to one end of a wire. It puts a signal on the wire, which can be wirelessly detected by the tracker probe. The probe is then used to follow the route, or locate the other end, of the wire.

Optical Time-Domain Reflectometer
An optical time-domain reflectometer (OTDR) is used to test fiber-optic cables (see Figure 12-19). The OTDR can estimate the length of the fiber cable and overall attenuation, plus it can locate joints or breaks in the cable. It uses pulses of light, with the strength and delay of returning pulses, to locate problems in the fiber path.

PART V

Figure 12-19
Optical time-
domain
reflectometer

Figure 12-20
Wireless network
tester/analyzer

Wireless Network Analyzer

A wireless network tester/analyzer is used to troubleshoot wireless networks (see Figure 12-20). It reports on the frequency and field strength of a wireless signal and gathers additional information about the signal source, including the service set identifier (SSID) and channel of the access point. It can also determine whether the mode of the wireless device is ad hoc or infrastructure. The analyzer usually has a ping function to check connectivity and may support DHCP and wireless security features. A portable computer device with wireless analysis software may also fulfil this role.

NIC Loopback Tester

A NIC loopback tester is a device (see Figure 12-21) that sends a signal from the send port and receives the same signal on the receive port of a network interface card. The loopback test can determine whether an individual node is communicating properly by simulating a complete communication path. Loopback testers are available for both copper and fiber network interfaces.

Figure 12-21
NIC loopback
testers

Figure 12-22
Network protocol
analyzer

To test the NIC, plug the loopback tester into the physical network port. Start a network sniffer program such as Wireshark to monitor the outbound and inbound traffic separately. Then, send a packet using a utility program like ping. Watch the packet leave and return to the sending computer. Finally, compare the two packets in the sniffer. If the packets are the same, the test was successful.

Network Protocol Analyzer

Network protocol analyzer software can be useful in monitoring a system and tracing problems. However, you cannot always load protocol analyzer software onto an existing networked computer, especially if you are a "guest" in the system. In such cases, you may be able to use a hardware-based network protocol analyzer (see Figure 12-22). This device does not require a host operating system. It performs the same roles as a software-based analyzer, and it is designed to troubleshoot and monitor communications on a LAN. The analyzer shows you what protocols are present on the network at all layers of the OSI Model. A portable computer device loaded with network analysis software may also fulfil this role.

A protocol analyzer will always show a small percentage of errors at the Physical and Data Link layers, even if the network is working fine overall. Most of these errors will be malformed packets, meaning a packet that is too short or too long. Ethernet standards have defined a variable packet length between 64 and 1518 bytes. Any packets not within those boundaries are considered errors. Troubleshooting malformed packets is difficult. Normally, you have to disconnect physical devices from the network until the problem device is located.

Network Device Indicator Lights

Indicator lights on network devices provide immediate feedback such as connectivity, "activity," and "speed," as shown in Figure 12-23.

PART V

Figure 12-23
Indicator lights
on network
devices (Image:
A stockphoto /
Getty Images)

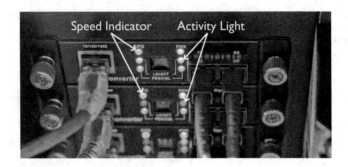

The specific color, patterns, and diagnostic meaning of the LED lights will vary by device and by manufacturer. Read the technical information provided by the manufacturer to translate the light colors and patterns into usable diagnostic information.

Internet Control Messaging Protocol

Internet Control Messaging Protocol (ICMP) is a TCP/IP protocol stack with some built-in utilities to help monitor the network and diagnose problems. It is a messaging service used to diagnose network connectivity issues. Because ICMP works at layer 3 of the OSI stack, it is not totally reliable. It cannot detect errors at the Transport layers and above. Still, it does provide valuable feedback about problems in the communication pathways.

Ping   *Ping* is a command-line interface (CLI) tool. In IPv4, ping uses ICMP messages to request a reply from another device, called an *echo request*. If the ping test is successful, the device responds with an *echo reply*. If there is at least one device with a functioning network connection, it can be used to send echo requests to other networked devices. This will help verify that those devices are connected to the network and responding to their assigned IP addresses or DNS names.

By default, when you send a ping request from a Windows system, it will issue four echo requests to the target system's IP address. Unix-based systems will continue to ping the target system's IP address until the command is terminated. In both Windows and Unix/Linux, it is possible to control the number of echo requests sent to a target system. Figure 12-24 shows a successful ping test that sent only one ping.

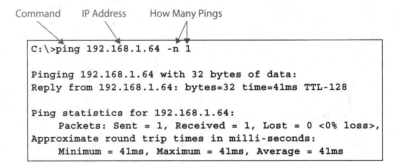

Figure 12-24   A ping test showing one ping sent to the IP address 192.168.1.64

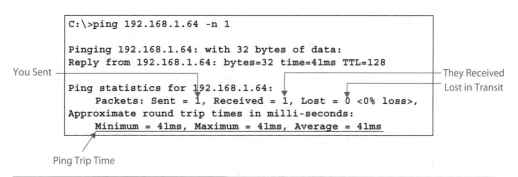

```
C:\>ping 192.168.1.64 -n 1

Pinging 192.168.1.64: with 32 bytes of data:
Reply from 192.168.1.64: bytes=32 time=41ms TTL=128

Ping statistics for 192.168.1.64:
    Packets: Sent = 1, Received = 1, Lost = 0 <0% loss>,
Approximate round trip times in milli-seconds:
    Minimum = 41ms, Maximum = 41ms, Average = 41ms
```

You Sent

They Received

Lost in Transit

Ping Trip Time

Figure 12-25 The results from a ping command show that the host is able to communicate with the network

Ping results show success, failure, and time in transit. This information can then be used to determine whether the target system is responding and, if so, how quickly. Latency is a by-product of distance and volume of all traffic on the network. Because latency changes constantly, there is no "normal" latency. However, excessive latency can be a sign of network congestion. When ping fails to connect with a target system, ping will return an error and show a loss.

Figure 12-25 shows a successful ping test. One ping sent, and a reply arrived within 41 milliseconds. This ping test confirms that these devices were able to communicate over the network.

Figure 12-26 shows the ping has failed. The request timed out, and the packet was lost. Either the device sending the ping or the device addressed is unable to communicate with the network.

If you want to test the NIC of the device you are currently using, you can ping the device's loopback address. Remember, IPv4 reserves 127.0.0.0 to 127.255.255.255 as loopback addresses. Any address in this range can be used, but most network devices automatically use 127.0.0.1 as their loopback address.

Traceroute Traceroute and Windows *tracert* are diagnostic tools for checking the network path between two nodes: the route. Traceroute, like ping, uses ICMP echo requests to map each intermediate hop. Traceroute works by incrementing the time-to-live (TTL) value of each ICMP request sent. The response received from each hop will include either

PART V

```
C:\>ping 192.168.0.254 -n 1

Pinging 192.168.0.254: with 32 bytes of data:
Request timed out.

Ping statistics for 192.168.0.254:
    Packets: Sent = 1, Received = 0, Lost = 1 <100% loss>,
```

Ping Error Indicators

Figure 12-26 A failed ping test may result in a "request timed out" error message

```
C:\>tracert 12.81.104.88

Tracing route to 12.81.104.88 over a maximum of 30 hops
                                                                  Time-in-Transit
    1    1 ms      3 ms      4 ms     home[192.168.1.254]
    2   24 ms     22 ms     25 ms     99-144-132-2.lightspeed.jcvlfl.sbcglobal.net [99
.144.132.2]
    3   24 m      23 ms     25 ms     99.132.13.24
    4    *         *         *        Request timed out.              Error
    5    *        23 ms     22 ms     99.132.13.57
    6   22 ms     23 ms     67 ms     70.159.204.198
    7   22 ms    103 ms     38 ms     12.81.50.46
    8   25 ms     25 ms     25 ms     12.82.104.88

Trace complete.
```

Figure 12-27 A successful traceroute test showing an error at one hop

an error or a time-in-transit value. Figure 12-27 shows traceroute in action, revealing the time-in-transit and error values. This implies that the intermediate point is administratively blocking the ICMP request. This is not an unusual response. There is no real error unless the trace fails completely.

Ipconfig/ifconfig

Ipconfig and *ifconfig* are similar CLI tools used for diagnosing IP configuration issues on the local computer system. Ipconfig is a Windows tool, and ifconfig is a Linux/Unix tool, each with additional features that can vary between operating systems. Either tool will reveal the device's IP configuration.

You can use these tools to reveal the DNS name, IP address(es), subnet mask, gateway, and other details about the network the device is currently using. Running ipconfig from the CLI by itself will result in displaying basic information about the IP configuration. The Linux/Unix command ifconfig actually allows you to change the configuration. Figure 12-28 shows basic information such as whether the device has an IP address, the network it belongs to, and the gateway used for communication.

```
C:\>ipconfig
Windows IP Configuration

Wireless LAN adapter Wireless Network Connection:

    Connection-specific DNS Suffix  . : gateway.2wire.net
    Link-local IPv6 Address   . . . . : fe80::ada4:7b47:79c3:8f6c%12
    IPv4 Address. . . . . . . . . . . : 192.168.1.86
    Subnet Mask . . . . . . . . . . . : 255.255.255.0
    Default Gateway . . . . . . . . . : 192.168.1.254
```

Figure 12-28 Basic IP configuration information

```
C:\>ipconfig /all
Windows IP Configuration

    Host Name . . . . . . . . . . . . . : JAFLT
    Primary Dns Suffix  . . . . . . . :
    Node Type . . . . . . . . . . . . : Hybrid
    IP Routing Enabled. . . . . . . . : No
    WINS Proxy Enabled. . . . . . . . : No
    DNS Suffix Search List. . . . . . : gateway.2wire.net

Wireless LAN adapter Wireless Network Connection:

    Connection-specific DNS Suffix  . : gateway.2wire.net
    Description . . . . . . . . . . . : Intel<R> Centrino<R> Advanced-N 6205
    Physical Address. . . . . . . . . : 08-11-96-EA-97-C4
    DHCP Enabled. . . . . . . . . . . : Yes
    Autoconfiguration Enabled . . . . : Yes
    Link-local IPv6 Address . . . . . : fe80::ada4:7b47:79c3:8f6c%12<Preferred>
    IPv4 Address. . . . . . . . . . . : 192.168.1.86<Preferred>
    Subnet Mask . . . . . . . . . . . : 255.255.255.0
    Lease Obtained. . . . . . . . . . : Wednesday, December 22, 2021 3:54:34 PM
    Lease Expires . . . . . . . . . . : Sunday, December 26, 2021 7:44:29 AM
    Default Gateway . . . . . . . . . : 192.168.1.254
    DHCP Server . . . . . . . . . . . : 192.168.1.254
    DHCPv6 IAID . . . . . . . . . . . : 302518678
    DHCPv6 Client DUID. . . . . . . . : 00-01-00-01-16-88-9C-F0-DC-0E-A1-2A-D1-E9
```

Figure 12-29   Ipconfig /all delivers DHCP server details and more

By using an additional command switch, /all, more information is available. Figure 12-29 shows ipconfig /all delivers the DHCP server, lease information, and more. This is called *verbose mode.*

You can use this output to determine whether the host node has an active IP address, to confirm that the IP address is in the correct network, and to determine whether the IP address is statically or dynamically assigned. If the IP address is set to dynamic and has an IP address of 0.0.0.0 or 169.254.X.X, then you know that the host is not talking to the DHCP server.

Address Resolution Protocol

The Address Resolution Protocol (ARP) command tool is the Data Link layer protocol that resolves IP addresses and Ethernet addresses. Every time a computer talks to another computer on the local network, a new entry is made in the ARP cache. Every time a computer talks on the network, the computer checks the ARP cache first.

ARP has an associated command-line tool, *arp*, which can be used to display the contents of the ARP cache and/or define static entries in an ARP table. To display the contents of the ARP cache, you enter the command *arp -a* from the CLI, as shown in Figure 12-30.

ARP is a convenient way to discover the MAC addresses of the devices on a network. If you know the IP address of a particular device but not the MAC address, you can use

Figure 12-30
ARP reveals the
IP addresses and
MAC addresses of
local hosts with
which a device
has recently
communicated

```
C:\>arp -a

Interface: 192.168.1.86 --- 0xc
   Internet Address        Physical Address       Type
   192.168.1.254           b0-e7-54-c9-4c-19      dynamic
   192.168.1.255           ff-ff-ff-ff-ff-ff      static
   224.0.0.22              01-00-5e-00-00-16      static
   224.0.0.252             01-00-5e-00-00-fc      static
   255.255.255.255         ff-ff-ff-ff-ff-ff      static
```

ARP to discover it. When a datagram is sent, if the receiver's address is not in the cache, a new ARP request is generated. You can force a new entry in the ARP cache by contacting the target device using any communication protocol. The simplest way is by sending a ping to the target system.

Troubleshooting any computer or network problem should always start with a series of questions. Some of the questions should be directed to the user and some to yourself. In most scenarios, you will deviate from a standard line of questioning only after the problem is better defined.

Anything beyond this basic level, you will likely need to refer to a network support professional. Everyone's troubleshooting processes will be different. There is no "best" order in which to check potential error sources, but performing the simplest checks first can potentially save you a lot of time.

Make sure you document your process and follow a logical path to resolving the problem. Know when a problem goes beyond your knowledge level, and feel free to escalate as necessary.

Chapter Review

In this chapter, you learned the basics of IT network operations, including how LANs, WANs, and VPNs work. The seven layers of the OSI Model, Ethernet, and IP addressing were described in detail. You learned about a range of networked AV technologies such as AVB, AES67, ST 2110, SDVoE, Dante, and Q-SYS.

Upon completion of this chapter, you should be able to do the following:

- Create a document detailing the network requirements of any AV equipment using the enterprise IT network
- Identify and define common networking components
- Name the layers of the OSI Model and describe the functions of each layer
- Identify common physical network connections and describe their capabilities
- Describe the function and capabilities of Ethernet technologies

- Compare AVB, ST 2110, SDVoE, AES67, Dante, RAVENNA, Q-SYS, SVSI, NAV, and DM NVX protocols
- Define the function and capabilities of Internet Protocol technologies
- Differentiate between TCP and UDP
- Identify network security technologies and their functions
- Identify some network troubleshooting tools and technologies

Now that you have an understanding of how networks and networked AV systems work, you are ready to take a look at AV control systems.

Review Questions

The following questions are based on the content covered in this chapter and are intended to help reinforce the knowledge you have assimilated. These questions are similar to the questions presented on the CTS-I exam. See Appendix E for more information on how to access the free online sample questions.

1. Which of the following should be included in a Network Device Inventory? (Choose all that apply.)

   A. Ports used by each device

   B. Device model numbers and firmware versions

   C. Physical dimensions of each device

   D. MAC address of each device

2. Which device provides a direct physical connection between multiple nodes?

   A. Bridge

   B. Router

   C. Switch

   D. Gateway

3. Which LAN topology connects devices to a central point?

   A. Star

   B. Mesh

   C. Bus

   D. Ring

4. In the OSI Model, Ethernet and switches are elements of _____.

   A. layer 7, the Application layer

   B. layer 4, the Transport layer

   C. layer 5, the Session layer

   D. layer 2, the Data Link layer

5. Crossover cables are used to connect _____.

    A. digital and analog devices

    B. devices configured to accept different Data Link layer protocols

    C. two devices without the use of an electronic crossover device

    D. a device's NIC to its MAC address

6. What are the limitations of a Wi-Fi network for streaming live HD video? (Choose all that apply.)

    A. Lower bandwidth than wired connections

    B. Expensive infrastructure

    C. Potential RF interference

    D. Restrictions on range

7. What does 1000Base-T refer to?

    A. Ethernet speed of 1Gbps

    B. IP speed of 1Gbps

    C. Ethernet speed of 100Mbps

    D. IP speed of 100Mbps

8. Which of these are layer 2 protocols commonly used to transport real-time media over a network? (Choose all that apply.)

    A. Dante

    B. UDP

    C. TCP

    D. AVB

9. Which of the following are parts of an IP address for a networked device? (Choose all that apply.)

    A. Subnet mask

    B. Bit rate

    C. Network identifier

    D. Host identifier

10. What CLI command would you use to discover the IP address of a device you are currently using?

    A. ping

    B. ipconfig

    C. arp

    D. loopback

Answers

1. **A, B, D.** A Network Device Inventory should include the required ports, model number and firmware version, and MAC address of each device, along with other network requirements.

2. **C.** A switch provides a direct physical connection between multiple nodes.

3. **A.** Star topology connects devices to a central point in a LAN.

4. **D.** In the OSI Model, Ethernet and switches are elements of layer 2, the Data Link layer.

5. **C.** Crossover cables are used to connect two devices without the use of an electronic crossover device such as a switch or router.

6. **A, C, D.** The limitations of a Wi-Fi network for streaming live HD video include lower bandwidth than wired connections, potential RF interference, and restrictions on range. Wi-Fi networks do not require expensive infrastructure.

7. **A.** A 1000Base-T Ethernet connection has a speed of 1Gbps.

8. **A.** AVB is a layer 2 protocol commonly used to deliver real-time media. Dante is a layer 3 real-time media delivery protocol. Although UDP is commonly used in real-time media transport, it is a layer 4 protocol.

9. **A, C, D.** Subnet mask, network identifier, and host identifier are parts of an IP address. Bit rate is related to speed and is not part of an IP address.

10. **B.** The ipconfig command can be used to discover the IP address of a device.

CHAPTER 13

Audiovisual Control Systems

In this chapter, you will learn about

- The capabilities of control systems
- Types of control systems
- Installing control components
- Verifying control systems

Control systems link audiovisual devices together to form functional audiovisual (AV) systems and connect AV systems to external systems such as lighting control, atmospheric effects, mechanical staging systems, HVAC, security systems, building management systems, fire detection systems, and pyrotechnic control systems. They are also the interface between the user and the integrated AV system.

AV control systems are used to integrate, automate, and simplify a series of operations to produce a specific AV system function or configuration. Control systems can be as complex as controlling dozens of interactive exhibits across a multi-space museum site or as simple as a remote control device that operates a consumer television set.

Control system installation follows preparation guidelines similar to audio and video systems. However, the complexity of the control system often requires that special attention be given to details such as the functional design, technical specifications, signal-flow drawings, network configuration, and Internet Protocol (IP) addressing information.

To install a control system correctly, AV installers will often need to consult with the software programmers, system designers, systems engineers, and IT professionals involved in the project.

Control Systems Overview

Control systems allow complex AV systems to be operated using simple interfaces. Complex AV systems are composed of many individual components, integrated to perform like a single system. A control system simplifies the complex individual functions and steps necessary to complete specific tasks, allowing users to focus on delivering the message.

The control system is the link between the end user, the electronics, and the operational environment. All systems of control have common traits such as communications from the controlling element to the remote elements and feedback from the remote elements to the controller. Diagrams, often called *panel layouts,* determine the control system user interface layout.

Control systems can interconnect and operate a wide range of systems and devices, including the following:

- AV components such as projectors, projection screens, monitors, video screen systems, vision mixers, video switchers, cameras, codecs, media servers, AV switch matrixes, image processors, audio systems, camera motion control systems, digital signage systems, and media players
- Environmental control systems such as lighting, curtains, draperies, occupancy and positional sensors, fire detection systems, light sensors, HVAC, and thermostats
- Production elements, including stage and house lighting, pyrotechnic systems, moving scenery, stage machinery, atmospheric and flame effects, front-of-house operations, flying systems, stage operations crew, motion sensors, signaling systems, and safety interlock systems
- Power supply systems, power sequencing systems, power source selection, and changeover systems

Some applications of control systems in commercial AV systems include the following:

- Dimming room lights and playing a media stream
- Selecting a media source for replay
- Setting playback volume levels
- Lowering or raising a projection or display screen
- Changing presentation modes in a huddle room
- Connecting multiple points for an audio or video conference
- Starting a playback, lighting, atmospherics, and animatronics sequence when a visitor enters an exhibit
- Providing notification when a digital signage screen goes offline and automatically resetting it

Audiovisual control systems are composed of both hardware and software elements. Examples of control systems include the following:

- Handheld TV remote control
- A dedicated room-control system with a touch-panel interface
- A voice-interactive, cloud-connected, smart-device-driven product presentation space

- Rack-mounted control systems for handling dozens of devices in a large auditorium
- Networked control systems tying together many dispersed locations and devices into one coherent solution

Control can be initiated by a user, an event, a sensor, or a programmed automatic process. It can be initiated by remote-control devices in the same room or from any point on the Internet. In addition, control of remote devices can be made available via touch-screen devices, physical sensors, cloud-connected smart media devices, and physical buttons.

Duty Check

This chapter relates directly to the following task on the CTS-I Exam Content Outline:

- Duty C, Task 7: Load Configuration and Control Programs

This task comprises 4 percent of the exam (about four questions). This chapter may also relate to other tasks.

Control System Functions

According to AVIXA best practice, a control system's behaviors are described in a *button-by-button* document. This document narrates the events that occur when someone "presses a button" (initiates a trigger event) on the control system interface. The events described are the button's functions and programs.

A *function* is an individual action. For example, if you select Lights On from a user interface, the lights in the room switch on. Common functions include the following:

- Raising or lowering a projection screen
- Commanding a switcher to change an input (see Chapter 7 for information about switchers)
- Setting volume levels
- Routing a signal from one device to another
- Powering on devices
- Opening or closing drapes
- Activating various functions on a single device (play, pause, and stop)

In a more complex system, a user may execute a programmed sequence of several functions from a single trigger action. Such a sequence is often referred to as a *macro*.

For example, a user might press a button marked Play Show. After the user presses the button, the control system may execute the following programmed sequence of functions:

- Power up the haze and smoke machines
- Power up the effects lighting system
- Power up the show replay system
- Power up the projectors, video processor, and video displays
- Power up the speaker arrays
- Power up the animatronics system
- Switch the audio and video inputs to the preshow players
- Display the preshow video content
- Play the preshow music
- Wait 10 minutes
- Lower house light levels in the room
- Fade out the preshow music and video
- Switch the video and audio inputs to the show replay system
- Trigger the show replay system
- Switch on external "Show in Progress" signs

In general, as the project requires more functions, the design complexity of the control system increases. When a system has been designed correctly, anyone should be able to operate it seamlessly. The operators' level of knowledge and experience can vary widely, so accommodating a wide range of users makes control-system design and programming a challenge.

User Interfaces

Although it is becoming increasingly common for control systems to trigger functions based on internal timers or external sensors, such as when a visitor steps on a pressure mat or enters the trigger zone on a camera image, control systems generally require some form of *human–machine interface* (HMI) to trigger control sequences.

A user interface (UI) might be a multi-button panel installed at the entry door of a presentation room, allowing the user to select a lighting level for the room. It might be a cloud-connected smart-speaker, a pocket-sized wireless transmitter, or an application installed on a personal digital device. Advanced graphical user interfaces (GUIs) on touch-sensitive displays and cloud-based control panels are widely used. The design and programming of the UI, and the total user experience (UX), are highly specialized skills, requiring advanced study and training.

The UI is the most critical element of the control system because its purpose is to enable end users to understand and operate the facility. The needs of the user and the type of experience will dictate the appropriate UI devices. Since the control system's users may not be technical, the UI must be designed to be reliable, predictable, and intuitive.

Control System Hardware

Control systems are usually made up of hardware, firmware, and software. The hardware consists of the system's central processing unit (CPU) and all the devices it connects to.

The common hardware components used in control systems are

- A central control unit where all of the processing and decision making take place. Other components generally check with the control unit for instructions when a command is sent or received. The control unit may be a dedicated CPU, an application or service running on a general-purpose computer, or a cloud-based processing unit.

- User interface points, including touch panels, push buttons, switches, remote controllers, movement and occupancy sensors, smart devices, and button panels.

- Wireless and IR transmitters and receivers.

- Interface components between the control system and other systems such as lighting, stage machinery, life safety, pyrotechnics, building management and automation, audio, flame and atmospheric effects, security, fire safety, and HVAC.

- Signal switchers and media controllers such as replay systems, media servers, streaming sources, cable decoders, off-air receivers, and optical-disc players.

- AV system devices, including display panels, digital signal processors, projectors, digital signage controllers, display controllers, AV switch matrixes, screen and projector lifts, drape controllers, and mixers. Some of these devices, such as projectors and audio amplifiers, may have an Internet of Things (IoT) control interface.

As noted earlier, hardware is just one of the components of a control system. Without firmware or software, the hardware cannot execute a command.

Firmware

Firmware is the built-in software that has been programmed into or stored in a hardware device. This firmware is usually stored in nonvolatile storage, such as flash memory, erasable programmable read-only memory (EPROM), or a similar component. The firmware tells the device how to operate: how to perform basic processes such as accessing memory and peripheral devices, how to load and run software, and how to communicate with external devices. On some projects you may have to install or update firmware; details on these tasks are provided in the *Installing Firmware* section later in this chapter.

Control System Software

Software is a general term for the instructions and data that are used to perform tasks on programmable devices, such as computers, controllers, and other smart devices with a processing capability. An AV control system generally must have some kind of operating firmware or software (operating system or OS) loaded to perform a function. You can install the hardware devices and connect up the system, but until the OS software/firmware is loaded, the system will not function.

A control system's operating software/firmware does not usually perform any sophisticated functions without an additional set of programmed or scripted action instructions. The manufacturer or author of the control system usually provides an *authoring program* to create scripts or instructions for functions and operations, or may provide a set of preprogrammed instructions for some standard functions.

In many cases, by the time the control system is ready for installation and testing, a programmer will have authored the scripts or functions required for the installation, as specified by the system designer. In other circumstances, the programming may be undertaken on site once the AV system has been installed and tested. The control system programmer may be an independent contractor, a member of the system integrator's staff, or a programmer working for the control system supplier.

Some devices in the system, such as controllers with a custom UI or an IoT interface, will require specialized creation or authoring software, which must interface with the main control system. For instance, to create a touch-panel layout and triggers, you probably need to have specific software.

Programmers develop the screens for the interfaces and the logic to operate the equipment. They must use the detailed systems drawings to know which device is connected to which port on a switcher, what audio channel is routed to which amplifier, the addresses of Internet sources, and so on. For each device, the proper commands must be obtained from the device manufacturer or from the control system supplier in the form of a device driver, a personality specification, or similar subprogram.

After creating a custom script or program for the system, the programmers/developers often put the program or script files through a process called *compiling*, where the human-readable instructions are converted into native machine-code instructions ready for loading and execution by the control system devices.

Some control systems have sufficient processing power to be able to *interpret* each human-readable instruction immediately prior to execution, thereby eliminating the step of compilation before execution. This allows the programmer to modify a script or program and run it immediately, which shortens the iterative write/test/debug/test/debug/test/debug, etc., process during system installation and integration.

You may be responsible for loading programs into the control systems on the job site. This takes some practice, but can be done by an experienced installer with some guidance from the programmer or control system manufacturer.

Software for a control system is often proprietary to the hardware. While some control systems are based on broadly used OSs like Microsoft Windows, Apple macOS, or various flavors of Unix and Linux, others utilize a dedicated operating system, although these are frequently based on customized versions of Linux. You will need to have a working knowledge of the OS, software, and hardware that are compatible with the system specified for your installation.

Types of Control Systems

Control systems communicate with the devices they control using a number of different communications protocols. The protocols discussed here are those most widely employed: serial data, IP, universal serial bus (USB), contact closure, infrared (IR), and radio frequency (RF). AV signal extension systems such as HDBaseT and AVX include serial data, USB, IR, and Ethernet in their suites of embedded signals, and so, too, do some AV over IP systems.

Serial Data

Before the wide availability and affordability of Ethernet networks, many AV systems components were controlled via serial data protocols. Serial digital control is a method of sending control messages between devices using binary digital codes, frequently over a wired connection. Although the control codes could be arbitrary sequences of 1s and 0s, it is common to use ASCII-encoded text sequences to make the messages human-readable for testing and debugging. Many AV system controllers still include a serial data port or two for communication with legacy equipment.

The main serial data communications that were used in AV control systems are described in the following sections.

RS-232

RS-232 is an unbalanced-line point-to-point protocol requiring two signal wires for unidirectional and three signal wires for bi-directional communications. It operates reliably over a maximum distance in the range of 15 to 20 meters (50 to 60 feet) at a maximum speed of 19.2kbps. At the much slower rate of 2.4kbps, the maximum reliable transmission distance is approximately 1 kilometer (3,300 feet). The most common connectors used for RS-232 are the D-subminiature DB9 and DB25 or screw terminations (Euroblock).

RS-422

RS-422 is a balanced-line four-wire, multipoint (ten devices) protocol designed for bi-directional communications at speeds up to 10Mbps over distances up to 12 meters (40 feet), and at 90kbps over distances up to 1,200 meters (4,000 feet). No connector type is specified, but D-subminiature DB9, 8P8C modular (RJ-45), and screw terminations (Euroblock) are in common use.

RS-485

RS-485 is a robust, balanced-line, multi-drop protocol (32 transceivers), using a point-to-point transmission-line architecture in a topology similar to composite video. RS-485 drivers are tri-state (0, or 1, or off-line states) allowing for high-speed transmissions between devices over 120 ohm, twisted-pair cable. Data speeds up to 10Mbps are achievable over short distances, but 2Mbps over 15 meters (50 feet) and 250kbps over 400 meters (1,300 feet) are common rates. No connector is specified, but D-subminiature DB9, RJ-11 (6P6C) and RJ-45 (8P8C) modular connectors, and screw terminations (Euroblock) are commonly used.

The DMX512 lighting control protocol is built on RS-485 for its communications, using the five-pin XLR (and sometimes the nonstandard three-pin XLR) connector.

USB

USB, among its many, many applications, is a serial data signaling system that is widely used for device control and interface in AV systems. While USB may be limited to transmission distances less than 5 meters (16 feet), there are USB-to-Ethernet adapter systems that use TCP/IP to allow USB control to be extended to any point on the Internet. Commercial off-the-shelf USB to RS-232, RS-422, and RS-485 converters can used to enable control equipment to communicate with devices fitted only with legacy serial communications interfaces.

Network Control

The Ethernet network has become the standard means of communicating between all types of AV devices, not only for control but also for content. Ethernet allows communication among control components, devices, input systems, output systems, applications, data, and the Internet. As mentioned in Chapter 12, while there remain a few AV applications that run over the data link layer (layer 2) of an Ethernet network, the vast majority of control applications use the TCP/IP networking protocol (layer 3 and above) as their underlying control architecture. This removes the limitation of operating only over Ethernet on LANs and allows the possibility of routing and communications, as well as control signals, across the entire Internet. As discussed in Chapter 12, placing control systems on networks requires advanced planning and coordination between AV integrators and IT departments.

Networked control solutions can be created that allow effective, real-time room support, loss prevention, inventory management, facility usage and scheduling, and web-centric presentation environments, to name only a few possibilities. Within this environment, the control scenario has shifted from the needs of one facility to enterprise-wide considerations.

Contact Closure

A *contact closure* is the simplest form of control communication. As a system output, this type of control communicates with a remote device by opening or closing an electrical contact that allows a current to flow in the device. It has the most basic protocol: On (contacts closed) or Off (contacts open).

Devices that can be operated by contact-closure control include motors; projection screens; drapes; shades; heating, ventilation, and air conditioning (HVAC); and power feeds to equipment. Operation typically requires a low-voltage interface such as a mechanical or solid-state relay to allow the low current of the control signal to operate a higher-current circuit on the device.

Contact closures can also provide inputs to the control system. In addition to reading a simple button press, a contact closure input can provide the status of sensors for such parameters as room occupancy, the position of a projector lift, a weight on a pressure mat, the position of a room-dividing panel, a light beam being interrupted, a fire alarm cut-off, or an audio over-level sensor. Connections for contact closure communications are usually screw terminals (Euroblock/Phoenix).

The main types of contact closure are

- **Momentary**   These contact closures activate only while pressed and return automatically to their "normal" or unactivated position on release. There are two possible normal positions:
  - **Normally open position**   When not activated, the contact is open or "off."
  - **Normally closed position**   When not activated, the contact is closed or "on."
- **Latching**   These contact closures remain in either the open or closed state until activated to switch to the opposite state. Latching contact closures may take the form of "push on/push off" or "toggle" devices.

Infrared Control

Infrared control systems use pulses of invisible IR light to send control signals between devices. The IR transmitter is usually an LED, and the receiver is usually a photo-transistor or similar light-sensitive device. Patterns of pulses corresponding to control instructions are used for a wide variety of functions. The most common IR controls are the lightweight remote controls used with consumer devices ranging from heating and cooling equipment to the entire spectrum of media players and display devices.

IR control has some significant limitations. As most objects, including walls and people, are opaque to IR light, there must be a clear line of sight between an IR transmitter and its receivers. Most IR systems are limited to an effective range of about 9 to 12 meters (30 to 40 feet), although very high-output LED transmitters can be used to extend beyond these limits. While signal repeaters can also be used, long IR paths are prone to being blocked, making the situation more suitable for a more penetrating radio-frequency communications link.

As IR receivers are light-sensitive devices, they can sometimes be overwhelmed by the IR light in bright sunshine or confused by the pulsed light from fluorescent and LED light sources. Also, IR is usually a one-way communication path, meaning that an IR-controlled device has no way to provide feedback on its status or confirm that it has received a command from the control system.

An AV control processor can also command IR-controlled devices by placing a wired LED IR emitter over the IR sensor of a receiving device. This type of IR emitter can also prevent the receiver from being affected by troublesome ambient light sources. Place an opaque mask over the IR sensor, leaving only a small opening. The IR emitter is then placed over the hole in the mask.

Radio-Frequency Control

RF control uses wireless signaling via radio waves to communicate between devices. Some common RF systems used for AV control systems are discussed in the following sections.

Wi-Fi

Wi-Fi, as discussed in Chapter 12, is a general data communications protocol used for computer networking, and is effectively the wireless extension to Ethernet. Because of its implementation across most spaces where AV systems are used and because of the wide availability of devices with integrated Wi-Fi capabilities, it is used for both control communications and data distribution. Wi-Fi is also used in some IoT implementations.

Zigbee

Zigbee is a relatively inexpensive, low-energy, low-speed wireless communications protocol that was designed for remote device communication and control over a small area such as a house or an office, with a maximum range of about 30 meters (100 feet) line of sight. It can, however, use mesh network relays to extend its coverage. It operates in the unlicensed industrial, scientific, and medical (ISM) 2.4GHz band worldwide and in the unlicensed ISM 900MHz, 780MHz, and 860MHz bands in some countries. It has been widely implemented in domestic systems for control of appliances and services and in industrial applications. Zigbee is the basis for some implementations of IoT device networking and control.

Bluetooth

Bluetooth is a low-energy data communications protocol designed for personal area networking over distances up to about 10 meters (30 feet) line of sight. It operates in the ISM 2.4GHz band. Bluetooth is widely implemented in computing and peripheral devices and is used for point-to-point control and data communications in portable audio equipment. Bluetooth is also used in some IoT implementations.

Ultra-Wideband

Ultra-Wideband (UWB) is a very low-power communications protocol that transmits an extremely wide bandwidth (500+Mhz) signal using time modulation for encoding the data. Its low-power ultrawide-bandwidth signals produce very little interference with nearby narrowband RF systems, while UWB signals themselves are highly immune to interference from other RF systems. Being time-modulated, UWB can be used to detect the distance between the transmitter and receiver, which enables proximity detection and triggering. UWB communications for personal devices have been developed in the 6 to 9GHz range.

LTE 5G

The fifth generation (5G) of the International Telecommunications Union's long-term evolution (LTE) cellular telephone network is a standard specifying a network that carries voice telephony and high-bandwidth data at speeds from 100Mbps to 20Gbps.

The available bandwidth is dependent on the frequency band, the cell sizes, and the number of devices on the network.

Low-band 5G operates in the 600 to 900MHz region, Mid-band 5G uses the 3 to 7GHz range, and High-band 5G uses the 24 to 40GHz and 64 to 86GHz regions. Frequencies used and bandwidths available vary by country and telecommunications carrier.

Each 5G device requires either a software or hardware subscriber identification module (SIM) and a current connection contract for a cellular network that services the installation site. Device connectivity for data and control in both mobile and static applications makes 5G an ideal channel for content distribution and control on personal devices, IoT devices, and smart devices where secure, reliable Wi-Fi access may not be available.

Spectrum Management

When using any type of RF control, you must verify the frequencies that are already in use in the environment you intend to control. With the near-universal use of Wi-Fi and Bluetooth on portable computing and personal communications devices, it is almost impossible to guarantee that the required communications channels will be available and free from interference to use for remote control. This can become critical at times when many people are present in the AV space. As mentioned in the discussion of wireless microphones in Chapter 9, the coordination of wireless frequencies allocated to the wireless technologies involved in an audiovisual installation is critical. A full-spectrum wireless frequency allocation plan is an important component of preparing for any AV installation.

Control System Cable Tips

Control system wiring is defined by specifications covering the installation and electrical properties of the signals. Control system wiring specifications determine how control signals go from point A to point B, how the hardware can operate or send feedback, the topology of the network, the location of power supplies, and the signal quality.

Here are some points to keep in mind while wiring together a control system:

- Read and understand all the manuals—RTFM (read the friendly manual).

- Check cable specifications for construction, combustion rating, voltage rating, jacket, gauge, and bundling.

- If others have run the cables, verify that there have been no surprise substitutions or intermediate joins, and verify the cable certifications if available.

- Label every cable methodically, and keep a clear list of the labels and their meanings.

- A multimeter can be useful for measuring and verifying input/output in contact closures.

- A multi-connector cable continuity and pin-out tester is a valuable tool to verify cable terminations.

- Be aware of signal quality versus length of a cable run. Running cable too far may mean reduced reliability of the system. Signal problems increase proportionally to the length of wire and number of connections.

- If you have trouble locating the IR receiver on a device, use a torch/flashlight to shine a light on the front of the device. The IR window will appear foggy or fuzzy.

- Confirm cable separation between signal types. Signal crosstalk problems are often random and intermittent.

- A mobile phone camera or other digital camera can detect whether IR is being transmitted. Point the camera at the IR transmitter; if the transmitter is working, a bright point of light on the IR emitter will appear in the image. If you borrow a professional camera for this task, ensure that no IR filters are fitted.

System wiring can be challenging, especially within a complex system with many devices and connections. Wiring problems can increase proportionally to the length of wire and number of connections within the control system. Wiring specifications for control systems can include wire length, gauge/size and diameter, voltage ratings, jacket type, and bundling requirements. Refer to your system drawing and documentation for these details when installing a control system.

Control System Signal-Flow Diagrams

Control system flow diagrams depict the general organization of the control system and the logical flow of control. They may or may not depict device locations, but will depict devices that are grouped for some functional purpose, such as an equipment rack. This varies by company.

As noted earlier, similar to audio and video system flow diagrams, control system flow diagrams show that signals generally flow from left to right, although this pattern is quickly disappearing because of the increase in the number of devices that need to be connected and because of the bi-directional nature of most control signals.

As shown in the control flow diagram in Figure 13-1, user interfaces are indicated on the left, the central processor is in the center, and controlled devices are on the right. Equipment is shown within and outside of the equipment rack. Cables that terminate at wall or floor boxes are shown, with their connectivity continuing to the device. Control signals from outside of the system such as LAN and WAN IP networks and RF feeds are also shown. Environmental devices such as lighting, HVAC, fire services, staging machinery, projector lifts, and shades are shown as controlled devices. Attention to power switching, control buses, and the organization of similar signals grouped together, such as IR, network, and serial data, provide information for the installers.

The process for creating a control system flow or function diagram is similar to the process for creating general system diagrams, such as audio or video "block" diagrams. The designer must determine how to represent control systems in drawings and how much detail is required in flow diagrams and in the schematic or wiring diagrams.

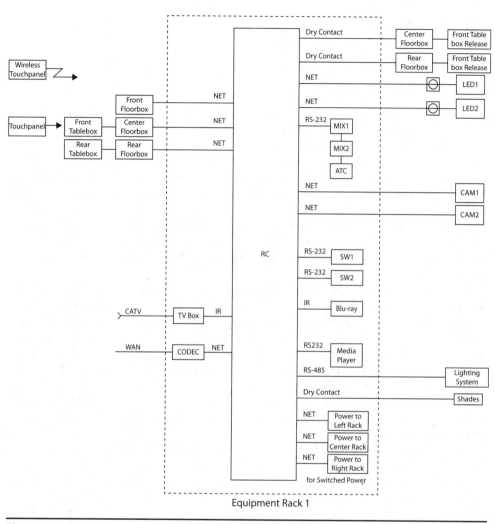

Figure 13-1   Example of a control system signal flow diagram

Installing Control Systems

Installers need to know which devices are being used and how they will connect to and interact with the system. Set aside some time to briefly discuss with the programmer the overall software design concepts.

The design plan should indicate the following:

- Required equipment
- Cable types

- Termination types
- Mounting types and methods
- Equipment address settings
- Button-by-button interface specifications

Open and inspect each piece of equipment; check for any damage and power on the device to verify that it is in good physical condition.

Use the software design plan to assign addresses and operating modes to the equipment as required. There are several ways to assign operating modes and addresses. Using a local HTTP (web browser) interface, manufacturer-specific configuration applications, dip switches, and jumpers are the common methods. Always RTFM for each piece of equipment in the system before attempting to configure it. After setting the addresses and operating modes, mark the settings on a label on the backside of the gear for future reference. Not only will this help if there are several pieces of the same model to install, but it makes it easy to determine what settings have been established on each device without searching through menus.

One hour of testing in the workshop before moving the equipment to the installation site may save many hours of troubleshooting in the field. Therefore, whenever possible, configure and test each device in the workshop. Verifying settings and performance of the system components in the workshop allows the technicians to interact with many of those involved in the project. User interface designs can be verified, troubleshooting is less time consuming, programmers have more time to fix any problems, and clients have a functional control system quicker.

If you must assemble and configure the control system at the client site, you should, at a minimum, set a brief meeting time with the designer and software programmer for review or questions.

Control system installation follows similar preparation guidelines as audio and video systems. However, the complexity of control systems requires special attention be given to any design layouts, signal flow diagrams, IP addressing information, and other technical documentation.

While installing a control system, strive to develop a positive relationship with the software programmers, system designer, and IT professionals involved in the project. Effective interaction with these people helps to assure that the control system is installed and configured correctly.

Installing Firmware

The control system must have the correct firmware and software properly installed and configured to achieve the required performance. From time to time, new firmware is made available to remove existing firmware bugs, accommodate changed hardware in new models of the equipment, add new system functionality, and improve operating performance.

At the workshop, confirm that the equipment has the latest firmware installed. Use the equipment manual to check the version numbers, consult the manufacturer's website for the current version, and update as necessary. It is better to upgrade firmware in the workshop than at the customer's site, since the required resources are probably more accessible in the office than in the field.

The control programs are the unique action scripts written for a specific install project. There may be separate scripts for the controller and devices such as switchers, DSPs, touch screens and control panels; each of these scripts must be installed on their respective devices. The manufacturers of the equipment may have proprietary software for loading the scripts onto the devices. Always refer to the software manual for the installation process: RTFM.

Make a special effort to ask the software programmer if there are any "hidden" or service pages, buttons, or modes in the device UIs that allow the technician to configure passwords, startup sequences, and other technical settings. Such service and technical interfaces will be included in the documentation provided by the designer. Asking about these technician interfaces can save you time configuring the system.

Verifying the Control System

Verifying the control system is an extremely important task. You should receive a list of what a control system is supposed to do from the control programmers. Once you have this list, you will need to press every button and test every trigger in the control system, one at a time. Check every single option to make sure it does what it is supposed to do. If the button or interface does not do what it is supposed to do, you will need to note that it is not right. Keep a detailed list of malfunctioning items for the system designer, programmers, and any involved members of the project management team. They can look through the list, examine the original specifications for the system, and make a decision if it is actually "broken." If there is a problem, you or your project manager may need to fill out a change order.

Inspect and configure each control system and controlled device before installation of the AV system. If you have time, connect the system together as completely as possible and verify its operation.

Use the design plan to confirm the locations of all the control devices in the room. Then, go through a final system check. Here is a checklist:

- Verify that all cables are connected correctly according to the drawings. Modify drawings or connections as needed.

- Confirm that termination pin-outs are correct.

- Power up the control processor and connected devices.

- Verify that all required device selection buttons in the UI are available. For example, if the button for a media player is activated, the switcher selects the correct input.

- Verify that the correct buttons appear for the operation of each device selected. For example, if the button for a media player is touched, the appropriate controls, such as "play" and "pause," should appear.

PART V

- Confirm that each device's UI action buttons perform the specified action in accordance with the button-by-button specification.
- Test all buttons, control switches, motion sensors, infrared sensors, screen mechanisms, projector lifts, and other devices that interface with the system. This process can take several hours.
- Document all irregularities and anomalies, and report them to the programmers immediately.
- Confirm lighting presets in accordance with design specs and applicable lighting standards.
- Confirm audio presets in accordance with design specs and applicable listening standards.
- Confirm video presets in accordance with design specs and viewing standards.
- Test streaming connections and cloud data sources.
- Test conference connections for audio- and videoconferencing.

The installer must document any changes requested by the client; however, do not commit to any changes without clearance from your project manager.

Troubleshooting Control Systems

Because control systems are often complex and involve interaction between equipment from different manufacturers, they are likely to be a source of problems.

Take a look at a couple of problematic scenarios, possible causes, and solutions presented here.

Problem 1   A selection is made on a handheld remote and nothing happens.

Possible causes and solutions:

- Are the batteries in the transmitter dead? Change or recharge the batteries.
- Is the transmitter actually transmitting? Check IR devices with a digital camera. Check the RF receivers for traffic or signal indications.
- Is the addressing (unique identification) of the transmitter, receiver, or controlled component incorrect? Correctly address all components.
- If the control system has a CPU and control panels, have the appropriate operating software and action scripts been loaded into these devices and successfully initialized? If not, load the program software and action scripts and initialize all devices.
- Is the RF or IR receiver located inside the rack greatly reducing the range at which it can read the transmitter's signal? Place the transmitter in close proximity to the receiver to see whether this improves its effectiveness.
- Is there a programming error? If this is the case, you will need to coordinate with the programmers to troubleshoot.

Problem 2 A device in the control system is not working or is working erratically. For example, a projector will not change inputs on command.

Possible causes and solutions:

- Is the cable carrying control information to the projector connected to the wrong output port on the control system? Check the appropriate port connection and make the change.

- Is the cable correctly connected; is it wired incorrectly or damaged? Check the pin-outs at both ends of the cable. Verify continuity of all active conductors and verify that there are no shorts between conductors created by staples, solder drips, cable ties, or other causes.

- Is there a programming error? If you suspect an error in programming, make every effort to confirm that it is a programming error and not an operational or configuration error, as calling back programmers can be time consuming and expensive (and potentially embarrassing if you made a wrong diagnosis). If it is a programming error, you will need to work with the programmers to troubleshoot it.

- Is the communications protocol for the device you are trying to control correctly configured? Verify the protocol in the device and operating system manuals—RTFM.

- Is the control port on the receiving device enabled?

- Have you followed all of the manufacturers' instructions? RTFM.

Talking Control with the Customer

When dealing with the customer and you encounter a discrepancy between what a UI control does and what the documentation says it should do, only write down what the button did. Do *not* say "This function is broken" or "This control should do. . ." to the customer.

Let's say you have installed a new control system. When you push "play," the control system checks to see whether the projection screen is down and the projector is on. It then lowers the screen, turns on the projector, waits two minutes, and starts the movie. Before you check it on your documentation, your customer says, "That button should also dim the lights and adjust the volume." If you say, "Yes, you are right," without knowing what your company has agreed to do, your company may have to program all of the customer's new requests, whether they agreed and intended to or not.

Performance Verification Standard

As you learned in Chapter 11, the ANSI/AVIXA 10:2013 Standard for Audiovisual Systems Performance Verification provides a framework and supporting processes for determining elements of an audiovisual system that need verification. As shown in Figure 13-2, this standard has a section dedicated to control performance verification.

PART V

9.5 Control Performance Reference Verification Items

Item Number	Item Name	Pre-integration	Systems Integration	Post-integration	Closeout
CON-100	Control System Communications		×	×	
CON-101	Interfacing and Control of External Devices and Systems		×	×	
CON-102	Mobile Device Integration		×	×	
CON-103	System Response to Emergency Conditions		×	×	
CON-104	Control System Automated Functions			×	
CON-105	Control System User Interface Performance			×	
CON-106	Control System Response Time			×	

Figure 13-2 Control performance verification items from the ANSI/AVIXA 10:2013 Standard for Audiovisual Systems Performance Verification

The tasks on the checklist are

- **100** Verify that all control communications are tested from endpoint to endpoint via the appropriate midpoint(s) for operation and functionality as defined in the project documentation.

- **101** Verify that AV control system interfaces to and from control systems provided by others conform to requirements as defined in the project documentation.

- **102** Verify that mobile devices that are to be supported are integrated and operating as defined in the project documentation.

- **103** Verify that any required response of the installed audiovisual system(s) in the event of a life safety or similar emergency operates in accordance with local regulations and as defined in the project documentation. This item specifically excludes sound system response to an emergency condition, which is covered under item AP-100, Emergency Muting.

- **104** Verify that all time-dependent or automated functions executed by the control system conform to requirements as defined in project documentation.

- **105** Verify that the control system is implemented in a manner consistent with the requirements as defined in the project documentation.

- **106** Verify that the control system provides the user response time and maximum latency defined in the project documentation. This verification item should require a metric to be verified.

Verifying the performance of control systems according to established standards will help you assure overall system performance. You will learn about audio and video system verification in the chapters that follow.

Chapter Review

In this chapter you studied how control systems work and how to set them up for networked AV equipment.

Upon completion of this section, you should be able to do the following:

- Identify the documentation used when installing a control system
- Identify the process for installing control program hardware and software
- Verify that the control system works as expected

Remember that the primary purpose of control systems is to make complex AV systems more accessible.

Control systems allow users to focus on communicating a message. Any control system that is connected to a network can become a very powerful tool for device management and delivery of content. If control systems are built according to the client's needs, using input from IT professionals and proper selection of devices, the final control system will be easy to use.

Review Questions

The following questions are based on the content covered in this chapter and are intended to help reinforce the knowledge you have assimilated. These questions are similar to the questions presented on the CTS-I exam. See Appendix E for more information on how to access the free online sample questions.

1. What is a control system?

   A. A strategy for managing workgroups

   B. An authorization policy

   C. A system for managing a network

   D. An interface between a human and a machine or system

2. Which of the following can control systems operate remotely? (Choose all that apply.)

   A. Power

   B. Drapes/curtains

   C. Display monitors

   D. Lights

3. Which of the following are elements of a control system? (Choose all that apply.)

   A. Firmware

   B. Hardware

   C. Software

   D. Spyware

4. What is a macro?

   A. A function activated by a button

   B. A set or sequence of functions activated by a single button

   C. Multiple functions activated by separate buttons

   D. A single function activated by a switch

5. Which of the following is a compiled program?

   A. Machine code

   B. DVD titles

   C. RAW codec

   D. Control script

6. Which type of control system provides device operation by opening or closing an electrical current or voltage loop?

   A. IR control

   B. Contact closure

   C. RF control

   D. Serial control

7. Which of the following are limitations of an IR control system? (Choose all that apply.)

   A. It does not provide feedback that commands have been executed.

   B. It requires direct line of sight to the device's control point.

   C. It may be subject to interference from ambient light.

   D. It always requires batteries.

8. Which of the following is true of an IP control system?

   A. It can be connected to equipment over the LAN.

   B. It works with all models of equipment.

   C. It is slower than serial control systems.

   D. It runs machine code only.

9. Why is it better to test control systems in the workshop? (Choose all that apply.)

   A. It allows more time to fix problems.

   B. It may provide better access to help from internal teams.

   C. It allows time for longer breaks.

   D. It prevents customers from asking questions.

10. What should you do after setting the control address of a device?

   A. Mark the equipment with this information

   B. Erase this information from the system documentation

   C. Give the customer an encrypted list of these settings

   D. Back them up by posting them on your blog

Answers

1. D. A control system is the interface between a human and a machine or system.

2. A, B, C, D. Control systems can operate power and drapes/curtains and display monitors and lights remotely.

3. A, B, C. Firmware, hardware, and software are elements of a control system.

4. B. A macro is a set or sequence of functions activated by a single button.

5. A. A compiled program is machine code.

6. B. Contact closure control provides device operation by opening or closing an electrical current or voltage loop.

7. A, B, C. Only these three descriptions state the limitations of an IR control system.

8. A. An IP control system can be connected to and operate devices over the LAN.

9. A, B. It is better to test control systems in the workshop because you will have more time to fix problems and have access to help from internal teams.

10. A. After setting the control address of a device, you should mark the device with this information.

CHAPTER 14

Audio Gain and System Equalization

In this chapter, you will learn about

- The audio verification process
- Methods for setting system gain
- Mixer, microphone, and line input adjustments
- System equalization
- Measuring and testing loudspeaker performance

The audio system can be one of the most challenging aspects of an audiovisual (AV) system design. As an installer, it can be equally challenging to ensure a system provides an undistorted, high-quality signal with adequate levels and intelligibility to all specified destinations.

An audio system requires final adjustment after all the components have been installed so that it can produce the desired or specified volume and quality. It is part of your job to set the audio gain and system equalization (EQ) and to adjust the various signal-processing components for the system to sound as good as it possibly can.

You will need to verify the performance of many components in an audio system and at many points along the signal path. You will want to use the ANSI/AVIXA 10:2013 standard for audiovisual systems performance verification discussed in Chapter 11. The list of verification items found in that standard will make it easier for you to document the status of the AV system and to keep track of what you and your team have already inspected. The verification items for audio are listed later in this chapter.

You and your team will likely be testing the control, audio, and video systems at the same time. You will need to use some of the knowledge you gained from your study of Chapter 9, for example, how to calculate the change in a sound and signal level using decibel equations. In this chapter, you will focus on the setup and verification of audio systems.

Duty Check

This chapter relates directly to the following tasks on the CTS-I Exam Content Outline:

- Duty C, Task 8: Test the Audiovisual System
- Duty C, Task 9: Calibrate the Audiovisual System
- Duty E, Task 4: Repair Audiovisual Systems

These tasks comprise 11 percent of the exam (about 11 questions). This chapter may also relate to other tasks.

Audio-Testing Tools

You will need to use different test instruments to measure and verify various aspects of the audio signal along its path from a microphone or other audio source to the loud-speakers. Audio-testing tools range from handheld devices to computer applications; some are available for single-function testing and others have a suite of testing functions.

In Chapter 9 you learned how an impedance meter is used to identify issues such as short circuits and to gauge the total impedance or load of the loudspeaker system. In this chapter, you will learn about audio signal generators, sound pressure level (SPL) meters, and other test gear. What follows is a look at some of the primary tools that are widely used for performing audio testing, setup, and system verification.

Piezo Tweeter

A tweeter is a term often used for a loudspeaker, often horn or dome shaped, that produces audio frequencies in the range of 2kHz to 20kHz (which is considered to be the upper limit of human hearing). Special tweeters can deliver high frequencies up to 100kHz.

A piezo tweeter contains a piezo-electric crystal mechanically coupled to a sound dia-phragm. When a voltage is applied to the surface of a piezo-electric crystal, it flexes in proportion to the voltage applied, thus converting electrical energy into mechanical movement. Figure 14-1 shows a piezo tweeter connected to an XLR audio connector.

Figure 14-1
Piezo horn-tweeter connected to an XLR connector

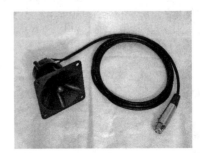

Audio Signal Generators

An audio signal generator is a test device that generates calibrated electronic waveforms in the audio frequency range for the testing or alignment of electronic circuits or systems.

Audio signal generators, such as the one shown in Figure 14-2, generate sine, square, triangle, or other waves at specific frequencies or combinations of frequencies. Many audio signal generators also generate *pink noise* (a quasi-random noise source characterized by an equal amount of energy per octave of frequency—it rolls off at 3dB per octave) and *white noise* (a quasi-random sound that has the same energy level at all frequencies). In addition to generating frequency sweeps, these devices can be used to test polarity.

Sine waves are used as steady references for setting signal levels and to reveal distortion caused by the equipment being evaluated. Common sine wave frequencies used for testing and verification include:

- **1kHz** Used for setting levels and setting system gain using the unity gain method
- **400Hz** Used in conjunction with a *piezo tweeter* to listen for clipping, as well as setting system gain using the optimization method

A wide range of professional-quality audio test signal generator applications are available for smart personal devices and laptop operating systems. Together with a quality audio interface (either in-built or an external adapter) and some time invested in calibrating the signal output, you may be able to configure an audio installation without requiring a dedicated, stand-alone signal generator.

A tone generator can generate a stable, constant signal (such as a 1kHz sine wave at 0dBu) that can be measured with a signal analyzer to establish the baseline measurements for setting gain. Whether you are setting unity gain or using the system optimization method, you will need at least a cable tester, signal generator, and signal measurement device.

Figure 14-2
An audio signal generator

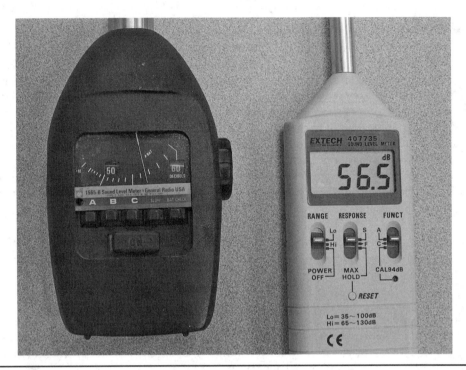

Figure 14-3 Meters for measuring sound pressure level

Sound Pressure Level Meters

SPL is a measurement of all the acoustic energy present in an environment. As discussed in Chapter 9, this is typically expressed in decibels (dB SPL), with reference to 0dB, the threshold of human hearing.

An SPL meter, as shown in Figure 14-3, gives a single-number measurement of the sound pressure at the measurement location. The meter consists of a calibrated microphone and the necessary circuitry to detect and display the sound level in decibels. Its function is simple: it converts the sound pressure levels in the air into corresponding electrical signals. These signals are measured and processed through filters, and the results displayed in decibels.

SPL Meter Classification

When selecting an SPL meter, you want to use one that can take readings as accurately as possible. The IEC 61672-1:2013 Electroacoustics - Sound Level Meters standard defines standards for sound measurement devices (also known as ANSI/ASA S1.4-2014 United States). Meters are classified by the accuracy of their measurements:

- **Type 1/Class 1** Precision-grade instruments for laboratory and field use, with measurement tolerance of ±0.7dB. These are intended for use in environmental applications, building acoustics, and road vehicle noise measurements.

- **Type 2/Class 2**   General-purpose-grade instruments for field use, with measurement tolerance of ±1.0dB. These are intended for use in measuring noise in the workplace, basic environmental applications, and motor sport measurements.

Devices that are not classified do not conform to a standard, so they are not reliable for measurement and testing purposes. For many audio purposes, a Class 2 meter is acceptable.

Always reference the project specifications for the SPL meter settings necessary for proper verification. If the verification requirements do not specify settings for the SPL meter, you can use SPL meter weighting guidelines to select one.

SPL Meter Weightings

You can apply weighting to the SPL meter measurement to correlate the meter readings with how people perceive loudness. As you saw in Chapter 9, the *equal loudness curve* shown in Figure 14-4 is a measure of sound pressure across the frequency spectrum of human hearing. This curve shows how loud different frequencies must actually be for the human ear to perceive them to be of equal loudness. The curve with dashes represents the threshold of human hearing: the minimum sound pressure required for an average human to hear the different frequencies within the audio spectrum. The x-axis of the graph shows the frequency in hertz, and the y-axis shows the SPL in decibels. From this graph you can see that at the threshold of human hearing, the 40Hz tone must be about

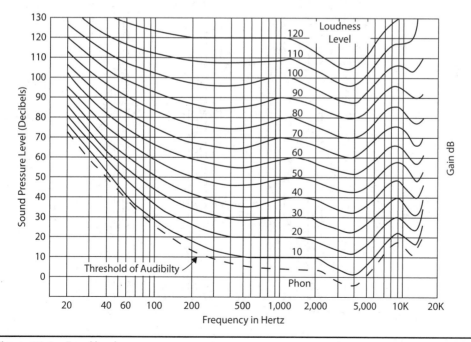

Figure 14-4   Equal loudness curve

50dB SPL louder before the human ear can perceive it equally as loud as a 1kHz tone. A 200Hz tone, however, would have to be only 15dB SPL louder to be perceived as equally as loud as a 1kHz tone. These SPLs use 0dB SPL (20µPa at 1kHz) as their reference.

The *weighting curves* represent standard filter contours designed to make the output of test instruments approximate the response of the human ear at various SPLs. An SPL meter will typically have three weighting settings: A-, C-, and Z- (or zero) weighting. All standards-compliant SPL meters are required to have the capacity to display levels with an A-weighting, while a C-weighting capability is also required in Class 1 meters.

- **A-weighting** This filter gives readings that closely reflect the response of the human ear to noise and its relative insensitivity to lower frequencies at lower listening levels.

- **C-weighting** This filter produces a more uniform response over the entire frequency range, but with –3 dB roll-off points at 31.5Hz and 8kHz.

- **Z- (or zero) weighting** This filter produces a flat frequency response (±1.5dB) between 10Hz and 20kHz.

You can see the differences in SPL weighting by examining the weighting curves in Figure 14-5. You can see that A-weighting discriminates against low-frequency energy. The A-weighting curve is almost the inverse of the equal loudness curve at a low listening level. The A-weighting curve reflects how the human ear perceives low-frequency energy as being at a lower SPL than it is in actuality.

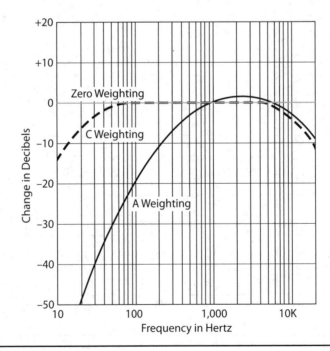

Figure 14-5 SPL weighting curves

A weighting is useful in situations with low listening levels, including most speech applications. As the listening level increases to 85 to 140dB SPL, the human ear's response flattens out (and starts to cause hearing loss at SPLs above 90dB). At that point you may then choose a C-weighted filter, whose low-frequency curve is far less steep.

Because they reflect the response of the human ear at low listening levels, A-weighted measurements are frequently used to quantify ambient background noise. While there are better metrics for quantifying background noise levels and their effects on the listener, an SPL measurement provides a simple one-number rating.

The ANSI/ASA S12.60 standard shows that maximum one-hour SPLs in learning spaces, including those from building services such as heating, ventilating, and air conditioning (HVAC), should not exceed 35dB SPL A-weighting.

The fact that a space should have a minimum 25dB signal-to-noise (S/N) ratio acoustically provides a goal of speech level of at least 60dB SPL A-weighted. However, a sound reinforcement system for speech may more typically operate in the range of 70 to 75dB SPL.

SPL Response Times

In addition to choosing a weighting for your SPL meter, you will need to select a response time. A fast response is used for capturing transient, momentary levels. A slow response will capture consistent noise levels. This is useful for averaging rapid fluctuations in sound pressure levels and more closely mimics the way your ears react.

Multimeter

Since electronic audio signal levels are typically measured in dBu or dBV, both of which are referenced to a known voltage, an alternating current (AC) voltmeter or multimeter can be used to measure or set signal levels. You can also use a multimeter for testing cable continuity; connector pinouts and polarity; measuring direct current (DC) voltage, current, and resistance; and to test equipment batteries.

When selecting a multimeter, look for a true root mean square (RMS) capability, which will enable you to measure the RMS (effective) value of an AC voltage or current. RMS, rather than peak voltage, is the most accurate means to compare voltage readings in audio applications. The multimeter must also have wide enough measurement bandwidth to accurately measure both the reference and the signal.

Measurement Microphone

Test and measurement microphones are used to accurately capture system performance. A measurement microphone should be omnidirectional and have a flat on-axis frequency response across the entire audio spectrum. Typical measurement microphones used for system alignment have a nominal diaphragm diameter of ½ inch (13 millimeters).

Oscilloscope

An oscilloscope is a sensitive test device that enables measurement and evaluation of electronic signals by displaying the signal as a time-referenced waveform. In audio system verification, an oscilloscope may be used to identify clipped signal waveforms when using the optimization method for setting system gain.

These devices are the basic tools for testing and evaluating the audio signal pathway and listening environment.

Setting System Gain

Setting the proper gain structure for an audio system is the foundation to providing a clean, undistorted signal. Knowing how to set gain correctly will help ensure optimal performance of the audio system you installed.

The two terms you will need to understand are

- **Gain**   The increase in the amplitude of a signal
- **Attenuation**   The reduction in the amplitude of a signal

As shown in Figure 14-6, there are many points within an audio system where signal levels can be adjusted. In fact, unity gain must be set at virtually every point along the audio signal chain.

Importance of Setting System Gain

Setting gain correctly provides an optimal signal-to-noise ratio for an AV system and helps to avoid signal distortion.

Under normal circumstances, the audio mixer's output level should be near the zero mark, as indicated by the mixer's output-level meter. This allows some headroom for moderate peaks in the signal to pass through undistorted. When peaks exceed the available processing headroom, the resulting problem is usually called *distortion clipping* because the waveform appears to be clipped off or flat at its peaks. Clipping occurs when a system or device can no longer provide an increase in output signal for a corresponding increase in input signal.

Figure 14-7 shows an oscilloscope display of an undistorted 1kHz sine wave and the same 1kHz sine wave in a clipped condition.

Most mixers will produce +18 to +24dBu output level without clipping. You should allow 10 to 20dB or so of headroom for emphatic talkers or loud passages in program material. This means the mixer should deliver approximately a 0dBu output level when the input sources are delivering their typical, intended output.

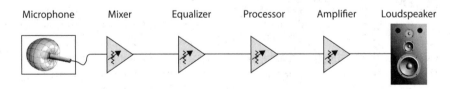

Figure 14-6   Points in an audio system where gain can be adjusted

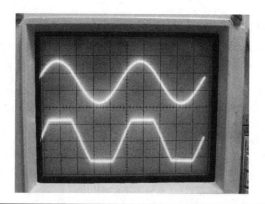

Figure 14-7 An undistorted 1kHz sine wave (top) and the same 1kHz sine wave in a clipped condition (bottom)

Proper setup of system gain results in improved signal-to-noise ratio by reducing noise and hiss, and avoiding clipping and distortion. It also improves dynamic range.

You will need to set the gain structure of each of these devices:

- Microphone preamplifiers
- Preamplifiers in mixer microphone inputs
- Line-level mixer inputs for external sources
- Audio mixer
- Processing devices (equalizer, compressor, limiter, delays, effects)
- Digital signal processors (DSPs) and loudspeaker processors
- Active/powered loudspeakers
- Amplifiers

You do not necessarily have to set gain structure in this sequence. You can set gain in the mixer first, for example, and then in the microphone preamplifiers. You just need to make sure you set gain in every item. One exception is that you should set the input of the power amplifiers or active/powered loudspeakers last.

Selecting a Method of Gain Structure

There are two methods for setting system gain: unity gain and the system optimization method. Both methods are relatively simple, and neither requires expensive equipment.

- **Unity gain** For a typical presentation room, conference room, or boardroom where you have installed professional audio components, unity gain will provide an adequate system electronic signal-to-noise ratio of around 60dB. Using the unity gain method, the strength of the output signal should equal the strength of the input. For example, if the first device you measure shows a signal-level output of 1.23V, you should be able to measure a 1.23V signal at the output of each device all the way to the power amplifier inputs.

- **System optimization** For a more critical listening environment such as a studio, broadcast facility, performing arts facility, or lecture hall, the system optimization method will provide the optimal signal-to-noise ratio allowed by the equipment in the signal path. This method takes more time and skill to complete than the unity gain method.

Let's take a closer look at each of these methods.

Unity Gain Method

Unity in the context of gain structure means "same thing in, same thing out." In other words, neither gain nor attenuation is applied to the signal. The decibel or voltage level at the output of the mixer should be at the same level along the rest of the signal path.

The unity gain approach commonly uses a 1kHz tone as a reference, which is measured by either a signal analyzer or a multimeter.

The benefits of the unity gain method are:

- It provides an adequate electronic signal-to-noise ratio in most basic audio systems.

- It is an easy and fast method that requires only a signal generator and multimeter.

The drawbacks of using unity gain are

- The system components may not be linear or uniform in their performance across the entire audio spectrum or dynamic range.

- Headroom and individual clipping levels vary from device to device, even between devices from the same manufacturer.

- A downstream device may clip and produce DC before the output of the mixer clips, possibly harming components or causing other problems.

Setting Unity Gain

As noted earlier, in unity gain, the mixer output meter should indicate near zero under normal operation.

The procedure for setting gain with the unity gain method is as follows:

1. First, define zero. Zero may mean 0VU at 1.23V, which is also +4dBu, the pro-audio line-level reference. Zero could also mean 0dBu (775mV) or even 0dBV (1V). To define zero, you will need to follow these steps:

   a. With the system power amplifiers off, set all mixer trims, faders, cross-point gains, and masters at their zero (unity) settings.

   b. Configure a signal generator to output a 1kHz sine wave at 0dBu (775mV).

   c. Apply this signal to a line-level input of the mixer and observe the mixer's output meter.

   d. Adjust the trim on the channel receiving the signal from the generator until the mixer's output meter shows zero.

2. Using a signal analyzer or an RMS multimeter, measure the output signal of the mixer. Normally, you should see something near 0dBu (775mV), +4dBu (1.23V), or 0dBV (1V).

3. Adjust the input trim until the analyzer or multimeter indicates the closest one of these reference levels while the mixer's output meter still indicates a zero output level. You will use this reference signal level to set unity gain through the rest of the system.

4. Adjust any filters and dynamic processors in the signal path so that they will *not* affect the reference signal. To do this, you will need to do the following:

   a. Make certain any filters, such as equalizers, are all set to zero, that is, their unity settings.

   b. Set compressors and limiters to their maximum threshold settings.

   c. Set downward expanders to their minimum thresholds.

   d. Turn off any other effects such as reverberation.

5. With the signal generator connected to a mixer input and the mixer output still reading zero, connect the analyzer or multimeter to the output of the next device in the signal path. Then adjust the gain on the device until the analyzer or multimeter shows the reference level established earlier.

During the gain setting process, you can check that the processors are set correctly by selecting and deselecting "bypass." If selecting and deselecting "bypass" does not alter the measured output of the device, the test signal is not being affected by any unintentional processing. Final settings for the dynamic processors will typically be set after the equalization process has been completed.

Continue this practice with each device in the signal path up to the point of the power amplifier or active/powered loudspeaker input.

TIP When setting system gain, keep in mind that finer adjustments are easier to execute at 0dBu, so drive the output of the mixer to 0dBu under normal conditions. When quiescent, a system with gain set correctly should not produce audible noise, such as hiss.

System Optimization Method

System optimization takes a different approach than the unity gain method. You still want to achieve mixer output meter indication of around zero under normal conditions. Unlike the unity gain method, however, the optimization method provides uniform headroom throughout the signal path. This method also maximizes the electronic signal-to-noise ratio of the audio system.

To set gain using the system optimization method, the mixer output is driven to clipping and then set just below clipping. Once the gain has been set on the mixer, the next device is connected, and its output is set the same way.

You can use an oscilloscope to look for the clipped waveform. You can also use a real-time analyzer to observe the harmonic energy associated with a clipped fundamental frequency, as shown in Figure 14-8.

In many instances, all you will need is an inexpensive piezo tweeter.

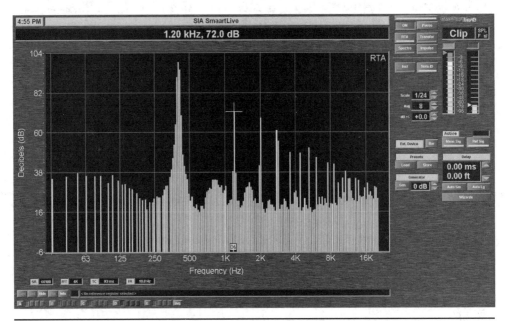

Figure 14-8 Harmonic energy associated with a clipped waveform shown on a signal analyzer

TIP To use a piezo tweeter, set a low-frequency sine wave, such as 400Hz, as the reference frequency in the system optimization method. Although the piezo will not faithfully reproduce a 400Hz fundamental frequency, it will easily pass the harmonic frequencies associated with clipping the 400Hz waveform. The piezo will be silent when the signal is a full sine wave, but it will make noise when the signal is clipped.

The Benefits of Using the Optimization Method

The benefits of using the optimization method for setting gain structure include:

- This method provides a much more robust signal-to-noise ratio than the unity gain method. This is because the normal system's signal level is much farther away from the devices' noise floors. Thus, the maximum signal-to-noise ratio can be achieved.

- Each line-level device has the same relative headroom as the mixer. The actual headroom (the amount of signal between 0dBu and the onset of clipping) will vary for each device.

There are, however, some drawbacks to using the optimization method:

- It requires more time and skill than the unity gain method.

- Some downstream devices, such as equalizers, can easily become overloaded, so passive attenuators (PADs) may be required on their inputs.

- Replacement of any system component requires system recalibration.

- As this optimization method requires finding the clipping point for each device and adjusting the level to just below that point, if the mixer is clipping, all the devices downstream will be clipping at the same signal level as the mixer.

Setting Gain Using the System Optimization Method

This is the method for setting system gain using the optimization method and a piezo tweeter:

1. As with the unity method, with the system power amplifiers off, set all mixer trims, faders, cross-point gains, and masters at zero (unity) settings.

2. Adjust any filters and dynamic processors in the signal path so that they will not affect the reference signal.

   - Verify all filters, such as equalizers, are set to zero (unity settings).

   - Set compressors and limiters to their maximum threshold settings.

   - Set downward expanders to their minimum thresholds.

Once again, you can check that the processors are set correctly by selecting and deselecting "bypass." This confirms that the test signal is not being affected by any unintentional processing. Final settings for the dynamic processors will typically be set after the equalization process has been completed.

3. Configure the signal generator to output a 400Hz sine wave at 0dBu.

4. Connect the signal generator to a line-level input of the mixer.

5. Connect the piezo to the output of the audio mixer. Since the 400Hz tone falls below the frequency passband of the piezo, it should be silent.

6. Increase the mixer's master output and listen. Once you hear noise from the piezo, you have exceeded the maximum level that the mixer will output without clipping and distortion.

7. Reduce the level until the piezo becomes quiet again. Since the transition between an unclipped and a clipped condition is very noticeable, a clean or clipped condition is easy to hear.

8. With the mixer's output level adjusted at just under the clipped level, connect the piezo to the output of the next device in the signal path.

9. Increase that device's gain until a clipped condition is indicated by the piezo, and slightly reduce the level until it is quiet again.

10. Continue this procedure with each device in the signal path up to the input of the power amplifier or active/powered loudspeaker.

11. After the procedure is complete, go back to the mixer and reduce the master output back to the unity setting.

Since the goal is to operate the mixer near its zero indication, the entire system has virtually the same headroom as the mixer, and the mixer's metering now indicates the system's condition.

Of course, in each step of the procedure just described, where the piezo tweeter is used to detect clipping, you can substitute an oscilloscope or a real-time analyzer to detect the clipping point.

Setting Gain Structure for Audio DSPs

DSPs may contain the various functions of all the devices normally found as separate components or devices within the rack of a conventional system.

Examples include but are not limited to the following:

- Microphone preamplifiers
- Equalizers
- Compressors/limiters
- Matrix routers and mixers

- Delays
- Feedback eliminators
- Loudspeaker controllers
- Echo cancellers

You should be familiar with two important terms when working with DSPs:

- **Software**   This is the operating firmware or software provided by the DSP manufacturer to perform the signal processing functions built into the device.
- **Configuration files**   These are the files that specify the signal flow settings, processing, and parameters within a DSP device. Initially configured by the system designer or system programmer, these settings may need to be fine-tuned (tweaked) by the installer at the job site for optimum performance.

Programming and configuring a DSP may require an external computer system running proprietary software to be connected to the DSP via a serial, universal serial bus (USB), or Ethernet interface. It is increasingly common for DSPs to include a web interface that allows configuration via a web browser from any device on the Internet. The owner's manual will indicate the software requirements and connection options that are available. If the unit has a front-panel display, some adjustments and settings may be made using menus via the front-panel interface.

Input signals in a DSP can be routed to either single or multiple outputs. Routing and gain settings are typically made through a cross-point matrix, where any or all inputs may be routed to any or all outputs.

Gain settings follow the same basic pattern as an analog system. Preamplifier gain settings bring the mic level up to line level, and all routing and processing are performed at the line level thereafter.

Usually, existing software configuration files are replaced (overwritten) when the installer uploads new configuration files. If possible, save backups of previous configuration files so that if a problem arises during the upload, you can revert to the previous settings.

After the system is tuned for optimum performance, the DSP configuration files should be saved and copied to a computer for backup, service, archival, and documentation purposes.

Mixer Input Adjustments

After you have set system gain, it's time to connect and adjust the microphones and program sources at the mixer inputs using the preamp/trim adjustments.

Microphone Input Adjustments

Microphones will have different sensitivities. That is, for a given sound-level input, the amount of output will vary depending on type, construction, and purpose.

To make microphone adjustments, you will need the following:

- The microphones
- An SPL meter
- A portable noise generator

The steps in the process for adjusting microphone levels consist of the following:

1. Set all channel equalizers, faders, cross-point gains, and masters to their zero (unity) settings.
2. Set each channel's microphone preamplifier all the way down to the minimum settings. The preamplifier could be labeled *Pre, Trim, Gain, Preamp,* or similar.
3. Connect each microphone to its appropriate mic input. If the microphone connected requires phantom power, ensure that phantom power is available on that channel.
4. Determine the appropriate level that each microphone is expected to receive. For normal speech applications, this is often about 65 to 70dB SPL.

NOTE Singers or enthusiastic presenters will be much louder than the conversational level of 65 to 70dB SPL—compensate accordingly.

5. Position a small acoustic noise generator outputting pink noise near the microphone.
6. Using an SPL meter positioned close to the capsule of the microphone, as shown in Figure 14-9, adjust the position and/or the level of the pink noise generator until the SPL meter reads approximately within the range of 65 to 70dB, which is the level for conversational speech.
7. With 65 to 70dB SPL at the microphone, adjust the preamp for the mixer's microphone channel until the mixer's output metering indicates 0.
8. Repeat this procedure for each microphone channel.

Often for a simple speech-only presenter system or a system that utilizes an automatic microphone mixer, only one microphone is active at any time.

Figure 14-9
An SPL meter
held close to
the capsule of
a gooseneck
microphone

If you are working on a system that has multiple microphones active simultaneously (for example, in a city council chamber, an interview, or a panel presentation), reduce the input trim for each microphone by 3dB for each doubling of the number of simultaneously active microphones. In other words, if you expect two microphones to be active simultaneously, adjust each preamp to read −3dB below the 0 indication on the mixer's output. If you expect four microphones to be active simultaneously, adjust each preamp to read −6dB below the 0 indication on the mixer's output.

Line Input Adjustments

In addition to the microphones being amplified up to line level, sources originating at line level will have to be adjusted. Many line sources in professional audio systems require little, if any, amplification before routing and processing. As consumer audio systems operate at a line level of 316mV, which is about 12dBu less than the pro-audio line level of 1.23V, they will require some amplification.

Line levels will vary depending on the equipment. Pro-audio line level is typically around +4dBu, and consumer line level is −10dBV (−7.79dBu). While these may be the levels you would expect, the actual levels will vary depending on the particular device and content. For example, laptop computers and smart personal devices have a wide variation in source line levels, although these outputs are usually easily adjustable on the device. The content from computer-based devices is often wildly variable in output level and quality and often requires continuous ongoing adjustment.

The process for making line input adjustments is as follows:

1. Set each channel's preamplifier all the way down to the minimum settings. The preamplifier could be labeled *Pre, Trim, Gain, Preamp,* or similar.

2. If possible, disengage any phantom power for line-level inputs.

3. Connect each line-level device to its appropriate line input. An audio mixer may have different inputs for mic and line level. For example, the mic input will usually be an XLR, while the line level uses a 6.5mm (1/4in.) jack connector. Other devices may use the same connection type, such as a captive screw (euroblock) or XLR, for both mic and line but offer a selection switch or software setting to select between mic and line. You may also see a PAD option, which passively attenuates the signal going into the mixer when connecting a line-level source.

4. Select or replay an output signal at 0dBu on the line-level source.

5. With the source active, adjust the preamp for the channel until the mixer's output metering indicates 0. If possible, use the actual source and program material intended for use with the system to make adjustments.

   If the actual source is unavailable, consult the manufacturer's information to determine the expected nominal output level and use a signal generator outputting that level as a substitute.

If you have followed the gain setting procedures outlined in this chapter, you should adjust the levels of the system components, from microphone or other mixer source inputs all the way to the power amplifier or active loudspeaker inputs, to provide optimal overall performance.

Power Amplifier Adjustments

The power amplifier is the last component you will adjust when setting gain. Up until this point, all of the gain adjustments have occurred with the power amplifiers turned off. Now it is time to turn them on and make sure that the amplifiers are sending enough power to drive the loudspeakers to the target SPL levels.

A common misunderstanding is that the adjustment on an amplifier is a gain or a wattage adjustment. The adjustments on a power amplifier are actually *input attenuators* (so you can't really turn an amp up to 11). A power amplifier, properly sized and specified for the application, should have no difficulty achieving the required SPL when the audio mixer output meter is near the 0 indication. In other words, when driven with the proper signal, which is somewhere between 750mV and 1.23V nominal, the power amplifier will produce enough power to drive the selected loudspeakers to specified levels without distortion.

The process for setting input attenuators on a power amplifier is as follows:

1. Drive the mixer with normal signal levels and output metering showing near 0.

2. Position an SPL meter at the listener location(s).

3. With the attenuators at minimum, turn the power amplifier on and adjust the attenuators until the SPL specified by the design is achieved.

The most common issue affecting sound systems is finding the input attenuators of the power amplifier all the way up (that is, in a "wide open" position) and the output of the devices earlier in the signal chain "turned down" (that is, less than unity gain). The common result is a system that produces audible hiss when it's turned on. Boosting a signal way up at one point and turning it way down at another to compensate reduces dynamic range. It also adversely affects the signal-to-noise ratio because the signal may now be closer to the noise floor.

 CAUTION   It is possible to damage equipment, particularly loudspeakers, by not powering the system components on or off in the correct order. To avoid such damage, follow these procedures:

- Powering on: Begin with devices along the signal path in sequence. You will typically start at the mixer and end with the power amplifiers.
- Powering off: First turn off all power amplifiers and then turn off all other electronics in reverse sequence along the signal chain toward the mixer.
- In brief: Amps on last. Amps off first.

Figure 14-10   A software-based graphic equalizer

Audio Equalization

As discussed in Chapter 9, EQs are frequency controls that allow you to boost (add gain) or cut (attenuate) a specific range of frequencies. An audio mixer typically provides high-, mid-, and low-frequency controls on individual channels. A digital signal processor may have several equalizers (such as the one shown in Figure 14-10) available to modify the signal of specific frequency ranges at various points along the signal path.

There are typically two places in the audio system where you will encounter an equalizer:

- The individual channel strip of an audio mixer
- Between the mixer and power amplifier

The individual input channel EQ provides tonal control of that input, whereas an equalizer found between the mixer and the power amplifier or powered loudspeaker is used to adjust overall system performance.

There are many different types of equalizers, ranging from simple tone controls to fully parametric equalizers. Some mixing consoles offer a combination of fixed and semi-parametric controls. Graphic and parametric can be separate devices, built in to audio mixers, or included as a function in a DSP unit.

Equalization Methods

There are several ways to equalize an audio system, including using a real-time analyzer (RTA) and using the dual-channel fast Fourier transform (FFT) method. First take a look at some of the terms you will need to understand.

PART V

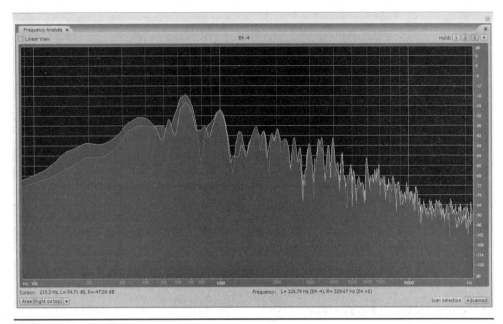

Figure 14-11 Analysis of a signal viewed in the frequency domain

System measurements can be performed in either the *frequency domain* or the *time domain*. Modern measurement methods use a complex mathematical function, called a *transform,* to convert a frequency-domain view into a time-domain view and vice versa. First, you need to understand the following terms:

- **Frequency domain** The view of a signal's amplitude in relation to its frequency. Figure 14-11 shows the amplitude of the signal at different frequencies. The energy can be divided into octaves or, more often, into fractions of octaves. It allows you to view a signal's spectral energy.

- **Time domain** The view of a signal's amplitude over a period of time (see Figure 14-12).

- **Transform function** The mathematical operation that maps information gathered in one domain, such as the frequency domain, and allows you to view it in a different domain, such the time domain.

- **Fourier transform** The mathematical filtering process that determines the spectral content of a time domain signal. The FFT is a simplified and easier-to-calculate version of the Fourier transform function that is used widely used in digital signal processing systems. An inverse FFT (IFFT) can be used to transform an amplitude-over-frequency map into an amplitude-over-time map.

Understanding the differences between the time and frequency domains means that you will be able to examine a measurement from multiple perspectives.

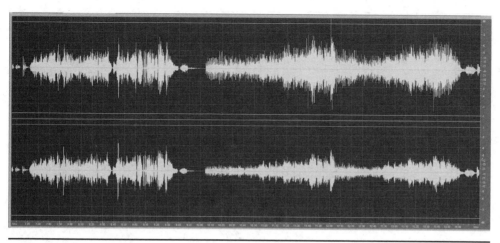

Figure 14-12 The same signal viewed in the time domain

 TIP One of the limitations of FFT is that the calculation is based on the assumption you are testing a linear, or distortion-free, system. This is often not true when testing loudspeakers in a room.

Single-Channel vs. Dual-Channel Measurements

Different tools for audio equalization offer different options for viewing measurements. Depending on the device you use, you may be able to choose to take single-channel or dual-channel measurements.

A single-channel measurement analyzes one input at a time, with no reference signal for comparison. The source signal used for testing is assumed to be of a particular type, such as pink noise.

Dual-channel measurements use two inputs for comparison: a reference test signal and a measured signal. The reference signal can be an external source or one generated by the testing device. The testing device will compare the measured signal to the reference signal, and the resultant output will be displayed as a sum, difference, coherence, or some other computed comparison. The result could be a number, a series of numbers, or a representative graph.

Room-Based Loudspeaker Measurements

When equalizing a system, you will typically make measurements in the near-field space and make some adjustments to compensate for loudspeaker frequency response or to meet a specified response curve. A smoother loudspeaker response can also help minimize feedback. No loudspeaker is perfect, so you can expect to make some adjustments.

When you measure a loudspeaker's response from any distance, you will generally also be measuring the acoustic properties of the room. The room will have frequency response

PART V

Figure 14-13
Sound energy
captured by a
microphone in
dB SPL over time

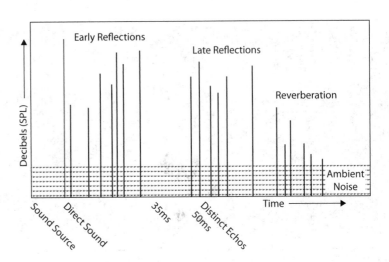

characteristics because of its size, its shape, and the surfaces and materials found in it. Unless you are an acoustician, you will not normally want to independently measure the response of the room.

As an AV professional, you will be concerned about the combined response from the loudspeaker and the entire acoustic environment so that you can apply appropriate adjustments to the equalizer. When you put a test microphone on a stand and place it in a room, the microphone picks up the energy coming from the loudspeaker together with the reflected and reverberant energy found in the room. Figure 14-13 shows acoustic energy picked up by a microphone over time. The microphone picks up the direct sound, early reflections, late reflections, and reverberations. It also detects the ambient noise present in the environment.

This combination of direct and reflected energy may, or more likely may not, represent the actual response of the loudspeaker. In fact, reflected energy off of the floor in front of the test microphone and other surfaces could result in a frequency response that looks like a comb-filtered response on the analyzer.

Tools for Loudspeaker Measurements

Common measuring tools, including RTAs and single-channel FFTs, cannot discern the difference between the direct sound energy from the loudspeaker and the sound energy coming back from the room. When using RTA and single-channel FFTs in large and live rooms, the majority of what is being captured by the measurement process is not what is actually coming out of the loudspeaker.

When measuring the response from the loudspeaker, you send a reference signal to the loudspeaker for testing. It takes time for that reference signal to make it from the loudspeaker to the test microphone's location. If you were able to take a measurement at only the precise moment the direct sound arrives, you could ignore the reflections and reverberation from the room. However, you cannot measure at only an instant in time.

Instead, you will need to start your measurement slightly before the direct sound arrives and end it slightly after the direct sound passes your microphone. This is called *time windowing*. You can set a time, or "window," around the arrival time of the direct sound.

Narrow time windows exclude most reflections from the room but also do not allow accurate measurements of lower frequencies. The window will need to remain open longer to make meaningful lower-frequency measurements.

Other factors in performing measurements include the type of reference signal used, the sampling rate, room modes, ambient noise, and the number of data points to include in each octave.

Real-Time Analyzer

You can use a real-time analyzer to equalize a system. This method provides you with a visual representation of the frequency bands that may need to be adjusted. It lets you make a significant adjustment to a system quickly. However, the measurements this method produces are "time blind," as an RTA cannot take measurements in the time domain. This means that room reflections and ambient noise almost certainly will contaminate your loudspeaker measurements. Therefore, you will need some technical knowledge to interpret the non-time-windowed measurement. The RTA method has no way to assure you that you are still within the critical distance of the sound field. For these reasons, many AV experts recommend you use other methods for system equalization.

The RTA Method

The goal of equalizing a system is to reduce the variation of acoustic audio-level output across the frequency range of a loudspeaker system. An equalized system should have a smooth response curve across the band of audio frequencies.

Loudspeakers are not linear devices. If a pink noise source is fed into a loudspeaker, the resulting output will not be flat. It will have peaks and valleys of varying levels across the frequency range of the loudspeaker. Correctly equalizing the system reduces, or flattens out, the variations.

As shown in Figure 14-14, an RTA measures energy levels at defined frequencies in a system. It then creates a visual representation of the SPL at each measured frequency. This is called a *frequency response* plot, and you can use it to visualize the acoustic energy within a given space.

It is important to note that the RTA test instrument measures the sound pressure arriving at the microphone, which will include the sound intended to be measured, plus ambient noise and room reflections. An RTA cannot differentiate between the direct sound from a loudspeaker and the constructive and destructive interference caused by the combined effect of direct and reflected energy.

RTAs are available as dedicated hardware devices and as a wide variety of software applications for virtually all platforms, from desktop and laptop operating systems to most portable, smart-device operating systems. An audio interface of sufficient accuracy and a measurement microphone is usually required to accompany a software application.

PART V

Figure 14-14
An RTA measures energy levels at defined frequencies

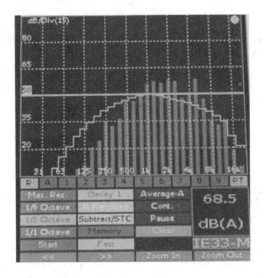

Preparation for the RTA Method

If you are using the RTA method to equalize an audio system, you must account for *room mode*. Also known as a *standing wave,* a room mode occurs between facing surfaces of an enclosure such as a room or loudspeaker cabinet, where the dimension between those parallel surfaces equals one-half wavelength (and its harmonics). The wave is thus reflected back on itself out of phase to produce interference patterns that create areas of maximum and minimum pressure.

If you choose to use the RTA method of equalizing a room-based audio system, follow these preparation procedures. You will need the following items:

- Real-time analyzer
- Measurement microphone
- Pink noise generator
- Appropriate connecting cables
- Microphone support system and mounting to reach to the loudspeaker locations

Before you measure, follow these steps:

1. Ensure that loudspeakers are not placed adjacent to acoustically reflective surfaces, as these may affect the input to the measurement microphone. As loudspeaker cabinets are often mounted on or near walls, it is important that the walls be at the rear of the loudspeakers' radiation patterns.

2. Turn off any source of noise such as fans, compressors, welders, dimmer racks, dimmable luminaires, computers, and HVAC systems.

3. Calculate the room modes using the formula in the next section, as these will affect the low-frequency performance of the loudspeaker system.

4. Set an equalization curve that is appropriate for the type of program content your client will use with the sound system.

Formula to Calculate Simplified Room Mode

Room modes will affect the low-frequency performance of a room. To discover where room modes will begin to be prevalent, use this simplified formula:

Mode Frequency (Hz) = 3 × (velocity of sound)/RSD

where

- The velocity of sound = 343m/sec (1,125ft/sec)
- RSD = The room's smallest dimension (in meters if using velocity in m/sec, or feet if using ft/sec)

Taking RTA Measurements

After you have completed the previous steps, you are ready to take measurements. Here is the process for taking measurements using an RTA device:

1. Find a position within the room that is on the same axis as the loudspeaker and that is within the listening position (audience area) of the space.

2. Measure and document the ambient noise of the room. Document the highest level by frequency band on your RTA.

3. Set the equalizer to the default position, zero, or no adjustment setting.

4. Insert the pink noise generator into the system at a point before the equalizer.

5. Make sure there are no other adjustments that may affect the frequency response of the system, such as a channel EQ.

6. Turn on the pink noise generator, and with the measurement microphone at the same place you measured the ambient noise, increase the generator's output level until it is 20dB higher than the measured ambient noise. Ignore frequencies where room modes are prevalent.

7. Place the measurement microphone close to the loudspeaker (no closer than 1 meter, or 3 feet). Note the sound level.

8. Move the microphone away from the loudspeaker and note where the sound levels no longer drop. Moving farther from this location should not reduce the sound pressure indicated on the meter. At this point, the intended signal and unintended signals (reflections and noise) should be equal.

9. From this point, move the microphone closer to the loudspeaker until the signal increases by 3dB. You are now receiving more direct or intended sounds from the loudspeaker and fewer unintended sounds. Note that moving the measurement microphone too close to the loudspeaker (into the near-field) will also produce unwanted effects. The microphone should be at a minimum of 1 meter (3 feet) away from the loudspeaker.

10. Fix the measurement microphone in this position.

Adjusting at the Equalizer

At this point, the RTA will display the curve of the signal from the loudspeaker and the unintended signals. You will need to make adjustments at the equalizer by following this process:

1. Use the equalizer adjustments to reduce the peaks of the frequency response. Best practice would be using the filters to attenuate, not to boost. This is sometimes referred to as the *cut-only* method.

2. Attempt to match the curve you have selected for your client's program type and shape your response to match the curve as smoothly as possible.

3. Do not use the RTA to try to compensate for room modes.

4. Since the RTA is not reliable for low-frequency measurements, use what you can hear to make adjustments to the low-frequency performance of the system.

5. Any adjustment more than 6dB may be trying to compensate for anomalies that cannot be equalized. In addition, if you make an electronic adjustment and that adjustment does not appear on the RTA, it is likely to be a problem in the time domain and cannot be equalized using the RTA display.

6. Readjust the gain at the equalizer to compensate for the energy removed during the equalization process.

Recommended Response Curves

When equalizing a signal using an RTA, your goal is to match the measured signal to a recommended curve. If one has not already been recommended to you, you can use the curves shown in the figures here as a reference. These curves are based on curves from *Altec Lansing Tech Notes TL-232A* by Tom Uzzle.

Figure 14-15 shows the recommended response curve for speech reinforcement systems.

Figure 14-16 shows the international standard response curve for cinema playback systems.

Figure 14-17 shows the recommended response curve for studio control room monitoring systems.

Figure 14-18 shows the recommended response curve for high-level rock music reinforcement systems.

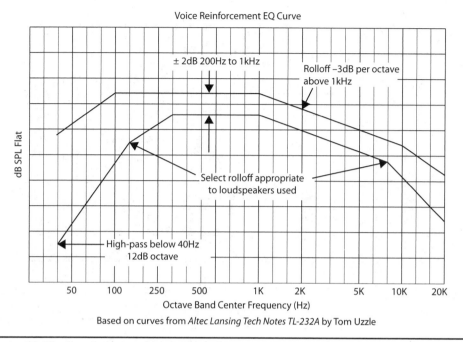

Figure 14-15 Response curve for speech reinforcement systems

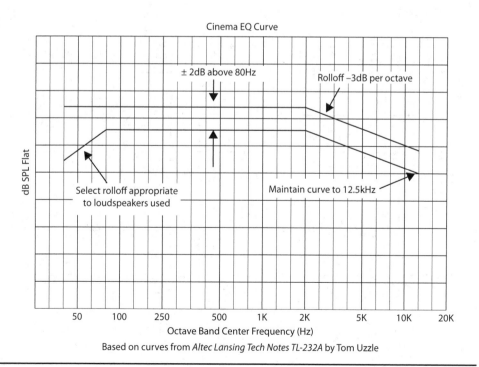

Figure 14-16 Response curve for cinema playback systems

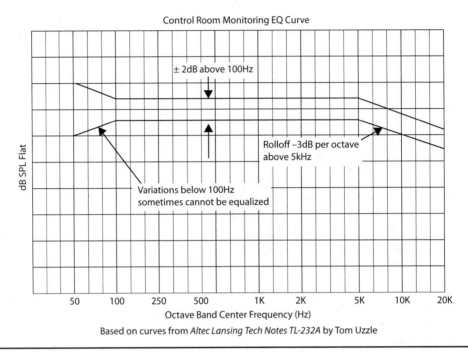

Figure 14-17   Response curve for studio control room monitoring

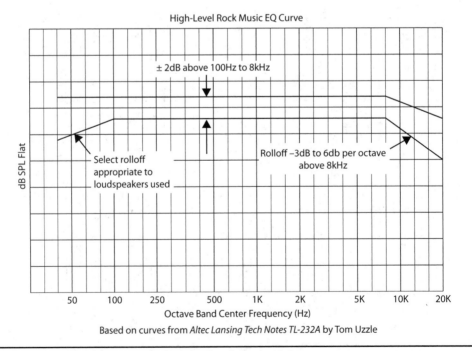

Figure 14-18   Response curve for high-level rock music reinforcement systems

Dual-Channel FFT

Using a dual-channel FFT analyzer is another popular method for equalizing an audio system, and if you utilize the time-windowing features when taking dual-channel FFT measurements, this method has some significant advantages over the RTA method. It allows you to view the sound spectrum in both the frequency and time domains. The time domain allows you to separate the energy coming from the loudspeaker from the reflected energy in the room, thereby enabling you to identify reflections in the room and isolate only the signals you want to measure. In addition, dual-channel FFT analyzers are designed to let you compare two signals. This means you can compare a reference signal, such as a sweeping sine wave, pink noise, or even program material, with the output signal to identify reflections in the room.

There are not many downsides to using an FFT analyzer for room equalization. The dual-channel FFT method is more complicated than the RTA method; hence, it will take more time to learn how to use it. With practice, however, you will likely find that the benefits of using this method are worth investing the effort.

Using Dual-Channel FFT

As noted earlier, the fast Fourier transform allows you to view a time domain signal (amplitude versus time) in the frequency domain. Using a test signal such as a sweeping sine wave or pink noise as a reference, it compares a measured signal against the reference signal in the time domain. Using time windowing allows you to separate the loudspeaker's direct sound from the reflected energy in the room.

For the Fourier transform to display this comparison, the measured signal is required to be at zero at both the start and end of the measurement window. To prevent truncation errors, the window function determines how the endpoints of the measurement transition to their zero value.

There are numerous types of window functions, but you should use the one recommended by the manufacturer of the measurement tool you will be using. Note that the window function is not the same as time windowing. A window is just a selected section of a measurement or waveform.

Transfer Function

A *transfer function* is the relationship between the input and output of a system or device. It is the result of what happens inside the system. Getting louder, greener, less noisy, more treble, brighter, being delayed, phase shifted, wetter, or colder are all examples of different systems' transfer functions. Dual-channel measurement devices can compare the input *to* a system with the output *from* the system, and thus provide you with information about the system's transfer function.

Tools for Using Dual-Channel FFT

To set audio system equalization using the dual-channel FFT method, you will need a dual-channel FFT analyzer; either dedicated hardware or as a software application.

PART V

You will need a signal source, such as a sine wave sweep, which ideally should contain all frequencies in the range you are measuring. Most FFT analyzers have a provision to generate suitable test signals.

Figure 14-19 shows a possible setup for dual-channel FFT audio signal measurement.

Dual-channel FFT analyzers sample the reference signal sent to the input of the sound system and sample the measurement made at the output of the sound system as picked up by a measurement microphone and compare these two signals. This comparison shows any changes in amplitude, phase, and time relative to the reference signal. The result is the system's transfer function. After delay and time windowing have been applied appropriately, you can compare the two signals, producing the transfer function of the loudspeaker. This also allows various test signals to be used as the reference signal, including music.

You will need a professional-quality, two-channel USB or FireWire audio interface to connect your reference signal and the measurement microphone(s) to the computer running the FFT analyzer software. The audio interface will require XLR microphone input(s) with phantom power capabilities.

You will also need a good-quality test and measurement microphone; cheaper ones are inconsistent. Many good-quality measurement microphones include a calibration chart showing the response characteristics for that particular device. The characteristics of a quality test microphone are listed earlier in this chapter.

If you are equalizing line-array loudspeakers, you will need to use three to four microphones, and you will need to average the measurements you take from each microphone position.

In addition to a microphone, you may need a stand capable of extending the microphone to heights of around 3 meters (10 feet) to get away from acoustic reflections from the floor or seating. For example, you could use a long boom-arm mic stand, a grip/staging/photographic C-stand, or a tall lighting stand, equipped with an adapter that fits the clip for the microphone.

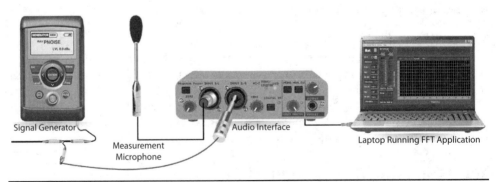

Signal Generator

Measurement
Microphone

Audio Interface

Laptop Running FFT Application

Figure 14-19   A dual-channel FFT setup

The Dual-Channel FFT Method

The process setting system EQ using the dual-channel FFT is as follows:

1. Verify that your system's gain structure has already been set.

2. Determine the lowest frequency you can make changes to with the normal measurement process, because standing waves produce nodes and antinodes at wavelengths similar to the room's dimensions. Calculate the lowest adjustable frequency using the simplified room mode formula.

3. Choose a loudspeaker or clustered group of loudspeakers to measure and adjust.

4. Turn off or mute any other loudspeakers in the space.

5. Check microphone placement. You do not want the microphone too close to the sound source (within the near-field) or too close to nearby boundaries, as it is difficult to provide a long enough time window for adequate frequency resolution. The microphone should be placed in the sound source's *far field*, which begins at a distance of three to ten times the largest dimension of the sound source. For a loudspeaker enclosure with the largest dimension of 1 meter, the far field would begin at a distance somewhere between 3 and 10 meters.

6. Connect the measurement microphone to one channel of the audio interface that is feeding the FFT analyzer.

7. Select a reference signal such as a swept sine wave or pink noise. Select an impulse response or energy time curve (ETC) measurement. Connect it to the other channel of the audio interface feeding the FFT analyzer.

8. Discover the time delay between the reference signal and measured signal. This can be done automatically through the FFT measurement device in most cases. If your test gear does not do this automatically, you will need to calculate time of flight, plus any latency introduced by the signal processors in the signal path. You will need this calculated number to synchronize the reference and measured signals for comparison.

9. Capture the signal from the loudspeaker with your measurement microphone.

10. The FFT analyzer will show you a visual representation of the signal. Examine the signal's time response. Note how close the earlier reflections are located relative to the direct signal.

11. Use the window function of your FFT analyzer to create a time window that excludes the earliest reflection.

 CAUTION Do not skip setting the time window. Some FFT systems will allow you to skip this stage, but setting a time window is one of the greatest benefits of using a dual-channel FFT analyzer. Without this step, you may as well use the RTA method.

Note that if the earliest reflection is close in time to the direct sound, the time window will limit the low-frequency resolution of the measurement. If this limit is not much higher than where the room modes dominate the low-frequency performance of the room, it will not affect the equalization process of the transfer function too much.

Next Steps

Once you have the results of the FFT analysis there are a number of further steps to consider to ensure that the results are both meaningful and useful:

- If the earliest reflection arrives less than 5 milliseconds after the direct sound, windowing may not allow enough of the entire frequency response to be analyzed. In this case, it may be helpful to put acoustically absorbent material on the surface near the loudspeaker that is creating the reflections or to move, or re-aim, the loudspeaker to substantially reduce these reflections.

- If the earliest reflection arrives *up to* 30 milliseconds after the direct sound, proceed to the Reading a Comb Filter instructions that follow.

- If the earliest reflection arrives *more than* 30 milliseconds after the direct sound, skip over the section on comb filters and follow the Adjusting for a Neutral Transfer Function instructions later in this chapter.

Reading a Comb Filter

If the earliest reflection arrives up to 30 milliseconds after the direct sound, you probably will not be able to create a time window that excludes all the reflections. This means you will likely need to allow for some *comb filtering* into your measurement. Comb filtering is a function of reflected sounds *interfering* with (adding to and subtracting from) the direct sound at certain frequencies.

In the comb-shaped graph shown in Figure 14-20, the *dips* are the frequencies where the pressure waves cancel each other out at the point of the measurement microphone. These frequencies measure as lower SPLs than what actually exist in the room. The *peaks* are the frequencies where pressure summing causes the measurement microphone to pick up higher SPLs than actually exist in the room.

Figure 14-20
A comb filter graph shows peaks and troughs in signal levels

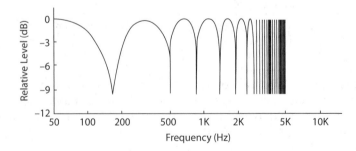

The teeth of the comb, or dips, can be 20 to 40dB, but summing at the peaks is limited to no more than 6dB above (double) the direct signal. The frequency at which the comb filter peaks and notches occur is highly location dependent. Moving the microphone even 50 millimeters (2 inches) will likely shift the peaks and notches to slightly different frequencies.

You cannot eliminate comb filtering unless you can change the room's acoustics or move the loudspeaker relative to room surfaces. Therefore, your job is to look through the comb filter and work out what the sound system is actually doing. The actual response will be 3 to 6dB below the summed peaks. You also need to ignore the notches, since they will be at different frequencies at different listener locations.

There are two ways to reduce the comb filter effect or potential for a comb filter response. The best method is to put the mic in a boundary microphone or pressure zone configuration where it is hard against a boundary surface (such as a wall or a panel of acoustically reflective material), which serves to block the arrival of interfering reflections from behind the microphone. The other method is to put in place acoustically absorbent material over the surfaces that produce the reflections.

The frequency of the first notch or node (peaks are called *antinodes*) in a comb filter occurs where the difference in distance traveled by the direct and the reflected versions of a wave from the sound source corresponds to half its wavelength. This places that reflected wave completely out of phase with the direct wave and causes its cancellation (destructive interference). The same effect occurs for harmonics of that wavelength, hence the spacing of the "teeth" of the comb.

Adjusting for a Neutral Transfer Function

If you used a 30-millisecond time window on the FFT and the earliest reflection arrives 30 milliseconds or more after the direct sound, and you have eliminated the reflections using time windowing, you can equalize the loudspeaker to produce a neutral transfer function or flat response (see Figure 14-21).

PART V

Figure 14-21

A flat or neutral frequency response

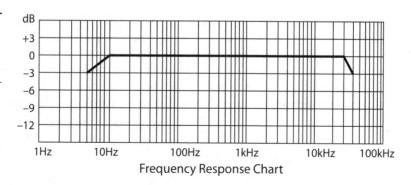

Frequency Response Chart

To adjust a response into a neutral transfer function:

1. View the signal on the FFT analyzer and examine the shape of the response. Identify any peaks or anomalies on the FFT response curve.

2. If there are large peaks or anomalies in the response, select a wide-bandwidth filter in the parametric EQ and focus on reducing one broad response anomaly at a time. Try to cut the response with EQ and avoid boosting wherever possible.

3. View the data at a few different resolutions. Move from 1/3rd to 1/6th to 1/24th octave resolutions, for example.

   - Progressively apply smaller and narrower (lower Q) parametric filters to the signal, smoothing out anomalies as you find them. There are diminishing returns as you get to narrower bandwidth filters and as you get to reduced filter depth. You may decide, for example, not to use filters narrower than 1/3rd octave or less than 1 or 2dB in depth.

   - For dips where it might be tempting to apply a boost filter, insert the boost EQ filter and listen to program material. If it sounds better with the filter, leave it in. If it sounds worse with the filter, remove it and do not try to make the response flat through this section of the loudspeaker's response.

4. Look at the room-ring modes. You can find them by looking at a 3D "waterfall" or spectrogram that shows frequency versus time, with the levels marked in colors.

   - Room-ring modes are more prominent in rooms with regular dimensions, concave surfaces, hard or live surfaces, and more reverberant larger rooms. The room-ring modes will persist longer than adjacent frequencies. They will be narrow-band in nature, so zooming in on an octave or two may help to reveal them. They will occur at the lowest frequencies produced by the loudspeakers and can extend up to 2 or 3kHz.

   - Attenuating room-ring modes by 3 to 9dB will improve the intelligibility of speech and clarity of music.

5. After you have balanced the overall levels and identified any room-ring modes, determine whether it is necessary to equalize the frequencies below the frequency where room modes dominate. Loudspeakers are generally consistent in their low-frequency response, so you may opt not to equalize in this range unless it is audibly necessary. There are two ways to equalize in this range. The first is to do it by ear using well-known source material. Alternatively, you can follow this process:

   a. Place a microphone just off the dust cap of the low-frequency or sub driver. Place a second microphone in the port of the cabinet. Make a summed-power-addition measurement of both mics.

   b. Next, calculate the surface area of all the port openings in the cabinet and the surface area of the driver's moving-piston parts. For example, an 18-inch (500-millimeter) woofer may have a 16-inch (400-millimeter) moving piston portion within the surround.

    c. Now compare the total port area with the total piston area of the speaker enclosure. If the total port area is one-half of the piston area, reduce the level of the port mic by 3dB. If it is one-quarter, reduce it by 6dB, and so on. This final response is the far-field response to be equalized flat.

    d. Depending on the frequency where room modes dominate, you may need do this for both a low-frequency driver in a full-range cabinet and the system subwoofer.

6. After you have flattened or smoothed the signal as close to a neutral response as it reasonably can be, your system is equalized. Next, you will need to finalize your measurements.

If you make an adjustment electronically but it does not respond acoustically, you have a problem that simply cannot be fixed through EQ adjustments. The issue may be related to the acoustics of the room or to the type of loudspeakers being used. If it appears at multiple locations within the room, it is likely to be a loudspeaker issue. If it appears at only one location in the room, it is likely to be a reflection issue. In either case, you have a non-minimum (summed) phase anomaly that cannot be fixed using equalization.

Finishing the Equalization Process

No matter what set of tools or method you use for equalizing an AV system, after you have flattened the signal to a neutral response, you will need to verify that the audio system has been equalized correctly. The steps you should follow are as follows:

1. Document all current system settings to retain a record of the starting point for this process.

2. Select program material that you know very well. This could be a favorite song or other audio recording. You should also use the type of program material that is typical for the system's intended use such as speech, musical instrumentals, or vocals.

3. To identify any distortion, turn on the program material to the maximum level at which you believe the system will be used. Distortion becomes more apparent at louder volumes.

4. Balance the amplifier levels and/or crossover levels by ear for best tonality. Use well-recorded, uncompressed program material that you have previously listened to with high-quality headphones.

5. Now listen to the program through the audio system you are verifying to check whether the program sounds the way it should.

6. Walk around the room with an RTA and listen. Does the audio system sound good in all areas of the room? If something sounds odd or off in the audio system, an RTA can assist you in pinpointing out-of-place frequency bands.

7. If the audio system does not sound good to you, use your ears to make some manual adjustments at specific frequencies using the cut-only method. You can use the RTA to help you identify which specific frequencies may need to be adjusted.

8. Repeat steps 2–7 for every loudspeaker or tightly packed/clustered group of loudspeakers in one part of a room at a time.

- When you have a group of three loudspeakers together as a cluster, treat that like a line array.
- Take several measurements at several points in space, at multiple listener positions.

9. After you have finished taking measurements, document your results. If you are using a software application, take screenshots or export the results of your measurements and put them in your project documentation.

Do not be tempted to continue changing the equalization after you have all the speakers turned on. Rather, you might try changing levels and delay times to improve results. Once you have equalized on-axis, if you change the equalization to accommodate off-axis, you may leave narrow peaks in the on-axis response that will be more prominent than dips off-axis. Do not use equalization in an attempt to fix coverage problems.

Performance Verification Standard

As you learned in Chapter 11, the ANSI/AVIXA 10:2013 Standard for Audiovisual Systems Performance Verification provides a framework and supporting processes for determining elements of an audiovisual system that need verification. As shown in Figure 14-22, this standard has a section dedicated to audio performance verification. These verification processes are described next.

Audio Performance Verification Items

What follows is the detailed expansion of each of the ANSI/AVIXA 10:2013 verification items listed in Figure 14-22.

- **AP-100**   Verify that any required muting or operational change of the installed sound system(s) has been made in accordance with local regulations in the event of a life safety or similar emergency.
- **AP-101**   Verify that loudspeaker zones are wired as defined in the project documentation.
- **AP-102**   Verify calibration of permanent audio system inputs such that the difference between any input signal level after the first common gain adjustment meets the requirements of the project documentation. This verification item should require a metric to be verified.
- **AP-103**   Verify that no audible noise caused by improper installation of any equipment provided in completed system(s) is present.
- **AP-104**   Verify that all audio routes are tested from endpoint to endpoint via the appropriate midpoint(s) for operation and routing as defined in the project documentation.

9.1 Audio System Performance Reference Verification Items

Item Number	Item Name	Pre-Integration	Systems Integration	Post-Integration	Closeout
AP-100	Emergency Muting	X	X	X	
AP-101	Loudspeaker Zoning	X	X		
AP-102	Alignment of Multiple Audio Source Levels		X	X	
AP-103	Audio Buzz and Rattles		X	X	
AP-104	Audio Routes		X	X	
AP-105	AV Room Reverberation Time		X	X	
AP-106	DSP Programming		X	X	
AP-107	Loudspeaker Physical Alignment		X	X	
AP-108	Loudspeaker Polarity		X	X	
AP-109	Loudspeaker Time Alignment		X	X	
AP-110	Phantom Power		X	X	
AP-111	Loudspeaker Transformer Tap Setting		X		
AP-112	Acoustical Ambient Noise			X	
AP-113	Assistive Listening Devices			X	
AP-114	Audio Coverage In Listener Areas			X	
AP-115	Audio Dynamics			X	
AP-116	Audio Levels Exceeds Background Noise Level			X	
AP-117	System Electronic Frequency Response			X	
AP-118	Audio System Equalization for Spectral Balance			X	
AP-119	Audio System Latency			X	
AP-120	Audio System Speech Reproduction at Listener Positions			X	
AP-121	Audio System Total Harmonic Distortion			X	
AP-122	Conferencing Audio Levels			X	
AP-123	Conferencing Echo Suppression Performance			X	
AP-124	Loudspeaker Impedance			X	
AP-125	Microphone Physical Alignment and Placement			X	
AP-126	Microphone Gain Before Feedback			X	
AP-127	Microphone Level Alignment			X	
AP-128	Multi-channel Loudspeaker System Output			X	
AP-129	Sound Masking			X	
AP-130	Audio Reinforcement System Headroom			X	
AP-131	Audio System Signal-to-Noise Ratio			X	

PART V

Figure 14-22 Audio performance verification items from the ANSI/AVIXA 10:2013 Standard for Audiovisual Systems Performance Verification

- **AP-105** Verify reverberation time meets the requirement defined in the project documentation. This verification item shall require a metric to be verified.
- **AP-106** Verify that all DSP-based products have been programmed as defined in the project documentation.
- **AP-107** Verify that loudspeakers are placed and aimed as defined in the project documentation.
- **AP-108** Verify that all loudspeakers have correct polarity as defined in the project documentation.
- **AP-109** Verify that loudspeaker time alignment performs as defined in the project documentation. This verification item shall require a metric to be verified.
- **AP-110** Verify that phantom power is provided at the correct voltage and correct locations as defined in the project documentation.
- **AP-111** Verify the loudspeaker transformer tap setting in constant voltage systems is as defined in the project documentation.
- **AP-112** Verify that the background acoustic noise levels within audiovisual spaces are within the required limits as detailed in the project documentation. This test is specifically related to ambient noise levels and not audio system quiescent noise, which is tested separately. This verification item shall require a metric to be verified.
- **AP-113** Verify that all devices that are part of the assistive listening system have been tested as a complete end-to-end personal listening system. Verify that the assistive listening system complies with regulatory requirements and adheres to project documentation. This verification item should require a metric to be verified.
- **AP-114** Verify that coverage of the audio systems in listener areas meets the performance requirements as defined in the project documentation. ANSI/AVIXA A102.01:2017, Audio Coverage Uniformity in Enclosed Listener Areas, should be used. Perform separate tests for all independent systems within the project, including but not limited to program sound, speech reinforcement, and show relay. This verification item should require a metric to be verified.
- **AP-115** Verify use of audio dynamics, including but not limited to noise compensation, automatic gain control, gating, feedback suppression, compression, limiting, delays, and levelers, meets the requirements defined in the project documentation.
- **AP-116** Verify that the audio level provided by the installed audio system exceeds the background noise level as defined in the project documentation. This verification item shall require a metric to be verified.
- **AP-117** Verify that the electronic frequency response of the audio system is as defined in the project documentation. This verification item shall require a metric to be verified.
- **AP-118** Verify that the audio system equalization is in accordance with the acoustic response curves as defined in the project documentation. This verification item shall require a metric to be verified.

- **AP-119**   Verify that audio system latency meets requirements defined in the project documentation. This verification item shall require a metric to be verified.

- **AP-120**   Verify that the audio system provides speech reproduction (intelligibility) as defined in the project documentation. This verification item should require a metric to be verified.

- **AP-121**   Verify that the total harmonic distortion of the installed audio system is as defined in the project documentation. This verification item shall require a metric to be verified.

- **AP-122**   Verify that in a conferencing audio application the incoming and outgoing audio levels are checked and adjusted in the system as defined in the project documentation. This verification item should require a metric to be verified.

- **AP-123**   Verify that a system with conferencing capability performs at nominal operating levels in a full duplex mode with echo and latency performance as defined in the project documentation.

- **AP-124**   Verify that all loudspeaker circuits have the correct impedance as defined in the project documentation. This verification item should require a metric to be verified.

- **AP-125**   Verify proper alignment and placement of microphones in the system as defined in the project documentation.

- **AP-126**   Verify that the speech reinforcement system is operating without feedback and at audio levels as defined in the project documentation. This verification item shall require a metric to be verified.

- **AP-127**   Verify calibration of microphone inputs so that the difference between any input signal level after the first common gain adjustment meets the requirements of the project documentation. This verification item should require a metric to be verified.

- **AP-128**   Verify that the audio outputs of a multichannel loudspeaker system are assigned correctly to designated outputs as defined in the project documentation.

- **AP-129**   Verify that audio system sound-pressure levels and equalization are adjusted to the level of sound masking as defined in the project documentation. This verification item shall require a metric to be verified.

- **AP-130**   Verify that the audio system is capable of performing above nominal operating levels without distortion as defined in the project documentation. This verification item shall require a metric to be verified.

- **AP-131**   Verify audio system electrical signal-to-noise ratio meets the minimum levels defined in the project documentation. This verification item should require a metric to be verified.

You may find some of these verification processes applicable to an installation you are undertaking, or you may find some processes that can form the basis of your own verification procedures. Verifying the performance of audio systems according to established standards will help you assure overall system performance.

Chapter Review

In this chapter, you studied how to adjust and verify the performance of the audio system.

Upon completion of this section, you should be able to do the following:

- Set the gain structure for audio system sources, including microphones and program audio sources
- Differentiate between the two most common methods of setting system gain
- Identify the equipment necessary for setting gain
- Set unity gain in an audio system
- Use the system optimization method to set gain in an audio system
- Adjust input levels in an audio DSP so the correct signal-to-noise ratio is achieved
- Adjust mixer input levels in an audio system so the correct signal-to-noise ratio is achieved
- Use the RTA method to set EQ
- Use the dual-channel FFT method to set EQ

Review Questions

The following questions are based on the content covered in this chapter and are intended to help reinforce the knowledge you have assimilated. These questions are similar to the questions presented on the CTS-I exam. See Appendix E for more information on how to access the free online sample questions.

1. A limitation of using the RTA method to adjust equalization is _____.
    A. that it cannot differentiate between the direct loudspeaker sound and the interference from combined direct and reflected energy
    B. that an RTA is not available as a smartphone app, as computer software, or as dedicated hardware
    C. that it cannot differentiate between high and low frequencies
    D. that it is not legal in some jurisdictions

2. The advantages of using a dual-channel FFT analyzer to adjust equalization include ____. (Choose all that apply.)
    A. viewing the sound spectrum in both the frequency and time domain
    B. identifying reflections in the room and isolating only the signals you want to measure
    C. being less complicated than the RTA method
    D. the ability to be optimized for different types of music

3. A reference microphone for setting equalization of a speaker system should _____. (Choose all that apply.)

   A. be the same brand and type used in the system as designed

   B. be omnidirectional

   C. have a flat, on-axis response

   D. have a nonreflective surface

4. After adjusting system equalization using the best tools available, you should _____. (Choose all that apply.)

   A. balance amp and EQ levels by ear for best tonality

   B. check the system by listening to familiar audio content

   C. check the system for distortion by listening at the highest expected volume

   D. change equalization off-axis to fix coverage problems

5. Which of the following are features of an audio signal generator? (Choose all that apply.)

   A. Generates sine wave and other signals used for setting system levels

   B. Usually has both 1kHz and 400Hz signal outputs

   C. With a properly calibrated signal, can determine signal bit rate

   D. Can be helpful in measuring distortion

6. Which of the following are associated with sound pressure levels? (Choose all that apply.)

   A. Is typically measured in decibels (dB SPL)

   B. Is the ultimate method for quantifying background noise levels

   C. Eliminates distortion in speech reinforcement

   D. Can most accurately be measured with a Class 1 precision SPL meter

7. In what applications is an A-weighted SPL most useful?

   A. In large concert halls and other Class A venues

   B. In environments with low listening levels such as speech applications

   C. At high-volume levels greater than 85dB SPL

   D. When using a DSP-based system

8. For what reasons is it important to set system gain? (Choose all that apply.)

   A. To minimize distortion

   B. To maximize clipping

   C. To maximize S/N ratio

   D. To minimize the gain structure

9. In which situations or environments is the system optimization method of gain adjustment preferable to the unity gain method? (Choose all that apply.)

   A. In performing arts centers and lecture halls

   B. For installations with short deadlines

   C. For systems with 750mV input signal levels

   D. Where clipping levels will not vary within the system

10. Which testing tool can detect clipped waveforms? (Choose all that apply.)

   A. Piezo tweeter

   B. Oscilloscope

   C. Reference microphone

   D. Cable tracker

Answers

1. **A.** The RTA method cannot differentiate between the direct loudspeaker sound and the interference from combined direct and reflected energy.

2. **A, B.** A dual-channel FFT analyzer shows the sound spectrum in both the frequency and time domains and enables you to identify reflections in the room and isolate only the signals you want to measure.

3. **B, C.** A reference microphone for setting equalization of a speaker system should be omnidirectional and have a flat on-axis response.

4. **A, B, C.** After technically adjusting system equalization using the best tools available, you should balance amp and EQ levels by ear for best tonality, check the system by listening to audio content you are familiar with, and check for distortion by listening to the audio at the highest expected volume.

5. **A, B, D.** An audio signal generator will generate a sine wave and other signals used for setting system levels. It usually has both 1kHz and 400kHz signal outputs and can be helpful in measuring distortion.

6. **A, D.** Sound pressure levels are typically measured in decibels and can most accurately be measured with a Class 1 precision SPL meter.

7. **B.** An A-weighted SPL is most useful in environments with low listening levels (20 to 55dB SPL).

8. **A, C.** Setting audio system gain will help minimize distortion and maximize signal-to-noise (S/N) ratio.

9. **A.** The system optimization method of gain adjustment is best applicable at performing arts centers and lecture halls.

10. **A, B.** Clipped waveforms can be detected with an oscilloscope, a piezo tweeter, or a 400Hz sine wave on a signal analyzer.

CHAPTER 15

Video System Setup and Verification

In this chapter, you will learn about

- The video verification checklist
- Building an EDID strategy
- Digital rights management and HDCP compliance
- Verifying and inspecting the video signal path
- Verifying video sources, displays, and audio/video sync

Ensuring appropriate signal levels across all video equipment, adjusting video cameras and displays, and checking audio/video synchronization are all part of the video verification process.

Chapter 10 covers video signal types, transport systems, and installing video components. In this chapter you will learn about factors that affect the quality of the video signal from source to display and how to address some of the concerns presented by digital video and networked audiovisual (AV) systems. You will have to test and verify several items, and an efficient way to keep track is to utilize the Audiovisual Performance Verification Checklist.

Duty Check

This chapter relates directly to the following tasks on the CTS-I Exam Content Outline:

- Duty C, Task 8: Test the Audiovisual System
- Duty C, Task 9: Calibrate the Audiovisual System
- Duty E, Task 3: Address Needed Field Modifications
- Duty E, Task 5: Maintain Audiovisual Systems

These tasks comprise 15 percent of the exam (about 15 questions). This chapter may also relate to other tasks.

Video Verification

As you learned in Chapter 11, the ANSI/AVIXA 10:2013 Standard for Audiovisual Systems Performance Verification provides a framework and supporting processes for determining elements of an audiovisual system that need verification. For video, you will use the Video System performance verification items and the Audio/Video System performance verification items, which are listed in full detail at the end of this chapter. You may find some of these verification processes applicable to an installation you are undertaking, or you may find some processes that can form the basis of your own verification procedures.

At this point, it is a good idea to review Chapter 10 and master basics such as digital video signal types, bandwidth considerations, and high-definition video formats.

Digital video and networked AV require knowledge of extended display identification data (EDID), digital rights management (DRM), and high-bandwidth digital content protection (HDCP).

Introduction to EDID

Extended display identification data was originally developed for use between analog displays and computer video devices, but has since made its way into Digital Video Interface (DVI), High-Definition Multimedia Interface (HDMI), and DisplayPort. EDID has since been further extended in capabilities to become Enhanced EDID (E-EDID), and more recently by DisplayID.

Displays and video sources need to negotiate their highest common resolutions so they can display the image at the best available quality. EDID is a method for source and display (sometimes called *sink*) devices to communicate this information, eliminating the need to configure the system manually. An EDID data exchange is the process where a sink device describes its capabilities such as native resolution, color space information, and audio type (mono or stereo) to a source device.

Each generation and version of EDID consists of a standard data structure defined by the Video Electronics Standard Association (VESA). Without the acknowledged handshake between display and source devices, the video system could provide unreliable or suboptimal video. Building an EDID strategy will be vital to your success as an AV installer.

EDID Packets

During the handshake process, the sink (display) device sends packets of EDID data to the source. The data packets carry the following information:

- Product information
- EDID version number
- Display parameters
- Color characteristics
- Audio characteristics
- Timing information for audio and video sync
- Extension flags

How EDID Works

Figure 15-1 shows how the EDID negotiation works, and the steps are detailed after the figure. The process begins with the activation of a *hot plug* link. The hot plug signal is an always-on, 5-volt line from the video source device that triggers the EDID negotiation process when a sink device is connected.

The EDID sequence works as follows:

1. On startup, an EDID-enabled device will use *hot plug detection* (HPD) to see whether the device is on.

2. The sink device returns a signal alerting the source that it received the HPD signal.

3. The source sends a request on the display data channel (DDC) to the sink for its EDID information.

4. EDID is transmitted from sink to source over the DDC.

5. The source sends video in its nearest possible approximation to the sink's preferred resolution, refresh rate, and color space. The source's selection can be manually overridden in some cases.

Figure 15-1
EDID process between source and sink devices

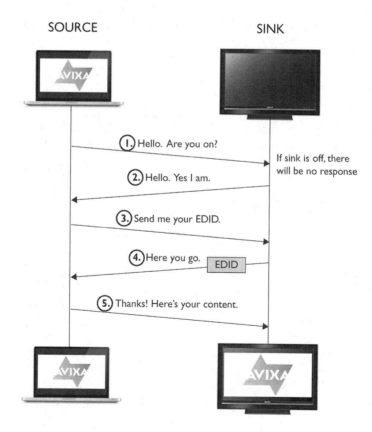

6. If the sink's EDID contains extension blocks, the source will then request the blocks from the sink. Extension blocks can be compatible timings relevant to digital video, as well as supported audio formats, speaker allocation, color space, bit depth, gamma, and if present, lip-sync delay.

NOTE Not all digital video extension technologies handle the hot plug detection and display data channel correctly. If EDID is required for a video display system to operate, it is important to verify the EDID compatibility of any video extension system specified in the installation.

EDID Table

An EDID table is a list of video resolutions and frame rates supported by a sink/display device. The DVI EDID data structure defines data in a 128-byte table; EDID for HDMI connections uses 256 bytes followed by additional 128-byte blocks, while DisplayID may include multiple 256-byte blocks of capability data. Table 15-1 shows some of the display capabilities of a specific monitor.

As an AV professional, what should you do when a user attempts to connect their legacy device to a new system? After all, your job is to make that device work and look as good as possible with the new system you have installed. Managing EDID will help you accomplish this goal.

Display Parameters:	
Video Input Definition	Digital Signal
DFP1X Compatible Interface	True
Max. Horizontal Image Size	600mm
Max. Vertical Image Size	340mm
Max. Display Size	27.2 inches
Gamma/Color and Established Timings:	
Display Gamma	2.2
Red	x = 0.653, y = 0.336
Green	x = 0.295, y = 0.64
Blue	x = 0.146, y = 0.042
White	x = 0.313, y = 0.329
Established Timings:	
800 × 600 @ 60Hz (VESA)	
640 × 480 @ 75Hz (VESA)	
640 × 480 @ 60Hz (IBM, VGA)	

Table 15-1 Capabilities of a Specific Display Screen Extracted from Its EDID Table *(continued)*

Established Timings *(cont.)*:

720 × 400 @ 70Hz (IBM, VGA)

1280 × 1024 @ 75Hz (VESA)

1024 × 768 @ 75Hz (VESA)

1024 × 768 @ 60Hz (VESA)

800 × 600 @ 75Hz (VESA)

Standard Timing:	
Standard Timings n°	4
X Resolution	1152
Y Resolution	864
Vertical Frequency	75
Standard Timings n°	5
X Resolution	1600
Y Resolution	1200
Vertical Frequency	60
Standard Timings n°	6
X Resolution	1280
Y Resolution	1024
Vertical Frequency	60
Preferred Detailed Timing:	
Pixel Clock	241.5MHz
Horizontal Active	2560 pixels
Horizontal Blanking	160 pixels
Horizontal Sync Offset	48 pixels
Horizontal Sync Pulse Width	32 pixels
Horizontal Border	0 pixels
Horizontal Size	597mm
Vertical Active	1440 lines
Vertical Blanking	41 lines
Vertical Sync Offset	3 lines
Vertical Sync Pulse Width	5 lines
Vertical Border	0 lines
Vertical Size	336mm

Table 15-1   Capabilities of a Specific Display Screen Extracted from Its EDID Table

EDID Troubleshooting

You may encounter disruptions in the EDID conversation. The best troubleshooting method is prevention. Compiling and maintaining an EDID data table for all devices in an installation can avoid many EDID problems. This table will help you track all of the expected resolutions and aspect ratios for every input and output.

If you need to troubleshoot a problem that you think may be an EDID issue, follow the usual fault locating principles:

1. Identify the symptoms:

- Did the system ever work correctly?
- When did the display system last work?
- When did it fail?
- Did any other equipment or anything related to this system change between the time it last worked and when it failed?

2. Elaborate the symptoms:

- Confirm that every device is plugged in and powered on.
- Make only one change at a time as you search for the problem.
- Note each change and its effect.

3. List probable faulty functions   Identify potential sources of the problem.

4. Localize the faulty function   Simplify the system by eliminating the equipment that you know is not the source of the error.

5. Analyze the failure   Substitute the suspect devices or components with devices that you know work correctly. You can also use EDID test equipment that can emulate sources and sinks.

EDID Tools

There are a range of tools you can use to discover and troubleshoot EDID problems:

- **Software**   Many software applications can be used to read, analyze, and modify EDID information. Some are available as tools to complement EDID hardware devices, but there is also a selection of freeware and public domain applications that run on Linux, Windows, or macOS platforms. With these you can connect a computer up to a display and read the EDID from displays and/or sources to identify whether there may be a problem. Some of these applications will also allow editing of the EDID information to test devices or to resolve a problem. As an example, the Linux "read-edid" command can extract a device's EDID information and help diagnose whether something is wrong.

- **EDID readers or extractors**   EDID readers and analyzers are available as special-purpose handheld devices or as additional functions on video test signal generators and analyzers.

- **Emulators**   An EDID emulator or processor may be a useful solution if you cannot get the handshake process to function properly. It takes the place of the sink device's EDID output and forces one or only a few EDID choices.

Resolution Issues

If a computer cannot read the EDID from the sink, it may default to its standard resolution. If the user subsequently attempts to manually set the system resolution to match the display, some graphics cards may enforce the default lower resolution and create a misaligned (size, aspect, centering) output without actually changing the video resolution.

If a computer is connected to multiple displays but can read the EDID from only one display, it may send an output that is mismatched to the other displays.

Some devices in the signal path between the source and the sink, such as switchers, matrix switchers, distribution amplifiers, video signal processors, and signal extender systems, have factory-set default resolutions. If the equipment has such settings, you will need to configure the device to pass EDID information from the sink to the source and vice versa. This may involve setting the EDID to match preset information about the capabilities of a sink device.

Note that if you make the presets something the source does not recognize, you may not get any image, or the image will have a very low resolution. For source components such as Blu-ray Discs (BD), be aware that some BD players will send a low-resolution 480p output that is compatible with many, but not all, older display devices.

No Handshake, No Picture

Many sources fail to output video if the handshake fails, but computer devices typically will send an output at a default lower resolution to ensure the user can still work with their computer. If this is the case, you may still see a picture from the PC source, but it will be of a lower-than-optimal resolution.

Some source devices will not output a video signal unless the display's EDID data gives confirmation that it can properly display the signal. If there is no signal output from the source, the problem could be that the display's EDID data is not being transferred to the source. If the hot plug detection pin cannot detect another device, the initiator of the conversation will interpret the sink as disconnected and cease the EDID communication. Potential problems with hot plug detection can involve the source not being able to supply sufficient voltage due to voltage drop in a long cable run, a bad (resistive) connection, or a digital video component such as a switcher or splitter intercepting the hot plug detect line.

Switching Sources

When switching sources, the changeover can be very slow, and there can be total picture loss during the switching process. This may be related to non-synchronous (crash) switching between sources and the delay required to resynch the frame (vertical sync) signal on the new source, or it may be due to a delay in EDID negotiations with the new source.

When EDID sources are not receiving hot plug signals, they generally conclude that the sink device is disconnected. Some low-quality direct switchers disconnect and reconnect signals when switching between devices. If an EDID connection is broken, when it is reconnected, the negotiation begins anew.

Managing EDID Solutions

Some approaches for dealing with EDID problems include the following:

- Always test the continuity of all circuits in cables.
- Only use cable lengths that fall within manufacturer guidelines.
- Determine whether a device has a default setting. If that setting is not optimal for the installation, research what it is optimal and how to change it.
- Always use EDID-capable signal extension, distribution, and switching devices.
- If the AV design specifies displays with different aspect ratios and resolutions connected to one source, select the highest common resolution on the EDID emulator.

Building an EDID Strategy

The goal of an EDID strategy is to allow a display to present the signal at its native resolution without scaling. If the system has displays of different resolutions or aspect ratios, without a proper EDID strategy, the resolution and aspect of the output will be unreliable and may vary when switching between sources.

All display devices in a video system should ideally have the same aspect ratio, which will prevent many EDID problems. Many source devices, including most computer systems, have a default output with a 16:9 aspect ratio, so projectors and other display components should match this. In projects where there are mismatches between source and sink capabilities, EDID processing management is required to retain consistent results during source and display switching. If there is an EDID strategy in place, your client's AV system should work for a long time.

Display devices in fixed installations (particularly projectors) are generally not upgraded as frequently as the computers, which means that those display devices may quickly become outdated.

To ensure that the system you are installing is EDID compliant, use a device that includes an EDID emulator as a sink. It can be set to a specific aspect ratio and native resolution so that the source outputs a consistent aspect ratio and resolution. You can then set the EDID for each connected sink device.

Digital Rights Management

Your customers may want to stream content that they did not create, such as content from video streaming services, music streaming services, or materials from a local media library. Unlicensed distribution of content can violate copyright laws.

Make sure your customers are aware of potential licensing issues related to the content they want to play. It may be necessary to negotiate a bulk license with a content service provider.

If you fail to obtain the proper licenses to play content, you are not just risking the legal repercussions of copyright infringement; you may be risking the system's ability to function at all. Publishers and copyright owners use DRM technologies to control access to and usage of digital data or hardware. DRM protocols within devices determine whether content can be allowed to enter a piece of equipment. Actual legal enforcement of DRM policies varies by country. It is also potentially illegal to circumvent DRM encryption.

High-Bandwidth Digital Content Protection

HDCP is a form of encryption developed by Intel to control digital audio and video content. If the content source requires HDCP, then all devices that want to receive that content must support it.

HDCP is merely a way of authorizing playback. The actual AV signals are carried on other wires in the cable. HDCP is used authorize the transmission of encrypted or non-encrypted content across a wide range of sources of digital content.

For example, there is a pause when you power up a Blu-ray player while the AV system is verifying that every device in the system is HDCP compliant. The Blu-ray player first checks whether the disc content is HDCP compliant. If so, the player proceeds to conduct a series of handshake exchanges to verify that all other devices in the display chain are also HDCP compliant before making the content available for display.

HDCP Interfaces

HDCP 2.x is capable of working over the following interfaces:

- DVI
- HDMI
- DisplayPort
- HDBaseT
- Mobile High-Definition Link (MHL)
- USB-C
- TCP/IP

Some Apple devices may block content if HDCP is not present, even if the content is not HDCP-protected.

How HDCP Works

HDCP's authentication process determines whether all devices in the display system have been licensed to send, receive, or pass HDCP content. No content will be shared until this entire process is completed. If there is a failure at any point in the process, the whole process has to restart.

PART V

The authentication steps are

1. **Device authentication and key exchange**, where the source verifies that the sink is authorized to receive HDCP-protected content and then exchanges device encryption keys

2. **Locality check**, where the source and sink verify that they are within the same location by confirming that the round-trip time for messages is less than 20ms

3. **Session key exchange**, where the source and sink exchange session encryption keys so the content can be shown

4. **Authentication with repeaters**, which allows repeaters in the system to pass HDCP content

When a switcher, splitter, or repeater is placed between a source and sink/display device to route signals, the inserted device's input becomes a sink/display, and its output becomes a new source and continues the video signal chain.

HDCP Device Authentication and Key Exchange

The authentication process is designed so the source verifies that the sink is authorized to receive HDCP-protected content. Each device must go through this process.

1. The source initiates the authentication process by sending a signal requesting the receiver to return its unique ID key.

2. The receiver sends its unique key to the source. The source must receive the sink's unique key within 100ms, or the device is not compliant with HDCP 2.x specification.

3. The source checks that the receiver's key contains a specific identifier that is given only to authorized HDCP adopters. If the key is missing, the process is aborted.

4. The source then sends its master key information to the receiver.

5. The receiver's software verifies the master key and uses it to compute new values that are returned to the source.

6. The source then verifies the receiver's calculated values. If the calculated values are not received within the appropriate amount of time, the authentication process is aborted.

HDCP Locality Check

The locality check verifies that the source and receiver are in the same locality, for example, in a nearby meeting room, auditorium, or classroom, and not a couple of cities away.

1. The source sends a message to the receiver containing a random number and sets a timer for 7 milliseconds. Both the source and receiver will have the same algorithm that takes the random number and generates a new one.

2. The receiver gets the random number, generates the new one, and sends that back to the source before the 7-millisecond timer expires.

3. The source verifies that the number it calculated and the one the receiver returned are the same. If they match, the authentication process continues. If they do not match or the timer expires, the authentication process for the locality check begins again. The locality check will be repeated for a total of 1,024 tries before it aborts the authentication process and everything must start over again.

HDCP Session Key Exchange

Once the source and receiver have passed the device authentication and locality check, the process moves on to the individual session using the following steps:

1. The source device generates a session key and sends it to the receiver with a message to pause for at least 200 milliseconds before using it.

2. The source pauses and then begins sending content encrypted using the session key. Each HDCP session has its own key and therefore unique encryption.

Once the session commences, the HDCP devices must reauthenticate periodically as the content is transferred. Several system renewability messages (SRMs) must be exchanged. If during SRM exchange it is discovered that the system has been compromised, the source will stop sending content.

HDCP and Switchers

All of the HDCP processes explained so far have assumed that the AV system you are installing has a single video source for a single video sink/display. In these systems, the number of keys and how they are exchanged will be handled between the two devices. However, when multiple video sources, displays, processors, and a switcher are added, HDCP key management becomes more complex.

Some switchers maintain key exchange and encryption sessions continuously, so the communication will not need to be restarted. These switchers can act as a source or sink to pass the protected and encrypted HDCP data to its destination. You will need to verify that the switcher can handle HDCP authentication from the manufacturer's documentation.

HDCP Authentication with Repeaters

An HDCP repeater is a device that can receive HDCP signals and transmit them on to another device, such as a switcher, display device, processor, or distribution amplifier. In a system with repeaters, as shown in Figure 15-2, the HDCP authentication process occurs after the locality check and device authentication have taken place between all the devices in the system. The session key exchange has not begun.

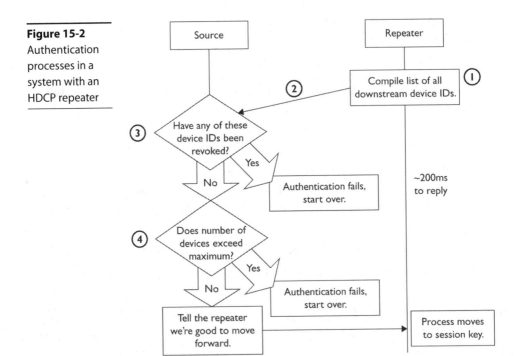

Figure 15-2
Authentication processes in a system with an HDCP repeater

The HDCP authentication process in a system with a repeater follows these steps:

1. The repeater compiles a list of IDs from all its connected downstream devices.

2. The repeater sends the list of IDs and number of devices to the source and sets a 200-millisecond timer.

3. The source reads the list and compares it to a list of revoked licenses in the media. Each new HDCP device or media has an updated list of revoked license numbers provided by Digital Content Protection, LLC. If any of the downstream devices are on the revoked list, the authentication process fails.

4. The source then counts how many devices are downstream. If the number of devices is less than the maximum of 32, the authentication process moves forward.

HDCP Device Limits

HDCP 2.x supports up to 32 connected devices for each transmitter and a maximum of four repeater levels for each transmitter. Figure 15-3 illustrates an example of connection topology for HDCP devices.

Although HDCP 2.x supports up to 32 connected devices, in practice, the number of sink/display devices allowable from a single HDCP-protected media source is typically much more limited and based upon the number of keys allowed by the source.

Figure 15-3
Depth of two
repeater levels
and device
count of six
in HDCP 2.x
topology

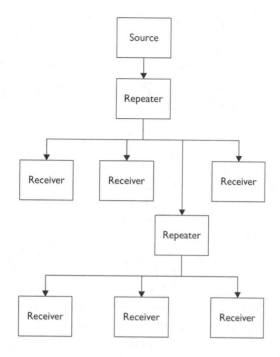

HDCP Troubleshooting

When HDCP authentication fails to initiate correctly, there can be a range of symptoms that do not directly implicate HDCP failure as their cause. These may include an image appearing for a few seconds at the start of a session and then disappearing, a green screen, white noise, or a blank screen, any of which could be caused by a broken cable, a bad connector, or dozens of other problems.

An HDMI signal analyzer or signal generator/analyzer is one of the few tools that will let you track and analyze the HDCP negotiation process, providing you insight into where the negotiations have gone wrong and a clue as to how you might remedy that issue. If such an analyzer is not available, then the signal flow troubleshooting strategy is likely to be the most productive. Start at the source end of the signal path with a known working display device and follow the signal path, testing at each point until you find the fault.

Verifying the Video Signal Path

When verifying a video system, you first should check the wiring and cabling of the video equipment.

Once all connections have been completed, verify that the wiring methods and cable pathways are as specified in the system design. This is typically a visual inspection process because you will need to compare the technical drawings, such as a signal flow drawing, against what you discover in the installation.

PART V

As with audio system wiring, during the inspection process, you need to look for the following:

- Correctly terminated connectors. Look for defects such as bent pins or frayed shields.
- Correctly connected cables. They should be attached to terminals as indicated on your system drawings.
- Cabling in the pathways that are properly organized in the walls, raceways, and the ceiling. Also, check for proper bend radius at all junctions and pull boxes.

As you work through this, record any errors or discrepancies for immediate repair. Permanent changes that occurred during installation should be documented and submitted to the AV system engineer or system designer for updating the as-built documentation.

Signal Extenders

Due to the resistive and reactive properties of all cables, long cable runs can cause significant signal attenuation. The high-frequency components in video signals limit the effective length of DVI, HDMI, and DisplayPort cable runs to a theoretical maximum of 5 meters (16 feet), which is often not sufficient for a professional AV system. One of the solutions to this problem is the use of an extender technology, which not only reaches farther than a direct video connection but is often cheaper than a long run of quality digital video cable.

Passive Extenders

Passive extenders (see Figure 15-4) generally use *balun* technology to interface between the digital video cable and one or two twisted-pair network cables such as Cat 6+ to extend the video signal over distances up to about 50 meters (165 feet). Short for *bal*anced-to-*un*balanced, a balun is a transformer used to connect between a balanced circuit and an unbalanced circuit. In an extender application, one balun is required at the point of connection between the video signal and the twisted-pair network cable, and a second balun is required at the far end of the twisted-pair extension cable to match the video signal to the receiving device.

Figure 15-4
A pair of HDMI baluns

Some extension systems use a single twisted-pair network cable, while others use two cables. As baluns are less than 100 percent efficient and there is signal attenuation in the extension cables, the specified distance limitations of extension systems must be observed. It is important that the extender devices at each end of the extension system are of completely compatible make and model, as there is no standard for pinouts between manufacturers and product ranges.

Active Extenders

Active extenders contain processing circuitry to boost and regenerate the video signals before transmission down the extension line and to regenerate them again at the receiving end. This enables active extenders to operate over greater distances than passive extenders. These extenders may use one or two twisted-pair network cables (Cat 6+) or one or two optical fiber cables as their extension medium. HDMI and DisplayPort signals may easily be extended for distances up to about 100 meters (330 feet) with wired extensions and up to 30km (19 miles) with fiber-optic extensions.

Active extenders require power to drive the processing electronics in both the transmitting and receiving devices. Some wire-connected systems send power down the cable(s) from the transmitter to the receiver, while others, including fiber-based systems, require a power supply at both ends. As with passive extenders, it is important that the transmitters and receivers are of completely compatible make and model, as there is no standard for pinouts and signal levels between manufacturers and product ranges. HDBaseT is a widely used active extender technology, with a full ecosystem of HDMI extension and distribution devices available.

It is also a common practice in event, large venue, and broadcast applications to use standard video conversion devices to convert HDMI or DisplayPort to XX-SDI, run the XX-SDI over coax or fiber to the destination, and then convert the XX-SDI back to HDMI or DisplayPort for display. This is a useful solution for locally originating content, but replay of copy-protected material can be problematic, as XX-SDI does not support HDCP.

Network Extenders

As discussed in Chapter 12, a range of AV over IP technologies can be used to extend video signal delivery to any point on a TCP/IP network.

 NOTE Active extenders reclock the signal and output a duplicate of the original signal. Passive extenders change the physical medium but do not reclock the signal.

Verifying Video Sources

In Chapter 10, you learned about the installation of video components, including video cameras and projectors. After the installation, you will need to fine-tune camera adjustments and complete projector calibration.

PART V

The location, intensity, and shadow quality of light sources in the picture area play a major part in the quality of the images picked up by cameras. When low-contrast lighting is used, images often look washed out and flat, while appropriately designed lighting will create images with contrast, dimension, and good exposure. The design of the lighting for video camera image capture is an integral part of any AV system design, and checking alignment, light levels, coverage, color temperature, color rendering, and contrast ratios should form part of the verification of the installation. The system design documents should include details of all lighting alignment parameters. It is quite common for a lighting designer, gaffer, director of photography, or lighting technician to be brought in for final alignments and level setting before system handover on projects with many luminaires and complex requirements.

Camera Adjustments

Clients tend to use cameras for videoconferences and to record and stream activities in their facilities. Similar to other AV devices, cameras should be set up for the environment in which they operate. Sometimes, cameras are connected and powered on, and that is the extent of the setup. However, a professional AV installer should make some standard adjustments to a camera to ensure that the images produced are of adequate quality.

Focus

The focus adjustment allows the camera to deliver a sharp, clearly visible image of the subject. To set the focus:

1. Have a subject stand in the general location where the camera will normally be capturing images.

2. Zoom the lens in tightly to the subject and adjust the main focus until the subject image appears sharp and crisp. A focus chart (see Figure 15-5) can make the process easier; the contrasting black and white sections allow for very accurate focusing.

3. Zoom back out to frame the shot for the best coverage.

Figure 15-5
A sample
focus chart

Back-focus

Lenses must stay in focus as they are zoomed from wide-angle shots to narrow fields of view. Adjusting the back-focus on the lens will keep zoomed-out images focused. To adjust back-focus:

1. Locate the back-focus setting on the camera lens. If the camera has this setting, it will likely be a manual, hardware-dependent adjustment near the focus setting.

2. At the target area, zoom into the previously used focus chart, and focus as before.

3. Zoom all the way out and adjust the back-focus until the shot is again in focus.

Repeat this process until the lens stays in focus throughout its zoom range. If possible, lock the setting in place.

Pan, Tilt, Zoom, and Focus Presets

In system designs that include preset scenes or setups for cameras with remote robotic pan/tilt/zoom (PTZ) capabilities, each preset should be set up as specified in the design and the preset labeled with the name, such as "wide shot," "presenter closeup," "white board," or "full stage," as indicated in the design documents. Fixed-position cameras with remote zoom and focus preset capabilities should be set up in a similar way.

Some more advanced robotic camera systems may include capabilities for remote dolly/pedestal position, camera elevation, and live moves. These systems will generally be set up in collaboration with the end user during system training prior to sign-off.

Iris Settings (Exposure)

The iris in the camera's lens controls how much light passes through the lens to reach the camera's imaging device(s). Some adjustment to the iris will usually be required to correctly set the exposure of the image. If light levels change in the area covered by the camera, the aperture (opening) of the iris will require adjustment to maintain the correct exposure of the image. This adjustment is generally found on the body of the camera, as shown in Figure 15-6.

Figure 15-6 Typical controls on a professional AV camera. Courtesy of Panasonic Corporation.

The automatic iris adjustment found on many AV cameras adjusts the aperture of the iris based on the output of the camera's image pickup device(s). Auto-iris responses may be based on the light falling on the center of the pickup (center weighted) or on the average of the light across the entire pickup area. Some camera's auto-iris systems can be switched between several exposure modes. Most lenses will also have a manually operated iris mode, which allows an operator, either local or remote, to set the lens aperture for the desired exposure.

There is a strong interaction in the exposure of an image between the level of light on the subject of the image and the level of light on the background behind the subject (often a wall):

- If the subject is substantially brighter than the background, it is possible for the auto iris to expose the subject correctly, while the background will be less prominent.

- If the subject is much darker than the background, the auto-iris systems may expose for the level of the background and leave the subject underexposed. However, if the auto-iris system is center-weighted, it may instead expose the central subject correctly and overexpose the background. Neither of these results are likely to be satisfactory pictures.

- If the subject and the background are of similar brightness, the auto-iris system will be able to get a good exposure of the subject, but the image may lack sufficient contrast to emphasize the subject and allow it to stand out clearly from the background.

If the system design includes adequate lighting control, you will be able to expose a camera with an auto iris using the following simple procedure.

With the background lighting at its minimum level, set the subject lighting (including fill lighting and backlighting) at close to maximum intensity, and allow the auto iris to expose the camera. Now, adjust the level of the background lighting to make the background visible, but not dominant, in the image. If there is no simple dimming control, it is possible to adjust the intensity of the background lighting by switching some of it off or on or by inserting neutral density filters/meshes or diffusion material in the luminaires illuminating the background.

The balance between subject and background lighting for a good exposure will vary substantially depending on the color and brightness of the subject material. If the subject is a person, the reflectivity of their skin will greatly affect the balance. You will most likely have to adjust the lighting to account for significant changes between darker and lighter skin tones of subjects. If the lighting control system has preset level memory, it is advisable to set up a number of lighting states to accommodate the range of skin tones that may be present in the subject area.

Backlight Adjustment

Some cameras have a *backlight adjustment* mode, which surprisingly has nothing to do with the lighting focused on the back and edges of a subject to give separation from the background. The backlight adjustment is a compensation control for when a subject is seen against a bright background, such as a daylit window, which usually tricks the auto-iris system into underexposing the less-lit subject.

When activated, the backlight adjustment attempts to compensate for the over-bright background by increasing the aperture of the camera above the auto-iris setting to bring the subject up to a better exposure. This inevitably overexposes the background in the process of solving the subject's exposure. Use of the backlight adjustment is, at best, a temporary measure to be used under duress. The solution to the problem is to improve the balance of light in the image by either reducing the light in the background (such as closing the curtain or blind over the window) or increasing the light level on the subject. Light-reducing window tinting can be used as a permanent solution to allow a subject to be seen against a brightly daylit exterior.

Automatic Gain Control

In situations where insufficient light is available to properly expose an image, it is possible to lift the brightness of the image by increasing the output of the camera's imaging device(s) via the video gain control functions. As with any amplification process, increasing gain also increases the noise in the signal, which results in random sparkles in the video output. Engaging the auto gain control (AGC), if available, will usually result in images that are of sufficient brightness, but will often be quite noisy. The best solution is to make sure that the subject is in an adequately lit area, which may be as simple as moving the subject into an existing pool of light or swinging a couple of lights around to cover the subject area.

Shutter Adjustments

The camera's shutter system controls the amount of light that the imaging system must process. In some cameras this is achieved by varying the size or speed of an actual rotating mechanical shutter that sits in the optical train before the imaging device(s). In other cameras the shutter control is a setting that varies the time between successive readings of the data from the imaging device(s). If a lot of light is entering the camera from a very bright scene, an automatic shutter system will reduce the amount of time the shutter is open, which may cause any fast movements in the scene to appear stuttered or jerky. The best solution to jerkiness caused by the shutter is to use some means other than the shutter to reduce the exposure of the imaging device(s). These solutions include reducing image gain, inserting a neutral density filter, or possibly closing down the iris.

White Balancing

Many cameras have an automatic white balance function. This circuit looks for bright objects in the image, assumes that they are white, and adjusts the color levels in the camera's output to produce a white signal for those bright objects. It is therefore critical that all automatic white balancing is performed using a matte white reference object, such as a piece of white cardstock or paper, illuminated by the same lighting as the subject. If the automatic white balance feature is off, manual white balancing is required. This is often performed using a grayscale chart to fine-tune the color balance across the camera's full dynamic range.

To automatically white-balance a video camera:

1. Place a large piece of white paper at the object location in the lighting that will be used for the subject.

2. Zoom into the white paper until it fills as much of the image as possible.

3. Press the white balance button. This will set the color reference levels for the current lighting conditions.

An automatic white balance is the best attempt the camera's electronics can make under limited conditions, so it is unwise to expect that two cameras capturing images of the same object will completely match, even if they are of the same model and have the same lens. Where color matching between multiple cameras is important, expert human intervention is usually required.

Some cameras have a number of preset color balance settings that may be selected to accommodate known color temperature conditions.

Verifying Display Components

Displays should be set up and adjusted to produce the best image based on the environment in which they are to be viewed. The viewing environment can have a major effect on the quality of the displayed image. Viewing a movie in the theater, watching a presentation in a boardroom, viewing the IMAG on a live event, and viewing digital signage in bright daylight are very different viewing experiences.

Display setup requires knowledge of signal generators and the purposes of common test patterns. A knowledgeable technician can make the necessary adjustments based on the viewing environment.

Identify Display Parameters

The first step is to identify the parameters of the display device, which could be a projector, an LED screen, or a flat-panel display. Next, determine the input signal types that will be used on the display. The signal will probably be HDMI, SDI, or DisplayPort, although DVI and RGBHV may still occasionally be used in existing systems.

Determine the aspect ratio of the displayed image. The most common ratio is 16:9, although other ratios may be used in cinematic; video-wall; and multi-panel, multi-module, or multi-projector systems. This information for each display device may be in the owner's manual, or it can be calculated by dividing the width by the height of the displayed image. As a guide, a ratio of width to height of 1.77:1 is equivalent to 16:9.

It is possible to display an image that was designed for a legacy 4:3 display on a 16:9 display, but depending on the display system settings, the image will be either horizontally stretched to fit, or have vertical black bars on either side of the image. Similarly, images intended for wide screen (2.39:1) cinematic display will usually display with horizontal black bars at top and bottom. It is usually preferrable to set up the display to show the entire image unstretched with black bars than with the image either cropped or distorted by stretching. If these settings are not defined in the system design documents, you should consult with the client or end user.

Figure 15-7
Test pattern to
set contrast

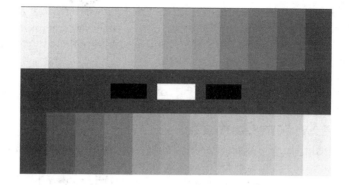

Set Contrast

The test pattern shown in Figure 15-7 is used to set contrast. The white rectangle in the center is super white (step 255), or the brightest item on the screen. The two very black rectangles beside it are super black (step 0), and the black background that surrounds the white and blacker rectangles is normal video black (step 16). The goal is to strike a balance between all the shades in the image. Make adjustments until all the bars are clearly distinct and visible. Ensure the white bars look white and the black bars look black. If the display's grayscale transfer function (gamma) is set correctly, the bars should appear in even, gradual steps.

If the black sections of the image are too dark, increase the brightness control. If the image sections are too bright, increase the contrast. It may take several adjustments to balance the image and ensure that all variations of white and gray can be clearly seen.

The procedure for adjusting brightness and contrast is as follows:

1. Display a grayscale test pattern.
2. Adjust the contrast level down and then increase the level while watching the white rectangle in the center of the screen. Adjust up until the white does not get any whiter.
3. Adjust brightness until you cannot see any difference between black and super black.
4. Repeat the adjustments until both settings are achieved.

Once complete, the gray bars should gradually increase/decrease in value in a linear fashion. Repeat these steps until all bars appear in even, gradual steps.

Set Chroma Level

Color bars are test patterns that provide a standard image for color alignment of displays. They can be used to set a display's chroma level (color saturation) and, where present, hue shift. Hue shift control is generally only found on systems that display images in the US NTSC format, which is prone to color phase errors.

Adjust the display for accurate monochrome brightness and contrast before color adjustments.

A commonly used test pattern is the HD Society of Motion Picture and Television Engineers (SMPTE) version. Alternatively, you may come across the Association of Radio Industries and Businesses (ARIB) version. You can use these interchangeably, and the process for using them is almost identical.

Check Appendix D for the link to the AVIXA video library, which includes a video explaining the process of setting the contrast, brightness, and chroma.

Image System Contrast

Image quality can be assessed using criteria such as contrast, luminance, color rendition, resolution, video motion rendition, image uniformity, and even how glossy a screen is. However, contrast remains the fundamental metric to determine image quality. Taking the viewing environment into consideration, the difference between system black and the brightest possible image is the system *contrast ratio*.

AVIXA developed the ANSI standard ANSI/AVIXA 3M-2011, Projected Image System Contrast Ratio, which set out how to measure contrast ratios and what contrast ratios are suitable for different viewing requirements. That standard has since been updated and revised and will shortly be replaced by ANSI/AVIXA V201.01:202X Image System Contrast Ratio (ISCR), a standard that extends beyond projection to cover system contrast for viewing all images, including both projection and direct-viewed displays. The complete standards are available from the Standards section of the AVIXA website at www.avixa.org, but here is a brief summary of the four viewing requirement categories and their required minimum contrast ratios:

- *Passive viewing* is where the content does not require assimilation and retention of detail, but the general intent is to be understood (e.g., noncritical or informal viewing of video and data). This requires a minimum contrast ratio of 7:1.

- *Basic decision-making* (BDM) requires that a viewer can make decisions from the displayed image but that comprehending the informational content is not dependent upon being able to resolve every element detail (e.g., information displays, presentations containing detailed images, classrooms, boardrooms, multipurpose rooms, product illustrations). This requires a minimum contrast ratio of 15:1.

- *Analytical decision-making* (ADM) is where the viewer is fully analytically engaged with making decisions based on the details of the content right down to the pixel level (e.g., medical imaging, architectural/engineering drawings, fine arts, forensic evidence, photographic image inspection). This requires a minimum contrast ratio of 50:1.

- *Full-motion video* is where the viewer is able to discern key elements present in the full-motion video, including detail provided by the cinematographer or videographer necessary to support the story line and intent (e.g., home theater, business screening room, live event production, broadcast postproduction). This requires a minimum contrast ratio of 80:1.

Figure 15-8
Contrast ratio
should be
measured from
five locations
within the
viewing area

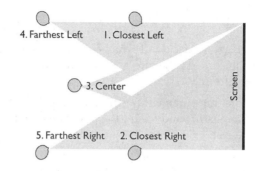

The standards include a simple measurement procedure to verify that the system conforms to the desired contrast ratio of the viewing task. First, you must identify the five measurement locations, as shown in Figure 15-8. The five locations are to be recorded on a viewing area plan.

- **Viewing location 1**   Viewing location closest to the image and farthest to the left in the plan view, as indicated in Figure 15-8.
- **Viewing location 2**   Viewing location closest to the screen and farthest to the right in the plan view, as indicated in Figure 15-8.
- **Viewing location 3**   Viewing location at the central point of viewing locations 1, 2, 4, and 5, as indicated in Figure 15-8. In the case where this central viewing location is obstructed (such as by a conference table), the measurement location will be the first available viewing location on the screen center line behind the obstruction.
- **Viewing location 4**   Viewing location farthest from the screen and farthest to the left in the plan view, as indicated in Figure 15-8.
- **Viewing location 5**   Viewing location farthest from the screen and farthest to the right in the plan view, as indicated in Figure 15-8.

After you have identified the viewing locations, measure the contrast ratio of the system using a 16-zone black-and-white checkerboard (intra-frame) pattern, as shown in Figure 15-9. You will also need a luminance meter or spot photometer with up-to-date calibration.

Figure 15-9
Checkerboard
pattern used
to measure
contrast ratio

PART V

The procedure for verifying image contrast ratio is as follows:

1. Display a 16-zone black-and-white checkerboard test pattern on the projection screen, as illustrated in Figure 15-9, under conditions that represent the actual viewing environment.

2. From the first measurement position identified on the viewing area plan (viewing location 1), measure and record the luminance values at the center of each of the eight white rectangles.

3. From the same measurement position, measure and record the luminance values at the center of each of the eight black rectangles.

4. Calculate the average of the eight white measurements and the average of the eight black measurements.

5. Divide the resulting average white value by the average black value to obtain the contrast ratio at that measurement position.

6. Repeat the contrast measurement procedure at each of the five measurement positions identified on the viewing area plan.

7. Record the resulting contrast ratios for each of the measurement positions on the viewing area plan.

   Contrast ratio = average maximum luminance / average minimum luminance

If the contrast ratio meets or exceeds the minimum laid out in the ISCR standard at all five viewing locations, then the system conforms to the standard. If the contrast ratio of one but no more than four measurement (viewing) locations falls below the required ratio for the identified viewing category by no more than 10 percent, the system partially conforms. If the contrast ratio at any one of the measured locations falls below the identified viewing category by more than 10 percent, the system fails to conform to the ISCR standard.

Audio/Video Sync

In many AV systems, the audio and video signals may take significantly different paths to get from the source to the point at which the user will experience them, such as in a seat within a venue, at a remote location, or distribution of recorded program material. The audio and video signals may also undergo very different numbers of processing steps, each with their own inherent latency.

In systems where associated video and audio signals may be transported or processed separately and are then subsequently combined for transmission or display, one of those signals could be delayed more than the other, creating synchronization errors sometimes known as *lip-sync errors*. These errors are typically corrected by applying delay to the least delayed signal so that it aligns with the most delayed signal. For a good user experience, the signals should arrive in synchronization at the point at which the user will experience them, regardless of whether that point is local or remote.

Sync Standards

As part of the metric selection process, the following documentation should be referred to in order of precedence to define the metric required for testing:

- ITU-R BT.1359-1
- EBU Recommendation R37
- ATSC IS-19

In the absence of defined requirements in the project documentation, the recommended maximum interval by which the audio should lead the video is 40 milliseconds and by which audio should lag behind the video is 60 milliseconds throughout an entire signal chain. This equates to approximately two or three frames on video systems.

For television applications, the recommended maximum synchronization error is audio leading by 15 milliseconds or lagging by 45 milliseconds.

A "blip-and-flash" test is a common method for measuring the synchronization of audio and video signals. A test signal consisting of one full frame of white video, accompanied by an audio tone of the same time length, is required. An alternative signal may be an image changing from full-frame white to full-frame black at regular intervals, with each change accompanied by a brief click or tone burst. In the absence of an appropriate signal generator, this can be produced by a computer running a slideshow. The signal at the point of reception can then be analyzed using a dual-channel digital storage oscilloscope to record the timings at which the audio and video signals are received.

If the sync alignment is not being measured, then a subjective reception test using the same test signal may be used, in addition to replaying video material that clearly depicts a person, such as a news presenter, speaking directly to camera. The subjective effect of any delay can then be evaluated for a pass/fail assessment.

Tests should be undertaken at the inputs of recording devices, at the inputs to transmission devices such as videoconferencing codecs and broadcast circuits, and locally within a presentation space at both near-screen and farthest-viewer positions.

Verifying Audio/Video Sync

After the AV system has been installed and all recording devices, presentation equipment, and link equipment/circuits are operational, the procedure for measuring the time alignment of the audio and video signals is as follows:

1. Set up a blip-and-flash source.
2. Measure delay (synchronization error) between audio and video at nearest and farthest viewer positions within the room.

PART V

3. Measure delay at inputs to all recording devices.

4. Measure delay at inputs to all transmission devices, such as codecs, network streaming interfaces, and broadcast circuits.

5. Note if the measured delay is within the specifications as stated in the project documentation or within +15 to –45 milliseconds (audio/video) in the absence of other information.

Alternative subjective tests may be undertaken using source video with no synchronization error, where a person is seen speaking. Validate subjective synchronization delay at locations listed earlier.

Correcting Audio/Video Sync Errors

Downstream video processing can cause delay. Some displays have built-in video processors, which, when used in conjunction with audio processors, scalers, and other processors, will cause delay in the signal.

Unfortunately, there is no clear way to predict how much, if any, delay will occur in an AV system. The latency introduced depends on the digital signal processors within the system, and the specific amount varies by manufacturer and device type. In some cases, you may be able to depend on the HDMI lip-sync feature, which will automatically compensate for delays in the video signal. In these cases, EDID can be used as a tool for correcting the sync issues. HDMI uses EDID to communicate delay information to upstream devices. During negotiation it will measure the delay introduced and send that information to the sink for automatic correction. This may be sufficient for systems with a single display, but with systems that involve multiple displays, when EDID detects different devices with different delay values, it is unlikely to be able to derive a single delay correction that will work for all sink devices.

In these cases, the integrator can manually introduce delay into an audio distribution system. Delay is a function typically found within digital signal processors (DSPs), switchers, or as a stand-alone, dedicated device.

Delay allows you to extract the audio at the beginning of the signal chain and send it right to the infrastructure. If you are responsible for selecting DSPs for a system, you should consider including a DSP that has the capacity to compensate for delay. Then, if you do encounter this problem, you can compensate for it by using a *blip-and-flash* test reel and adjusting the delay until the signals are back in sync.

Video Verification Checklists

The Video System Performance Verification Items and the Audio/Video System Performance Verification Items from the ANSI/AVIXA 10:2013 Audio Systems Performance Verification standard are listed here in Figure 15-10 and Figure 15-11, followed by a description of each list item. You may find some of these verification processes applicable to an installation you are undertaking, or you may find some processes that can form the basis of your own verification procedures. Verifying the performance of video systems according to established standards will help you assure overall system performance.

9.2 Video System Performance Reference Verification Items

Item Number	Item Name	Pre-integration	Systems Integration	Post-integration	Closeout
VP-100	EDID Management Plan	X	X		
VP-101	HDCP Management Plan	X	X		
VP-102	Projected Display Physical Alignment		X	X	
VP-103	Video System Pixel Failure Tolerance		X	X	
VP-104	Image Geometry		X		
VP-105	Displayed Image Performance	X	X	X	
VP-106	Colorimetry			X	
VP-107	Multiple Resolution Performance of Video Displays			X	
VP-108	Projected Display Brightness Uniformity			X	
VP-109	Projected Image Contrast Ratio			X	
VP-110	Test Video Routes			X	
VP-111	Video Camera Image and Operation			X	

Figure 15-10 Video performance verification items from the ANSI/AVIXA 10:2013 Standard for Audiovisual Systems Performance Verification

- **VP-100** Verify that the EDID (Extended Display Identification Data) management plan has been implemented as defined in the project documentation.

- **VP-101** Verify that the HDCP (High-bandwidth Digital Content Protection) management plan has been implemented as defined in the project documentation.

9.3 Audio/Video System Performance Reference Verification Items

Item Number	Item Name	Pre-integration	Systems Integration	Post-integration	Closeout
AVP-100	Emergency Communications		X	X	
AVP-101	Genlocking (Video Synchronization)		X	X	
AVP-102	Audio and Video Recording			X	
AVP-103	Audio/Video Sync			X	
AVP-104	Radio Frequency Television Distribution			X	
AVP-105	Source Testing			X	

Figure 15-11 Audio/video performance verification items from the ANSI/AVIXA 10:2013 Standard for Audiovisual Systems Performance Verification

- **VP-102**   Verify that the combined installation of projector and screen provides a displayed image that is correctly aligned to the active projection screen surface without misalignment unless an alternative condition is specified in the project documentation.

- **VP-103**   Verify that all displayed images do not have pixel failures (bright or dead pixels) that exceed the requirements of the project documentation or the manufacturer's specifications. This verification item shall require a metric to be verified.

- **VP-104**   Verify that all displayed images are correctly focused, have the correct image geometry and are free from distortion (e.g., stretching, keystone, barrel/pincushion). Any requirements for projection mapping or image shaping to unusual surfaces should be validated in accordance with the requirements of the project documentation.

- **VP-105**   Verify that the components of the displayed image system(s) (projection or direct-view) perform(s) as required with relation to image size, viewing angles, sight lines, viewer locations and/or any other requirements as defined in the project documentation.

- **VP-106**   Verify calibration of all video displays to ensure they display colors uniformly to a common reference standard as defined in the project documentation.

- **VP-107**   Verify that the system(s) accurately displays all resolutions required by project documentation on all displays within the system (i.e., no pixel shift, no geometric distortion, no artifacts from scaling, letterboxing, pillarboxing, or windowboxing).

- **VP-108**   Verify that the combined installation of projector and screen provides a display to the viewer that meets the requirements of the project documentation. This verification item shall require a metric to be verified.

- **VP-109**   Verify that the system conforms to the appropriate viewing category as defined in the project documentation. The testing methodology in ANSI/AVIXA V201.01:202X (and 3M-2011) shall be used. The projected image contrast ratio shall be measured for all projected images within the system. This verification item shall require a metric to be verified.

- **VP-110**   Verify that all video routes are tested from endpoint to endpoint via the appropriate midpoint(s) for operation and routing required by the project documentation.

- **VP-111**   Verify that cameras, lenses, and pan/tilt systems operate as defined in the project documentation. Inspect the camera image through the full lens operation.

- **AVP-100**   Verify that emergency communications systems properly receive inputs and information from other systems (including but not limited to life safety systems, security systems, and weather notifications), deliver appropriate notifications to target audiences, comply with regulatory requirements, and adhere to requirements defined in the project documentation.

- **AVP-101** Verify that the video synchronization of the system is performing as defined in the project documentation.

- **AVP-102** Verify that audio and video signals are being routed to the recording device(s) and that the recording device(s) is operating correctly, as defined in the project documentation.

- **AVP-103** Verify that audio/video synchronization is maintained to ensure the proper time alignment of signals during playback at the point of user experience or transmission as defined in the project documentation. This verification item should require a metric to be verified.

- **AVP-104** Verify that the radio frequency and satellite intermediate frequency distribution systems provide all services to all endpoints as defined in the project documentation. This verification item shall require a metric to be verified.

- **AVP-105** Verify that the signal produced by a source typical of what will be used in normal operation of the system is routed through the system to applicable endpoints and produces the performance as defined in the project documentation. A test generator shall not be used for this verification item.

Chapter Review

Verification that video components and the overall AV system perform as per specification is an important deliverable in an installation project.

Upon completion of this chapter, you should be able to do the following:

- Identify and address problems in the video system
- Identify HDCP concerns
- Verify that an AV system has an EDID strategy

After you ensure that the AV system you have installed works as expected, you are ready to perform system closeout, which you will learn about in Chapter 16.

Review Questions

The following questions are based on the content covered in this chapter and are intended to help reinforce the knowledge you have assimilated. These questions are similar to the questions presented on the CTS-I exam. See Appendix E for more information on how to access the free online sample questions.

1. Which of the following provides a method of ensuring the best image quality between source and sink/display?

   A. HDCP

   B. EDID

   C. HDMI

   D. Hot plug

2. During the handshake process, data from the EDID table is sent to _____.

   A. source devices

   B. sink devices

3. Which of the following information does the EDID packet contain? (Choose all that apply.)

   A. Timing for audio and video sync

   B. Display parameters

   C. Extension flags

   D. Color characteristics

4. What are possible reasons for no picture on the screen? (Choose all that apply.)

   A. The display device is not plugged in.

   B. The source device is unable to read EDID data packets.

   C. The hot plug pin is unable to detect a display device.

   D. A switcher is intercepting the hot plug detection.

5. Which of the following is a form of encryption to protect copyrighted content?

   A. VESA

   B. EDID

   C. HDMI

   D. HDCP

6. Which of the following are HDCP authentication steps? (Choose all that apply.)

   A. Locality check

   B. Key exchange

   C. Audio and video sync check

   D. Session exchange

7. Which of the following does HDCP 2.x support? (Choose all that apply.)

   A. 128 connected devices for each transmitter

   B. 4 layers or repeater levels for each transmitter

   C. 7 layers or repeater levels for each transmitter

   D. 32 connected devices for each transmitter

8. HDCP is capable of working over which of the following interfaces? (Choose all that apply.)

   A. DisplayPort

   B. HDMI

   C. MHL

   D. HDBaseT

9. According to the ANSI/AVIXA V201.01:202X Image System Contrast Ratio (ISCR) standard, what is the minimum contrast ratio of basic decision-making tasks?

   A. 7:1

   B. 15:1

   C. 50:1

   D. 80:1

10. What type of video is best used to verify audio/video synchronization?

    A. Landscapes with orchestra music

    B. News anchors speaking

    C. Team sports

    D. A *blip-and-flash* test video

Answers

1. **B.** Extended display identification data (EDID) is a method of ensuring the best image quality between video source and display.

2. **A.** Data from the EDID table in the display device is sent to the source device.

3. **A, B, C, D.** EDID packets contain timing for audio/video sync, display parameters, extension flags, and color characteristics.

4. **A, B, C, D.** If the display device is not plugged in, the source is unable to read EDID data packets, the hot plug pin is unable to detect a display advice, or the switcher intercepts the hot plug detection, then a display device may not show an image on the screen.

5. **D.** High-bandwidth digital content protection (HDCP).

6. **A, B, D.** Locality check, key exchange, and session exchange are HDCP authentication steps.

7. **B, D.** HDCP 2.x supports up to 32 connected devices for each transmitter and a maximum of four repeater levels for each transmitter.

8. **A, B, C, D.** HDCP is capable of working over all of these video interfaces.

9. **B.** 15:1 is the minimum contrast ratio for images used in basic decision-making tasks.

10. **D.** A *blip-and-flash* test will reveal audio or video latency.

PART V

CHAPTER 16

Conducting System Closeout

In this chapter, you will learn about

- The verification standard for system closeout
- Refuse management and cleanup
- Closeout documentation
- Customer training
- Obtaining customer sign-off

After verifying that the system is working, you must demonstrate to the client or the client's representative that it meets the performance specifications of the audiovisual (AV) design. To obtain customer sign-off, you will need to complete several tasks as part of the system closeout process.

You will need to complete a site cleanup process, which includes following proper refuse management. Handing over documentation to your client is an essential part of system closeout. If troubleshooting or requested changes remain undocumented or incomplete, the project remains open and could require additional staff hours to resolve. This could possibly delay payments and project conclusion. On some projects, your work may involve training the client's staff on how to operate the systems.

Review the verification items in the Audiovisual Performance Verification Standard again before you seek customer sign-off. After you obtain the sign-off, your handshake with your client indicates successful completion of the AV installation.

Duty Check

This chapter relates directly to the following tasks on the CTS-I Exam Content Outline:

- Duty D, Task 1: Demonstrate to Client or Client's Representative that the System Performs to Specifications
- Duty D, Task 2: Provide Training on System Operation
- Duty D, Task 3: Obtain Project Completion Sign Off from Client or Client's Representative
- Duty E, Task 1: Complete Progress Reports
- Duty E, Task 2: Coordinate with Other Contractors
- Duty E, Task 3: Address Needed Field Modifications

These tasks comprise 22 percent of the exam (about 22 questions). This chapter may also relate to other tasks.

Performing Site Cleanup

Before leaving the installation job site, clean up the space you worked in. Tidy all loose cables, vacuum the floors, wipe down equipment, replace any furniture or fittings that you moved, and straighten up anything that looks messy. A clean job site is a professional job site, which leads to future jobs and referrals.

TIP It may prove useful to comprehensively document the condition of the site at the completion of the AV cleanup by taking a set of photographs.

Refuse Management

Check with the builder/building contractor/general contractor to determine proper disposal methods. You may need to consider the following:

- Contractual disposal requirements.
- Contracted salvage rights.
- Where and how to dispose of materials.
- Safe and legal disposal of toxic and hazardous materials from old equipment, demolished structures, or wastage from the installation. Many jurisdictions have strict environmental stewardship requirements for the disposal of hazardous materials.

- Waste diversion: the amount of waste disposed of other than through destruction or in landfills, expressed in weight or volume. Examples of waste diversion include reuse and recycling.

- Use of chemicals and aerosols onsite in relation to volatile organic compounds (VOCs), usually measured in grams per liter (g/L).

NOTE Volatile organic compounds from many materials, including adhesives, sealants, paints, carpets, and particleboard, vaporize into harmful gases. Limiting VOC concentrations protects the health of both construction personnel and building occupants.

Prior to disposing of waste on a job site, the installer should check whether they have access to any onsite rubbish receptacles/dumpsters before using them. Some job sites may ask you to work with dedicated personnel for disposing work site debris, so coordinating with them will accelerate your cleanup procedures. If the arrangements are unclear, the safest and most responsible option may be to return the waste to your own base of operations for proper disposal.

Cleaning a Screen

The AV installation should look clean and fresh when your client walks in. Flat-panel display screens will pick up a lot of dust and dirt over the course of the installation, so you should clean them before inviting the client in.

Direct-view flat-screen panels are easy to scratch or damage. You will need to take care to clean them gently. To clean a flat-screen display:

1. Turn off the device; it is easier to see dust and dirt on a dark background.

2. Using a dry, soft cloth, preferably made of microfiber, gently wipe the screen.

3. If dust or oil remains, dampen the cloth with distilled water. Alternatively, some companies sell spray bottles of screen cleaner; you may choose to use this instead. If you are in doubt about the appropriate cleaning agent, refer to the product documentation or to the manufacturer.

4. Gently rewipe the screen. Move the cloth in a single direction, either up and down or side to side. This will minimize the appearance of any messy streaks.

Do not press hard on the screen while you are cleaning it; pressure can damage flat-panel displays.

Do not spray liquid directly onto the screen. If liquid gets inside the display electronics, it may cause damage.

Closeout Documentation

The importance of documentation at every stage of an AV project can scarcely be overstated. From the first meetings through installation and verification, documentation of every change is crucial for several reasons, including future maintenance of the system.

PART V

Closeout documentation includes all the information gathered during the project, as well as drawings of record, operational documentation, and possibly a completed punch/snag/problem list.

Verification Standard for System Closeout

Adhering to the standard ANSI/AVIXA 10:2013 Audiovisual Systems Performance Verification can help you smoothly close out your project. This standard requires you to meet with project stakeholders to determine what verification tests will be performed on the system and what metrics will be used to assess system performance. It includes guidance as to when in the installation process each step should be performed. The standard requires the installation team to turn in verification testing reports for each phase of construction.

When you use this standard to manage system verification, you are also forming a shared understanding with your client and other stakeholders of what constitutes a complete, fully functional system. Your performance and outcome expectations are explicitly stated and aligned early in the project. This reduces risk by identifying potential problems or disagreements early, thereby reducing the need for remedial work. The standard also helps both you and your client verify the project outcome by mandating documentation of each stage of construction and testing as it is completed. Both you and the client know when the system is ready for use.

The System and Record Documentation items from the ANSI/AVIXA 10:2013 Audio Systems Performance Verification standard are listed here in Figure 16-1, followed by a description of each list item. You may find some of these verification processes applicable to an installation you are undertaking, or you may find some processes that can form the basis of your own verification procedures.

- **DOC-100**   Verify that all equipment has been delivered as defined in the project documentation.
- **DOC-101**   Verify that samples of all equipment to be used as defined in the project documentation have been submitted for approval.
- **DOC-102**   Where samples of products have been required for approval, verify that the products that are delivered are the same and of the same quality.
- **DOC-103**   Verify that the correct and valid wireless frequency licensing permits have been obtained for legal operation of the system.
- **DOC-104**   Verify that any consultant's testing requirements defined in the project documentation have been performed and approved.
- **DOC-105**   Verify that any builder/building contractor/general contractor's testing requirements defined in the project documentation have been performed and approved.
- **DOC-106**   Verify that any integrator's testing requirements have been performed and approved as defined in the project documentation.

9.6   System and Record Documentation Reference Verification Items

Item Number	Item Name	Pre-integration	Systems Integration	Post-integration	Closeout
DOC-100	Final Inventory of AV Equipment	X	X	X	
DOC-101	Approval of Samples	X			
DOC-102	Delivered Product Against Samples		X		
DOC-103	Wireless Frequency Licensing		X		
DOC-104	Consultant's Testing			X	
DOC-105	General Contractor's Testing			X	
DOC-106	Integrator's Testing			X	
DOC-107	Manufacturer's Testing			X	
DOC-108	Owner's Testing			X	
DOC-109	Third-Party Testing			X	
DOC-110	Substantial/Practical Completion			X	
DOC-111	As-Built Drawings Complete				X
DOC-112	Audio System Test Reporting				X
DOC-113	Control System Test Reporting				X
DOC-114	Final Commissioning Report and System Turnover				X
DOC-115	Required Closeout Documentation				X
DOC-116	Software Licensing				X
DOC-117	User Manuals				X
DOC-118	Video System Test Reporting				X
DOC-119	Warranties				X
DOC-120	Final Acceptance				X

Figure 16-1   System and Record Documentation verification items from the ANSI/AVIXA 10:2013 Standard for Audiovisual Systems Performance Verification

- **DOC-107**   Verify that any manufacturer's testing requirements defined in the project documentation have been performed and approved.

- **DOC-108**   Verify that any owner's testing requirements defined in the project documentation have been performed and approved.

- **DOC-109**   Verify that any third-party testing requirements have been performed and approved as defined in the project documentation.

- **DOC-110**   Verify that a conditional acceptance of the project has been issued by the owner or owner's representative, acknowledging that the project or a designated portion is substantially/practically complete and ready for use by the owner; however, some requirements and/or deliverables defined in the project documentation may not be complete.

- **DOC-111**   Verify that a complete set of accurate as-built drawings indicating all AV devices, AV device locations, mounting details, system wiring and cabling interconnects, and all other details has been provided as defined in the project documentation.

PART V

- **DOC-112** Verify that the audio system test report has been completed and issued as defined in the project documentation.

- **DOC-113** Verify that the control system test report has been completed and issued as defined in the project documentation.

- **DOC-114** Verify that the final commissioning report has been completed, issued to the proper entity, and accepted as defined in the project documentation.

- **DOC-115** Verify that a complete set of as-built system documentation has been provided as defined in the project documentation.

- **DOC-116** Verify that the usage and ownership rights have been assigned as defined in the project documentation.

- **DOC-117** Verify that manufacturer's user manuals are delivered to the owner in a format defined in the project documentation (e.g., binders, PDFs), or dispose of the manuals in a responsible manner (recycling) if the owner specifies that they do not wish to receive the manuals.

- **DOC-118** Verify that the video system test report has been completed and issued as defined in the project documentation.

- **DOC-119** Verify that all warranties are activated and that all warranty details have been passed to the owner as defined in the project documentation.

- **DOC-120** Verify that a final acceptance of the project has been issued by the owner or owner's representative, acknowledging that the project is 100 percent complete, that all required deliverables, services, project-specific verification lists, testing, verification and sign-offs have been received, and that all requirements defined in the project documentation have been satisfied and completed.

A critical factor in ensuring that documentation deliverables are up to date is determining who is responsible for each document. You must understand both your specific role and the tasks involved in your team's closeout documentation. Knowing who is responsible for which items will enable you to successfully gather all of the relevant documentation for closeout.

Drawings of Record

After an installation has been completed and verified, the installation team should hand over comprehensive documentation, including the *drawings of record,* to the project manager.

Drawings and documents of record reflect changes made onsite to the original AV design and represent the actual installation as it exists at the point of handover. The original plans often cannot take into account product changes that may require changes on the job site. By carefully keeping track of changes to wiring, changes in model numbers of devices, audio and video signal modifications, and other system design elements, these drawings of record become valuable references during service calls and routine maintenance in the future.

Document every single change, even if it was just changing a connection point. Note the changes on the system plans. This step is vital not only in communicating system changes to all parties currently involved in the project but also to those who will work on changes to the system in the future.

Operational Documentation

Operational documentation should be included with a contract at the end of a project. Even if user documentation is not specified in a contract, training or job reference manuals should be prepared for the customer.

System documentation may include the following:

- System-specific operating instructions and manufacturer equipment manuals
- System design and block drawings
- Control system configuration, scripts, settings, and MAC and IP addresses
- DSP software and configuration files
- Network device inventory information
- Drawings of record
- Spreadsheet listing all equipment provided, including model numbers, firmware versions, serial numbers, and physical locations
- Description of recommended service needs and schedules of maintenance
- Warranty information (both manufacturer and AV system as a whole)
- Support telephone numbers and/or URLs

Other deliverables should include remote-control devices, adapters, cables, spare parts, and media (CDs, DVDs, flash, or cloud storage) with software programs and user documentation.

Best Practice for Equipment Manuals and Software Documentation

To ensure ease of system maintenance, organize system documentation as follows:

- Arrange equipment manuals in alphabetical order and group by function. For example, all display device manuals would be placed in one group.
- Leave a copy of the equipment operation documents with the equipment rack, the project manager, or the client's support personnel or end user.
- Create backups of all computer files associated with the equipment, especially if the files contain records of the system settings.

Punch/Snag/Problem List

Since the technician's role in the closeout process may be to answer questions about the installation and make corrections as they are discovered, a punch/snag/problem list is useful. It contains the corrections or changes that need to be made to conform an installation to the scope of the project.

Some of the items such a list may contain include

- Minor changes such as correction of illegible labels on equipment inputs
- Physical changes such as moving a wall-mounted connector to a higher location
- Functional changes such as adjusting a monitor
- Programming changes such as renaming input buttons on the control UI
- Technical changes such as rewiring a rack that was not properly dressed

Even if your client does not require this document, the use of such a list has become standard practice in the construction industry. In fact, it could be the document that protects you and your company from any future misunderstandings.

Be sure to have all your equipment, test gear, and tools available so you are able to make some changes on the spot. Spare connectors, wire ties, and cabling are also helpful. When all the work on the list cannot be completed immediately, make plans for follow-up and arrange to meet with the project manager to review the changes as soon as possible.

Customer Training

After you have completed all the procedures, work with the client to schedule a training session for the typical users of the system.

There may be several levels of training required for different groups of users. Each training session should be catered to the needs and interests of the audience.

Different training groups might include the following:

- Technical staff who will be responsible for operations in the future and require in-depth training on all systems
- Help desk or support staff who will be responsible for new-user or inexperienced-user hand-holding
- Power users who take full advantage of a system's functionality
- Casual users who make limited use of the system's functionality

You may also want to conduct in-depth "train the trainer" sessions with individuals from within the owner organization so that training can continue in the future.

First, study the system. If you are training people on the system's operations, you should be completely familiar with them yourself. Lack of familiarity or uncertainty on the part of the trainer can result in similar uncertainty in the trainees. More operational mistakes are likely if the trainer is uncertain or ineffective.

The client briefing should be informal. It is preferable to demonstrate operations by working with the newly installed system in front of the end users. Sometimes this will be done with the designer as part of the commissioning process. You, your project manager, or a sales team member may also brief the client.

Not all users require the complete and detailed operating procedures. Identifying the appropriate end users for extended training is most important.

It is also necessary to provide some level of documentation to customers describing how to operate a system. Detailed operating instructions for every piece of the AV system are only one portion of the total package.

End-user training should at least include the following:

- How to turn the system on and off
- How to configure and initiate a multi-site communications session
- How to switch between the various program sources
- How to start, pause, rewind, and stop the various program sources, including cloud streams
- How to turn on and mute microphones
- How to adjust the volume levels of audio sources
- How to use the control system interface
- How to adjust internal HVAC and lighting presets and settings

When creating this documentation, also keep in mind that presenters using the system may only have time to glance at instructions. They will only need simplified, bullet-point instructions.

You may at times need to connect your customer with the manufacturer or distributor for additional training or support during closeout. The technology manager in charge often requires more extensive training than you may be able to provide. Remember your steps for escalating questions or concerns to address this type of briefing.

During training, some end users may be disappointed that the system does not include functions or features that they desire, or may request the trainer to add functions or features to the system that lie outside the system's scope or design specifications. Such requests for changes to the system should be directed to the project manager for consideration through the formal change request process.

Best Practice for User Training

Create an outline of the topics you are going to cover. An outline will keep you on track and make sure you do not overlook important points. It will also engage the trainees in taking notes on points they need to remember.

It is best to conduct training as a separate scheduled event for the appropriate end users so that they are able to focus their attention on learning how to use the AV system.

Customer Sign-off

To obtain client sign-off, you should review the overall project. Before beginning this review process, you need to have completed the following tasks:

- You have completed a step-by-step approach to testing all the parts of the AV system.
- You have identified any problems or issues associated with the installation and setup of equipment and have addressed them prior to turning over the system to the customer.
- You have provided a means for your company to demonstrate to the customer that the installed system meets both design and performance specifications and standards.

Either at or following the training, ask the consultant or client to sign closeout documentation as specified. Having everything at the training provides an opportunity for sign-off with all parties present.

Chapter Review

In this chapter you studied the tasks that need to be completed before customer sign-off, which indicates successful completion of the project.

Upon completion of this chapter, you should be able to do the following:

- Identify the AV deliverables to the customer at the end of an installation project
- List the cleanup tasks that an installer must complete on a post-installation job site
- Conduct training for end users on AV equipment operation

Review Questions

The following questions are based on the content covered in this chapter and are intended to help reinforce the knowledge you have assimilated. These questions are similar to the questions presented on the CTS-I exam. See Appendix E for more information on how to access the free online sample questions.

1. Which of the following is an essential deliverable before sign-off?
   A. A punch/snag/problem list
   B. A clean site
   C. Referral solicitations
   D. Redundant cables

2. What should you do with equipment manuals and software documentation? (Choose all that apply.)

   A. Leave a copy with the AV rack

   B. Leave a copy in the storage closet

   C. Arrange them in alphabetical order

   D. Arrange them by function

3. What should be done with empty cardboard equipment boxes?

   A. Returned to the manufacturer

   B. Stored in a closet

   C. Discarded in the dumpster/waste container

   D. Discarded where indicated by the client

4. Which topics should *not* be covered in an end-user training session? (Choose all that apply.)

   A. How to turn the system on and off

   B. How to change the IP addresses of the devices

   C. How to switch between the various program sources

   D. How to use the control system interface

5. What is a benefit of using the verification standard for system closeout? (Choose all that apply.)

   A. It aligns outcome and performance expectations at an early stage in the project.

   B. It reduces project risk through early identification of problems, thereby reducing the likelihood of remedial work.

   C. It provides a verifiable outcome.

   D. It creates a common language between all parties.

Answers

1. **B.** Site cleanup is an essential deliverable before sign-off.

2. **A, C, D.** Best practices when compiling equipment manuals and software documentation include leaving a copy of the equipment manuals with the AV rack and arranging the manuals in alphabetical order and by function.

3. **D.** Empty cardboard equipment boxes should be discarded where indicated by the builder/building contractor/general contractor, or client.

4. **B.** End users should not be trained in reconfiguring the system. Training for these tasks should be given only to the technical staff.

5. **A, B, C, D.** All are benefits of using the verification standard for system closeout.

PART V

CHAPTER 17

Maintaining and Repairing Audiovisual Systems

In this chapter, you will learn about
- Troubleshooting audiovisual systems
- Devising a preventive maintenance schedule
- Creating a maintenance log
- Maintaining audiovisual tools

After completing and handing over an audiovisual (AV) project to the client, the project's working life begins in earnest, providing AV services and facilities to the end users. Most installation contracts include a period of warranty for the installed equipment and its configuration. Some also include an allocation of technical support hours during the bedding-down period for the new installation.

The AV installer's duties may include supporting the installation by diagnosing and resolving any equipment, configuration, or operational problems that may arise during the entire working life of the system. They may also be responsible for maintaining the system in optimal operating condition through a preventive maintenance program that preempts the faults and failures that can arise through the normal use and operation of the installed system.

It is important that both you and the client have a clear understanding of the extent of the contractual arrangements to provide warranty support, operational support of the facilities during the post-warranty life of the installation, preventive maintenance to reduce the likelihood of operational failures during the life of the installation, and modifications and upgrades to the installation that are outside the scope of the original system installation contract.

Your tools of installation trade include some potentially dangerous devices that can slow you down and cause you harm if not well maintained and some test and measurement devices that need loving care to remain serviceable, accurate, and fit for purpose.

Duty Check

This chapter relates directly to the following tasks on the CTS-I Exam Content Outline:

- Duty E, Task 4: Repair Audiovisual Systems
- Duty E, Task 5: Maintain Audiovisual Systems
- Duty E, Task 6: Maintain Tools and Equipment

These tasks comprise 10 percent of the exam (about ten questions). This chapter may also relate to other tasks.

Troubleshooting

There are going to be times when an AV system will not work correctly. When this happens, you will need to troubleshoot the systems to find where the problems are located before carrying out repairs or adjustments.

In a high-pressure situation, such as the sudden failure of the AV system during a presentation, conference, or live event, AV professionals are expected to calmly and quickly diagnose the problem and then either fix the problem or apply a workaround until permanent repairs can be made.

As you learned in Chapter 11, troubleshooting is a process for investigating, determining, and settling problems. The most important process in troubleshooting problems is to be systematic and logical. A reliable troubleshooting approach follows a process intended to clearly define the nature of the problem and narrow down the potential issues until the actual cause of the problem is identified. There are three steps involved in troubleshooting:

1. Symptom recognition and elaboration where normal and abnormal system performance is identified through consultation with the end user or their technical team.

2. Listing and localizing the faulty functions and identifying which systems and subsystems could produce the identified symptoms.

3. Analysis of the fault based on the information elicited in the previous steps to identify the nature and location of the fault. This process includes failure analysis to establish the probable causes of the fault to provide feedback about possibly preventing a recurrence.

Preparing to Troubleshoot

Prior to attempting to troubleshoot a system problem, you need to do the following:

- Take steps to familiarize (or refamiliarize) yourself with the AV systems and equipment, either as part of a preventive maintenance schedule or prior to an event. In many cases, an AV technician is under pressure to return a system to operation as quickly as possible, especially during an event. The troubleshooting and repair process will be more successful if you have already reviewed the system design and operation and know how individual devices should work, rather than learning about the system after you arrive on the scene.

- Have the project and/or system documentation available. System documentation will prove valuable if you need to examine drawings and trace component interconnections or determine appropriate system and component settings. And you may need equipment manuals to look up specific troubleshooting and repair procedures. All of these documents should be part of the project documentation provided during closeout and handover.

- Bring the proper tools and supplies. In many cases, you will need to test the output of devices, open the cases of individual components, and so on. This will require tools, such as the following:

  - A multimeter

  - Signal generators and signal analyzers

  - A laptop/notebook computer with a range of test software and signal adapters

  - Mechanical tools such as screwdrivers, cutters, and spanners to open equipment cases or remove components from racks

  - Tools to install new terminations and connections

  - A supply of standard connectors

  - A broad and inclusive selection of signal cables

  - A pair of headphones

- Understand the warranty and maintenance agreements. The service and warranty agreements covering the AV system and its components may dictate how repairs or replacements should be addressed. In cases where warranties cover the components and/or the overall system—whether they are manufacturers' warranties or the AV company's warranty—you should repair or replace components according to the terms of those agreements. In cases where the client is paying your company directly for service and repairs, you may need to explain the issues to the client and obtain approval prior to addressing more costly system faults.

PART VI

Selecting a Strategy

The next step in troubleshooting the problem is to apply the appropriate troubleshooting strategy. Strategies used to identify faults include

- **Swapping**   While this may not seem like a formal, step-by-step process, swapping the suspected faulty component with a known good component can help identify problems. Prior experience with similar systems and their problems may quickly lead the experienced technician to a faulty component. For example, swapping out a small loudspeaker that isn't working with another small loudspeaker that you know is working divides the problem into smaller parts and indicates whether the original loudspeaker is the cause.

- **Divide and conquer**   This strategy involves dividing a problem or system in half to define the half that is not working. Then divide the faulty half again and test for the failure point, repeating the process until you identify the source of the failure. In just a few steps, you can quickly identify a faulty component, even in a complex system.

- **Signal flow**   Start at one end of the signal path and continue to follow the signal path until you find the fault. With audio, you could begin at the source (such as the microphone or replay device) and continue all the way to the destination (such as the loudspeaker) until you find the faulty component (such as a cable, device, connection, or termination). Troubleshooting a video problem may begin at the sink device (the projector or display) and follow the signal path all the way back to the source. Of course, you could also do it the other way around.

Best Practice for Troubleshooting

The following are some basic troubleshooting best practices:

- Assume nothing. Anything that *can* go wrong *will* go wrong.
- Similar symptoms don't always have the same causes (see: assume nothing).
- Change only one thing at a time. If that doesn't resolve the problem, change it back.
- Test after each change.
- Use signal generators in order to provide a known reference to measure against.
- Use signal analyzers to obtain quantifiable information.
- Document your procedures and findings.

Troubleshooting in Live Events

In live event staging, there is often little time to implement a solution when a technical problem occurs. Troubleshooting starts before the gear leaves the warehouse by testing all equipment before dispatch and by determining what spare parts should be carried and which critical systems should be redundant.

On the site, the technical crew has tools, troubleshooting devices, and spare equipment specific to the systems involved. Troubleshooting checkpoints are established in each system for quick location of a signal or equipment failure.

When traveling or touring, a local AV supplier should be secured in advance as backup for equipment and services.

System Maintenance

If your company has been contracted for post-installation maintenance of a system, you may be required to provide ongoing support services. In a maintenance and repair role, your team will be required to maintain, troubleshoot, and repair AV equipment or systems.

Preventive Maintenance

Preventive maintenance is an equipment strategy that is based on inspection, replacement, servicing, or dismantling of AV equipment at a fixed interval, regardless of its condition at the time.

Preventive maintenance may include

- Checking hardware systems
- Checking and calibrating system levels and gains
- Checking signal cables and replacing bad cables or connectors
- Checking for available software and firmware updates and possibly installing them
- Cleaning critical system hardware
- Checking and adjusting power supplies
- Checking the status of system batteries and replacing them when necessary
- Checking mains power input cables
- Checking the status of light sources and replacing them where necessary
- Rebooting systems

Devising a Preventive Maintenance Schedule

The frequency and extent of preventive maintenance activities should be based primarily on the level of use of the installation and the types of equipment installed. Detailed information on the hours of use, and the types of uses, for each section of the installation is necessary to make an accurate estimate of the likely maintenance requirements. If equipment is in constant use, the schedule will require more frequent preventive maintenance, but on the other hand, site visits will probably need to be scheduled outside of normal operational hours to avoid disruptions.

Maintenance Requirements

Before considering the frequency of preventive maintenance activities, it is important to establish which parts of the installation require maintenance. Some devices may sit undisturbed in an equipment rack, suspended from a ceiling, or under a bench, quietly going about their business with no direct user intervention, while others may be moved about the installation several times each day and be connected to by dozens of users, plugging and unplugging a myriad of different personal devices.

Cooling Systems   Some devices are designed to operate entirely with passive cooling, relying on heat removal by radiation from heatsinks or by convection through ventilation holes or mesh. Passively cooled devices require regular but infrequent removal of dust and fibers that may accumulate on heatsinks and in ventilation apertures to avoid the device overheating and becoming unreliable.

Devices that are actively cooled with an internal forced air flow require two distinct maintenance interventions: filters and fans. As actively cooled equipment requires higher volumes of air movement than passive convection cooling, this equipment is usually fitted with dust and/or particle filters on the air intake to keep the air flow clean. Depending on the operating environment, these intake filters may require frequent cleaning or replacement to avoid overheating problems. Problematic environments for air intake filters include areas of high pedestrian traffic; food preparation areas with substantial amounts of smoke and/or grease vapors; rooms with chalkboards; places frequented by smokers; and venues such as night clubs, performance spaces, or exhibitions, where atmospheric effects including pyrotechnics, theatrical smoke, fog, and haze are used.

One factor to consider about system cooling health is the change of ambient environment that occurs when a facility is closed down at the end of an operational session. Devices that can comfortably stay within their safe operating temperature range during a normal operational session may find themselves in a space with minimal ventilation and cooling when all the humans pack up and go. If equipment is left after hours in full running mode, or even in warm standby, it may actually place more demand on the cooling system than when in normal operational use, and thus shorten the operational life of equipment.

Moving Parts   The other issue with forced-air cooling is the operating life of the fan or impeller devices forcing the air. Together with the phosphor wheels on some solid-state light sources and the color filter wheels on some single-DMD (single-DLP) projectors, these fans and impellers are among the very few moving parts found in most AV systems. Bearing failures on the drive motors may be the most likely point of failure on many pieces of gear, so some research into the estimated operational life of these motors should inform your estimates of service intervals.

Your ears, possibly augmented by an electronic stethoscope, are a very useful tool for assessing the health of motor bearings, as vibrations, whines, and grinding noises can usually be detected some time before the bearings fail completely.

Physical buttons and switches, as distinct from the virtual buttons on GUIs, are another moving part that can fail after repeated use, and checks should be factored into your preventive maintenance regimen.

Limited-Life Components   Some components of every AV system have a finite life, and any preventive maintenance regimen should factor the lifespan and replacement procedures for these components into the schedule and the budget.

Projector light sources are a familiar system component that requires regular checks of light output and lamp-hour counters. A preventive maintenance program should be based on the estimated life of the light source and the acceptable decrease in projector output or a specified lamp-hour count that will trigger a lamp replacement before failure.

Some projectors may have solid-state light sources with operational lives that should theoretically exceed the length of an equipment replacement cycle; however, solid-state light sources are particularly temperature sensitive and can fail, or be substantially degraded in either output or operational life, if the cooling system for the light source is compromised.

Batteries are a constant and lifelong maintenance issue with a wide range of AV devices. Although primary cells (single-use batteries) are an operational consumable rather than a maintenance issue, the vast number of portable, rechargeable devices in use is a significant preventive maintenance consideration. A preventive maintenance plan should take into account the entire inventory of rechargeable batteries in everything from wireless microphones, radio talkback units, and handheld radios, to collaboration dongles, handheld video cameras, audience response devices, laptop computers, and, of course, the forgotten army of uninterruptible power supplies (UPSs) lurking in the bottoms of racks and under consoles.

A battery maintenance strategy should take into account the usage hours of each battery-powered device together with the device manufacturer's specifications for the number of recharge cycles that the fitted batteries should be expected to perform. Batteries and their associated charging systems should be subject to periodic testing for voltages and the condition of the recharging connections on cables, charging stations, and the device's charging terminals.

UPS batteries are usually replaced at regular intervals based on the battery type and the manufacturer's documentation, although regular checks of UPS performance, in the controlled absence of an incoming mains supply, ensures that the UPS will indeed supply uninterrupted power when called upon to perform.

Cables and Connectors   It is widely acknowledged that the majority of electronic hardware failures are the result of faulty connections. Many of the cables and connectors used in AV systems are rated for only a few thousand (or sometimes even fewer) connect/disconnect cycles, and while this is rarely a significant limitation in a fixed and infrequently changing installation, it can become a critical source of unreliability and failures for systems being constantly moved and reconfigured. A preventive maintenance program should include regular visual inspections and continuity tests on the cables and connectors, on and between equipment subject to frequent connect/disconnect cycles, and the potential for physical mishandling.

Software and Firmware   Device firmware, operating systems, and AV system software are frequently updated—sometimes to add extra capabilities to devices and systems, but more often to remedy bugs and other issues that have arisen since the previous version was released. If a preventive maintenance schedule is to include the acquisition and installation of bug fixes, updates, and upgrades, it is important to establish guidelines for what type of updates are to be undertaken on each device in the system. Clients may prefer to leave some system elements in their current "if it ain't broke, don't fix it" state or may choose to keep elements up to date with the latest features, bug fixes, and security patches. It is not uncommon for clients to opt for stability and security with some critical systems, while wanting all the latest bells and whistles installed on others.

A preventive maintenance program may include periodic equipment shutdowns and restarts for systems that are otherwise left in continuous operation.

Fallback and Failover Systems   AV systems may include redundant duplicate (or even triplicate) components to provide fallback and failover contingencies for critical operations. A preventive maintenance program may include checking and testing the offline redundant devices and simulating, or deliberately creating, a controlled fault event that will verify the functioning of both the changeover system and the redundant equipment. The program may also include provision for rotating or exchanging the active and standby devices.

Creating a Maintenance Log

A maintenance log is essential on fixed installations. It should include every item examined and repaired and the service performed. Maintenance logs are easy to update, especially right after service has been performed.

Table 17-1 shows the types of data recorded, including the date and time the system was checked, who checked it, and the status of the system. It also includes problems that were discovered and what measures were taken to fix them. You should make it standard practice to record corrective measures that were taken to restore system performance or recommended actions and whether they are acted upon or not. From time to time equipment may need to be updated. For example, in a boardroom, a digital media player may replace a Blu-ray player. The date this took place should be documented, as well as how the signal was rerouted.

As you may have noticed in Table 17-1, the installers established a good relationship with their client, which opened the door for repeat business and referrals. When you do a great job installing a system, the client may choose to contract with your company to provide long-term maintenance on the equipment.

Date	Time	Initials	Work Performed
27/05/21	10 a.m.	AJB	Checked in with the onsite client before testing 1-year maintenance on Ballroom A. Tested for proper levels throughout the audio system using a meter and signal generator. All systems checked out OK. Noticed that fan in audio system rack was not operating. Recommended to client to call repair (800.123.4567) and speak with Tom for replacement. Before leaving at 3 p.m., secured all rooms and checked out with client.
16/06/21	8 p.m.	AJB	EMERGENCY CALL TODAY. Digital Signal Processor failed. Had to reference user's manual to get part number to replace the failed unit. Happened to have a spare device in the truck so replaced it. Reported to client status of problem. Will return to replace when part comes in. Left at 2 a.m.
20/07/21	8 a.m.	JAR	Normal tech not available today (AJB). Replaced DSP after notifying client. Had to get key from security guy to get access. Lucky break—DSP data configuration file was on USB backup flash drive! AJB left directions on how to upload crosspoints on DSP. Verified that DSP was operating within normal parameters. No problems Checked out with client at noon. Client said he was impressed with our work and has a friend who wants our company to do a retrofit install—told him I'd have sales dept. contact him.

Table 17-1 Type of Information in a Maintenance Log

Maintaining Your Tools

Keeping your installation, maintenance, test, and alignment equipment in safe working order is an essential process that underlies every aspect of AV technical activity. Every task you undertake requires that not only is every tool up to the task you require of it but that it will also be safe to use.

Staying Sharp

Many of your tools, from drill bits to jigsaws, from cable cutters to utility knives, and from wire strippers to angle grinders, require sharp and correctly aligned cutting edges to function effectively and safely. Some of these tools require regular sharpening and alignment, while others can be kept ready for use with replaceable cutting blades or discs. If a tool is losing its edge and underperforming, it is time to consider whether the tool is being used for tasks beyond its intended capabilities or whether it has simply become worn from use. Whichever is the case, the cutting capability should be restored by either resharpening and realignment or by replacing the disposable cutting edge. Not only does having a fully functional tool improve your efficiency at a task, but it also decreases the risk of the tool jumping or skipping in operation and putting you at risk of an injury.

It is your responsibility to have adequate tool replacement parts available to complete your tasks, to have an understanding of the processes required to acquire additional replacement parts, and to have tools resharpened, realigned, or replaced.

Staying Safe

No AV project, however critical or spectacular, is worth dying for. An important part of every project is to complete it to the highest possible standard—and then go home safely. There are a vast number of potentially dangerous tasks in every occupation, but acknowledging those risks and taking all the necessary steps to avoid dangerous outcomes is an important skill.

The requirement for the safe use of appropriate personal protective equipment (PPE) is discussed in Chapter 5 and mentioned in other sections where specific risks and safety practices are important. The use of the appropriate protective and safety practices should be an integral part of all work activities.

Preparation of your tools for each task should include checks for tool safety before each use. Visual checks of the condition of the tools should include verification that all safety guards are in place and operational, that all parts are present and undamaged, and that electrical supply plugs and cables are correctly terminated and in a safe condition.

Although many powered tools are driven by internal, low-voltage, rechargeable batteries that pose no electrical hazard, there are a substantial number of devices that require a mains supply for operation. It is good practice to inspect, and if indicated, test the wiring and insulation on mains-powered equipment before every use.

Many jurisdictions require regular inspection and testing of all mains-powered devices on a work site, including cables and distribution equipment. Most also require that labels be attached to the equipment to record when it was last tested and that records be maintained to track the safety history of each device. Building construction sites are often subjected to very stringent safety practices due to the high risks associated with building activities. Some jurisdictions categorize live event production as a construction activity from a safety perspective.

Staying Accurate

An AV installation and maintenance toolkit includes many devices used to measure, test, and align systems and equipment. These tools may include everything from a simple multimeter to sophisticated signal generators, from basic cable continuity testers to time domain reflectometers, and from light meters to color alignment systems. What they have in common is their capacity to provide a measurement or a standard signal that can be used to assess whether a device or system is functioning within the required parameters.

To provide a useful signal or measurement, the accuracy of the device must be known. The accuracy of the multimeter you use for basic continuity or pin-out testing on a microphone cable or a loudspeaker line is probably not important so long as the meter goes "beep" or lights up as an indication of low resistance. But if that multimeter is being used to check the voltage of a power supply or measure the voltage drop on a cable run, accuracy may become critical. The accuracy of the signal patterns produced by test signal generators are critical for the correct alignment of entire chains of AV equipment, so it is important that these signals are within the acceptable tolerances for their alignment and measurement tasks. The value of most test instruments is in their capacity to provide repeatable and accurate values.

Most test and measurement equipment is initially supplied with some documentation of its accuracy, reference information on the range of values that it can output or measure, and the range of operating conditions under which this accuracy can be expected. Professional-quality equipment will come with some form of certification of its calibration and an indication of how frequently it should be recalibrated.

Recalibration of test and measurement equipment is a complex and highly specialized procedure, usually undertaken by nationally or internationally accredited testing facilities. The calibration process involves the equipment being tested against certified standards for voltage, temperature, intensity, frequency, radiation, pressure, etc., and adjusted for an accurate response. Not only is such calibration quite expensive, it is rarely a fast process, meaning that equipment can be out of service for appreciable periods while it is being recalibrated.

While it is important to have all AV test equipment maintained in a state of accurate calibration, it is not usually necessary for every piece of equipment to be calibrated to laboratory accuracy. By regularly maintaining calibration of a single, high-accuracy device to laboratory standards, an organization can then use that device as a local master reference to check the accuracy of other equipment, either by directly comparing measurements against the reference device or by creating a local set of known standard values (resistances, voltages, signal patterns, etc.) to be used for calibration.

It is wise to consider which test and measurement equipment needs to be frequently and accurately calibrated for specific tasks. A multimeter that is reading low by 100 millivolts will usually not be a problem when checking if a power supply is producing an output, but a 100-millivolt discrepancy in a video test signal can be significant.

Chapter Review

In this chapter you studied the tasks involved in troubleshooting and maintaining AV installations and maintaining your tools and test equipment.

Upon completion of this chapter, you should be able to do the following:

- Explain the process for troubleshooting a problem in an AV system
- Create a maintenance schedule for an AV system
- Explain the processes for maintaining the safety and accuracy of AV tools and test equipment

This book provides a comprehensive study of the knowledge and skills listed in the CTS-I Exam Content Outline. This is a good time to try yourself out on the CTS-I online sample questions (see Appendix E), even if you have attempted them earlier.

Review Questions

The following questions are based on the content covered in this chapter and are intended to help reinforce the knowledge you have assimilated. These questions are similar to the questions presented on the CTS-I exam. See Appendix E for more information on how to access the free online sample questions.

1. Which of the following steps should be undertaken before attempting the troubleshooting process? (Choose all that apply.)

   A. Familiarize yourself with the AV systems and equipment

   B. Have the system documentation at hand

   C. Order a spare system controller

   D. Get an understanding of the warranty and maintenance agreements

2. What assumptions should you make about an audio system when there is no audio output? (Choose all that apply.)

   A. The gain on the main audio amplifier may be wound down to zero.

   B. The mixing desk channel for the microphone is switched to line input mode.

   C. The user is a complete idiot and did not switch the microphone on.

   D. Make no assumptions at all and select a troubleshooting strategy.

3. Which of the following should a maintenance log include? (Choose all that apply.)

   A. Devices replaced

   B. Description of problem

   C. Date of service

   D. Name of technician

4. Which of the following tasks would *not* be included in a preventive maintenance program? (Choose all that apply.)

   A. Checking the equalization and gain structure on the audio replay system

   B. Changing the preset frequencies on the handheld radio system

   C. Testing the failover systems for redundant spare equipment

   D. Checking the batteries in uninterruptable power supply systems

5. How often should your toolbox multimeter be sent to an accredited testing laboratory for recalibration?

   A. Before the commencement of each project.

   B. Every five years.

   C. Yearly.

   D. Never. It should be calibrated against your organization's highest-precision test instrument.

Answers

1. **A, B, D.** Before attempting fault finding on a system, you should familiarize yourself with the system, obtain all the system documents, and check what parts of the system are your responsibility to test and repair. Odds are that the problem will be a faulty cable connection and not the system controller.

2. **D.** Troubleshooting best practice is to make no assumptions at all about the nature of the fault. You should select a systematic strategy to identify the problem.

3. **A, B, C, D.** All the items listed should be included in a maintenance log.

4. **B.** Changing the frequencies on handheld radios is an operational matter. It does not affect the long-term operation or reliability of the handheld radios. All the other tasks are common inclusions in a preventive maintenance regimen.

5. **D.** Your toolbox multimeter is rarely required to perform measurements with a precision verified against the standards in an accredited calibration laboratory. It should be regularly calibrated against your organization's accurately calibrated, highest-precision test instrument.

PART VII

Appendixes and Glossary

Math Formulas Used in AV Installations

How long has it been since you solved a word problem or used a mathematical formula? Many Certified Technology Specialist – Installation (CTS-1) exam candidates have not done math in a formal setting in years. You may be familiar with the skills and tools in this appendix, but then again, you may need a refresher.

Using the Proper Order of Operations

Most audiovisual (AV) mathematical formulas use only the four common operators: add, subtract, multiply, and divide. However, some formulas require a solid foundation in the order of operations. The order of operations helps you correctly solve formulas by prioritizing which part of the formula to solve first. It is a way to rank the order in which you work your way through a formula. This section will review how to apply the proper order of operations.

This is the order of operations:

1. Any numbers within parentheses or brackets

2. Any exponents, indices, or orders

3. Any multiplication or division

4. Any addition or subtraction

If there are multiple operations with the same priority, then proceed from left to right: parentheses, exponents, multiplication, division, addition, and then subtraction. Several acronyms can help you remember the order of operations: PEMDAS, BEMDAS, BIDMAS, and BODMAS.

Steps to Solving Word Problems

All mathematical formulas summarize relationships between concepts. Word problems are designed to test how well an individual can apply that relationship to a new situation.

By following a few basic steps, you can turn a complicated word problem into a few straightforward steps. This section provides a structured approach to solving problems.

Within this structure, you will find many strategies for solving different types of problems. This strategy is based on *How to Solve It: A New Aspect of Mathematical Method* by G. Pólya (Princeton University Press, 2015).

Step 1: Understand the Problem

As is typical within the AV industry (and in general), the first step is to understand the problem you are trying to solve. Here are the tasks to complete for this step:

- Read the entire math problem.
- Identify your goal or unknown. What information are you trying to determine?
- Identify what you have been given. What data, numbers, or other information in the problem can help you determine the answer?
- Predict the answer if you can. What range of values would make sense as an answer?

Example: Calculate the current in a circuit where the voltage is 2 volts and the resistance is 8 ohms.

First, identify your goal. What are you trying to solve for? When you see the word *calculate,* generally the word that follows is your goal. Other words that identify the goal include *determine, find,* and *solve for.* Your goal in this problem is to calculate current.

An easy way to identify your given information is to find the numbers in the problem. In this example, the numbers are 2 and 8. Look for context clues or units to identify what those numbers represent. "The voltage is" identifies 2 as a voltage. The "ohms" after the 8 identifies 8 as the resistance.

Sometimes it's unclear what each number represents. In that case, drawing a diagram can help you make sense of what the problem is trying to say. You may want to make a chart of your given and unknown information for quick reference. For more complex problems, tables of given information can be extremely helpful.

Givens and Goal	Values
Current	?
Voltage	2V
Resistance	8Ω

Step 2: Create a Plan

The second step in this process is to translate the words in the problem into numbers you can enter into a formula. Begin with the following:

- Assign appropriate values to the goal and given information.
- Determine a formula that describes the relationships between your variables.

If there is a single formula that has all your variables in it, move on to the next step.

In some cases, you may be unable to identify a formula that will determine your goal based on your given information. If you're stuck, consider the following strategies:

- Use an intermediate formula to solve for the information you are missing.
- Use an outside reference, such as a chart or graph, to find information not listed in the problem.
- Use a strategy that has worked to solve similar projects in the past.
- Diagram the scenario described in the word problem. Use the diagram to keep track of the relationships between values. For instance, as listeners move farther away from a sound source, you know to expect a loss of sound pressure.

Example: Calculate the current in a circuit where the voltage is 2 volts and the resistance is 8 ohms.

Again, start by assigning variables to your given and unknown information. Note that some items are represented by different variables in different contexts. If you are having trouble determining which variables to use, consider drawing a diagram and labeling it with your given information, like in this chart:

Givens and Goal	Values	Variables
Current	?	I
Voltage	2V	V
Resistance	8Ω	R

Once you have assigned variables, think about the relationship between the information. There should be a formula that describes the relationship. For complex problems, you may need to use several formulas.

You may know formulas from memory. If you don't, look them up (we've provided several useful formulas at the end of this appendix). It is helpful to think about a time when you solved for the value previously or a similar problem you may have solved in the past. Try to think of a problem that used the same givens and unknowns.

For example, you might not remember how to solve for current using voltage and resistance. But if you remember how to solve for voltage using current and resistance, you can solve this problem. Voltage is equal to current multiplied by resistance.

Step 3: Execute Your Plan

The third step is to put your plan in action, as follows:

- Write the formula(s).
- Substitute the given information for the variables.
- Perform the calculation.
- Assign units to your final answer.

Example: Calculate the current in a circuit where the voltage is 2 volts and the resistance is 8 ohms.

Once you have determined the appropriate formula, write it down and then replace the placeholders with the numbers for this problem. People who skip this step are prone to making mistakes.

Formula: $V = I \times R$
Substitution: $2 = I \times 8$

To solve this equation, you need to get the I by itself. The 8 is currently being multiplied. To move it to the other side of the equation, perform the opposite mathematical function. In this case, that function is division.

$2/8 = (I \times 8)/8$
$2/8 = I$
$0.25 = I$

You need to assign units to your answer before it is final. Because I represents current, you would assign 0.25 the unit for current, which is amps (A).

$I = 0.25A$

Step 4: Check Your Answer

Your final step is to make sure the numbers you've calculated still make sense when translated back into words. Compare your answer to the scenario described in the problem. Is the result reasonable? Is it within the range you originally predicted?

For example, suppose that you are calculating the voltage present in a boardroom loudspeaker circuit. A result of 95 volts is probably a reasonable answer. A result of 50,000 volts indicates that you made a mistake in your calculations.

If you have an incorrect answer and use it to solve other parts of a process, it will result in cascading problems.

Example: Calculate the current in a circuit where the voltage is 2 volts and the resistance is 8 ohms.

In this example, is less than 1 amp a reasonable answer? Understanding the problem is essential here. A D battery has a voltage of 1.5 volts. When a D battery is attached to a circuit, there is not much current. So, the small number of 0.25 amps is a reasonable answer.

Rounding

Many of the results listed in workbook answer keys have been rounded to the nearest tenth. When solving multistep problems, you may be tempted to round at each step. The earlier or more often you round in a multistep problem, the less accurate your result will be. Round only your final result.

Some AV Math Formulas

This section presents some common math formulas that may be useful for AV professionals.

Aspect Ratio Formula

The formula for finding aspect ratio is as follows:

$$AR = W/H$$

where

- AR is the aspect ratio.
- W is the width of the displayed image.
- H is the height of the displayed image.

Note that W and H must be measured in the same units of length.

Screen Diagonal Formula (Pythagorean Theorem)

The diagonal of a screen can be calculated using the Pythagorean theorem.

$$A^2 + B^2 = C^2$$

where

- A is the height of a screen.
- B is the width of a screen.
- C is the diagonal length of a screen.

Estimating Throw Distance Formula

The formula for estimating throw distance is as follows:

$$Distance = Screen\ Width \times Throw\ Ratio$$

where

- *Distance* is the distance from the front of the lens to the closest point on the screen.
- *Screen Width* is the width of the projected image.
- *Throw Ratio* is the ratio of throw distance to image width.

Refer to the owner's manual of your projector and lens combination to find an accurate formula for your specific projector.

Decibel Formula for Distance

The formula for decibel changes in sound pressure level over distance is as follows:

$$dB = 20 \times \log(D_1/D_2)$$

where

- dB is the change in decibels.
- D_1 is the original or reference distance.
- D_2 is the new or measured distance.

The result of this calculation will be either positive or negative. If it is positive, the result is an increase, or gain. If it is negative, the result is a decrease, or loss.

Decibel Formula for Voltage

The formula for determining decibel changes for voltage is as follows:

$$dB = 20 \times \log(V_1/V_R)$$

where

- dB is the change in decibels.
- V_1 is the new or measured voltage.
- V_R is the original or reference voltage.

The result of this calculation will be either positive or negative. If it is positive, the result is an increase, or gain. If it is negative, the result is a decrease, or loss.

Decibel Formula for Power

The formula for calculating decibel changes for power is as follows:

$$dB = 10 \times \log(P_1/P_r)$$

where

- dB is the change in decibels.
- P_1 is the new or measured power measurement.
- P_r is the original or reference power measurement.

The result of this calculation will be either positive or negative. If it is positive, the result is an increase, or gain. If it is negative, the result is a decrease, or loss.

Current Formula (Ohm's Law)

The formula for calculating current using Ohm's law (see Figure A-1) is as follows:

$$I = V/R$$

where

- *I* is the current.
- *V* is the voltage.
- *R* is the resistance.

Figure A-1
Simple Ohm's law formula wheel

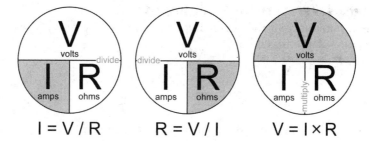

Power Formula

The formula to solve for power (see Figure A-2) is as follows:

$$P = I \times V$$

where

- *P* is the power.
- *I* is the current.
- *V* is the voltage.

Figure A-2
Simple power formula wheel

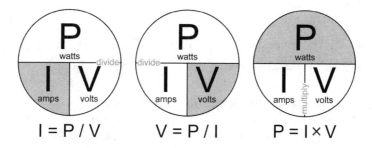

Series Circuit Impedance Formula

The formula for calculating the total impedance of a series loudspeaker circuit is as follows:

$$Z_T = Z_1 + Z_2 + Z_3 \ldots Z_N$$

where

- Z_T is the total impedance of the loudspeaker circuit.
- Z_x is the impedance of each loudspeaker.

Parallel Circuit Impedance Formula: Loudspeakers with the Same Impedance

The formula to find the circuit impedance for loudspeakers wired in parallel with the same impedance is as follows:

$$(Z_T) = Z_1/N$$

where

- Z_T is the total impedance of the loudspeaker system.
- Z_1 is the impedance of each loudspeaker.
- N is the number of loudspeakers in the circuit.

Parallel Circuit Impedance Formula: Loudspeakers with Different Impedances

The formula to find the circuit impedance for loudspeakers wired in parallel with differing impedance is as follows:

$$Z_T = \cfrac{1}{\cfrac{1}{Z_1} + \cfrac{1}{Z_2} + \cfrac{1}{Z_3} \ldots \cfrac{1}{Z_N}}$$

where

- Z_x is the impedance of each individual loudspeaker.
- Z_T is the total impedance of the loudspeaker circuit.

Series/Parallel Circuit Impedance Formulas

Two formulas are used to calculate the expected total impedance of a series/parallel circuit.

First, the series circuit impedance formula is used to calculate the impedance of each branch:

$$Z_T = Z_1 + Z_2 + Z_3 \ldots + Z_N$$

where

- Z_x is the impedance of each loudspeaker.
- Z_T is the total impedance of the branch.

Then the parallel circuit impedance formula is used to calculate the total impedance of the series/parallel circuit:

$$Z_T = \frac{1}{\dfrac{1}{Z_1} + \dfrac{1}{Z_2} + \dfrac{1}{Z_3} \cdots \dfrac{1}{Z_N}}$$

where

- Z_x is the total impedance of each branch.
- Z_T is the total impedance of the loudspeaker circuit.

Heat Load Formula (Btu)

The formula for calculating heat load in Btu is as follows:

Total Btu = $W_E \times 3.4$

where

- W_E is the total watts of all equipment used in the room.
- 3.4 is the conversion factor, where 1 watt of power generates 3.4 Btu of heat per hour.

This formula does not account for the heat load generated by amplifiers.

Heat Load Formula (kJ)

The formula for calculating heat load in kilojoules is as follows:

Total kJ = $W_E \times 3.6$

where

- W_E is the total watts of all equipment used in the room.
- 3.6 is the conversion factor, where 1 watt of power generates 3.6 kilojoules of heat per hour.

This formula does not account for the heat load generated by amplifiers.

Simplified Room Mode Calculation Formula

A simplified formula for discovering frequencies where room modes will be present is as follows:

$$\text{Mode Frequency (Hz)} = 3 \times (\textit{velocity of sound}) / \textit{RSD}$$

where

- The *velocity of sound* = 343m/sec (1,125ft/sec)
- *RSD* = The room's smallest dimension (in meters if using velocity in m/sec, or feet if using ft/sec)

AVIXA Standards

The American National Standards Institute (ANSI) is the official U.S. representative to the International Organization for Standardization (ISO). AVIXA's Certified Technology Specialist (CTS) certification exam, for which you are studying, is ANSI-accredited under the ISO and the ISO/IEC 17024:2012 Conformity Assessment—General Requirements for Bodies Operating Certification Schemes of Persons standard.

In addition, AVIXA is an ANSI-accredited Standards Developer (ASD), developing voluntary standards for the commercial AV industry. Accreditation by ANSI signifies that the processes used by standards development organizations (SDOs) to develop ANSI standards meet the ANSI's requirements for openness, balance, consensus, right to appeal, and due process. Subject matter experts work cooperatively to develop voluntary ANSI/AVIXA standards. There are currently more than 450 volunteers involved in task groups working on industry standards for audio, video, control, documentation, automation, and sustainability.

ANSI/AVIXA standards are not product-specific standards. They are system performance standards, management standards, documentation standards, and verification standards. These standards provide guidance for AV system performance. They take into account technology, physiology, architecture, and other variables in determining the best way to design, implement, and manage the performance of all types of AV systems. You can keep up with news about current standards, planned revisions, and those in development by visiting AVIXA's website (www.avixa.org/standards).

Don't be surprised if you see references to some of these standards on the CTS-I exam. It is important for CTS-I professionals to be able to understand and apply relevant standards. As a CTS-I–certified AV professional, you should consider standards when initiating, installing, verifying, and closing out AV system projects.

In this appendix, we offer a brief synopsis of the existing ANSI/AVIXA standards and those currently under development.

Published Standards, Recommended Practices, and Technical Reports

The following standards were available from AVIXA at the time of publishing.

A102.01:2017 Audio Coverage Uniformity in Listener Areas (ACU)   This standard defines measurement requirements and parameters for characterizing a sound system's coverage in listener areas. It provides performance classifications to describe the uniformity of coverage of a sound system's early arriving sound with the goal of achieving consistent sound pressure levels throughout defined listener areas. At the time of publication, this standard was under revision.

4:2012 Audiovisual Systems Energy Management   This standard defines processes and requirements for ongoing power-consumption management of audiovisual (AV) systems, using a tiered conformance approach. AV systems conforming to this standard will meet the defined requirements for automation, measurement, analysis, and training, which will vary based on each tier. At the time of publication, this standard was under revision.

10:2013 Audiovisual Systems Performance Verification   This standard provides a framework and supporting processes for determining elements of an audiovisual system that need to be verified, the timing of that verification within the project delivery cycle, a process for determining verification criteria/metrics, and reporting procedures. At the time of publication, this standard was under revision.

J-STD 710 – 2015 (CTA/AVIXA/CEDIA) Audio, Video, and Control Architectural Drawing Symbols   This standard defines architectural floor plan and reflected ceiling plan symbols for audio, video, and control systems, with associated technologies such as environmental control and communication networks. It also includes descriptions and guidelines for the use of these symbols.

F501.01:2015 Cable Labeling for Audiovisual Systems   This standard defines requirements for audiovisual system cable labeling for a variety of venues. It provides requirements to easily identify all power and signal paths in a completed audiovisual system to aid in operation, support, maintenance, and troubleshooting. At the time of publication, this standard was under revision.

V202.01:2016 Display Image Size for 2D Content in Audiovisual Systems   This standard determines required display image size and relative viewing positions based on user need. It can be used to design a new space or to assess or modify an existing space, from either drawings or the space itself. It applies to permanently installed and temporary systems. The standard does not apply to the performance or efficiency of any component. A free online calculator is available at www.avixa.org/discascalc (requires one-time registration).

RP-38-15:2018 (IES/AVIXA), Lighting Performance for Small-to-Medium-Sized Videoconferencing Rooms   This standard provides parameters and performance criteria for lighting small to medium-sized single-axis videoconferencing spaces (maximum of 25 participants), defined as one set of video displays and cameras oriented toward a group of seated participants, providing technical and practical requirements to assist practitioners who configure and specify lighting systems specific to videoconferencing projects.

3M-2011 Projected Image System Contrast Ratio　This standard defines projected image system contrast ratio and its measurement. Four contrast ratios based on four categories of content-viewing requirements are defined. Practical metrics to measure and validate the defined contrast ratios are provided. This standard has been revised and will be incorporated into *V201.01 Image System Contrast Ratio* when it is released.

F502.01:2018 Rack Building for Audiovisual Systems　This standard defines requirements for building AV equipment racks, which are defined as assembly of rack(s), mounting of AV equipment and accessories, cable management, and finishing.

F502.02:2020 Rack Design for Audiovisual Systems　This standard defines minimum requirements for the audiovisual rack planning and design process, including required process inputs and outputs. Key performance criteria validate the impact to internal and external integration with the facility requirements.

RP-C303.01:2018 Recommended Practices for Security in Networked Audiovisual Systems　This recommended practice provides guidance and current best practices for securing networked AV systems, including how to recognize risks and develop a risk mitigation management plan to address those risks.

2M-2010 Standard Guide for Audiovisual Systems Design and Coordination Processes　This standard provides a framework for the methods, procedures, tasks, and deliverables typically recommended or applied by industry professionals in the design and implementation of audiovisual communication systems. The framework enables clients and other design and construction team members to assess whether responsible parties are providing expected services. At the time of publication, this standard was under revision. The revision will define minimum documentation requirements for AV systems design.

TR-111.01 Unified Automation for Buildings　This technical report provides a detailed overview of the building automation environment and identifies the need for a unified set of standards to integrate multiple building systems, including but not limited to traditional building automation systems (BASs), into cohesive and functional systems and/or subsystems for increased benefits.

Standards in Development

The following standards were in development at the time of publishing and may now be completed and available.

V201.01 Image System Contrast Ratio　This standard defines contrast ratios based on user viewing requirements. It is designed to facilitate informed decision-making for any display, projector, and screen selection relative to location and purpose. It applies to permanently installed systems and live events, front and rear projection, and direct-view displays.

A103.01 Sound System Spectral Balance   This standard defines a measurement and verification process for sound system reproduction of spectral balance, also known as uniform frequency response, accomplished by documenting the frequency response from the sound system across a specified bandwidth within a low- to high-frequency range within the listening area.

A104.01 Sound System Dynamic Range   This standard provides a procedure to measure and classify the dynamic range, or signal-to-noise ratio, of early arriving sound from a sound system across a listener area.

UX701.01 User Experience Design for Audiovisual Systems   This standard defines processes that optimize user experience for AV-equipped spaces. Processes include user engagement, design, testing, deployment, and continuous refinement.

NOTE   AVIXA standards are constantly being reviewed and updated to keep them relevant and timely. To ensure that you are working with the current versions of standards, you should always check the standards section of the AVIXA website for updates and revisions.

Scan For AVIXA Standards

AVIXA AV Standards Clearinghouse

AVIXA is not the only provider of audiovisual (AV) standards. Other organizations around the world also develop AV standards. The process of researching these standards to determine which ones apply to your project can become a challenge.

In response to the need for an easier way to find AV-related standards, AVIXA offers the AV Standards Clearinghouse, an extensive, searchable spreadsheet of standards that are applicable to audiovisual technologies. This free resource was developed to help you spend less time researching and more time working toward delivering a high-quality, standards-compliant end product.

AVIXA's AV Standards Clearinghouse is a powerful tool that enables you to identify applicable standards by title, publication date, publisher, functional category (audio, video, control, other), interest category (integrator, tech manager, consultant, and so on), and regional applicability. The clearinghouse is updated periodically to include the latest standards.

The AV Standards Clearinghouse is not just a useful tool on the job. If you are focused on mastering certain areas of the CTS-I Exam Content Outline, the clearinghouse is a great resource for potential further study materials. Examples include standards for the following:

- Wiring the connectors used in audio systems
- Measuring jitter in digital systems
- Cable mapping documentation symbology
- Measuring projector performance
- Electrical safety precautions

You can find standards relating to almost any task on the CTS-I Exam Content Outline in the AV Standards Clearinghouse.

You can download the clearinghouse spreadsheet and the accompanying guide document from the AVIXA website at www.avixa.org/standards/av-standards-clearinghouse.

Video Link

AVIXA has produced several short videos that can help you prepare for the CTS-I exam, practice installation skills, and learn best practices. Most of these videos are less than ten minutes long. Use the following link, provided as both a URL and a QR code, for the current list of videos.

https://www.avixa.org/cts-supplemental-training-videos

About the Online Content

This book comes complete with TotalTester Online customizable practice exam software with 40+ sample exam questions from AVIXA.

System Requirements

The current and previous major versions of the following desktop browsers are recommended and supported: Chrome, Microsoft Edge, Firefox, and Safari. These browsers update frequently, and sometimes an update may cause compatibility issues with the TotalTester Online or other content hosted on the Training Hub. If you run into a problem using one of these browsers, please try using another until the problem is resolved.

Your Total Seminars Training Hub Account

To get access to the online content, you will need to create an account on the Total Seminars Training Hub. Registration is free, and you will be able to track all your online content using your account. You may also opt in if you wish to receive marketing information from McGraw Hill or Total Seminars, but this is not required for you to gain access to the online content.

Privacy Notice

McGraw Hill values your privacy. Please be sure to read the Privacy Notice available during registration to see how the information you have provided will be used. You may view our Corporate Customer Privacy Policy by visiting the McGraw Hill Privacy Center. Visit the **mheducation.com** site and click **Privacy** at the bottom of the page.

Single User License Terms and Conditions

Online access to the digital content included with this book is governed by the McGraw Hill License Agreement outlined next. By using this digital content, you agree to the terms of that license.

Access   To register and activate your Total Seminars Training Hub account, simply follow these easy steps.

1. Go to this URL: **hub.totalsem.com/mheclaim**

2. To register and create a new Training Hub account, enter your e-mail address, name, and password on the **Register** tab. No further personal information (such as credit card number) is required to create an account.

   If you already have a Total Seminars Training Hub account, enter your e-mail address and password on the **Log in** tab.

3. Enter your Product Key: **6bjc-2xb0-zrzh**

4. Click to accept the user license terms.

5. For new users, click the **Register and Claim** button to create your account. For existing users, click the **Log in and Claim** button.

   You will be taken to the Training Hub and have access to the content for this book.

Duration of License   Access to your online content through the Total Seminars Training Hub will expire one year from the date the publisher declares the book out of print.

   Your purchase of this McGraw Hill product, including its access code, through a retail store is subject to the refund policy of that store.

   The Content is a copyrighted work of McGraw Hill, and McGraw Hill reserves all rights in and to the Content. The Work is © 2021 by McGraw Hill.

Restrictions on Transfer   The user is receiving only a limited right to use the Content for the user's own internal and personal use, dependent on purchase and continued ownership of this book. The user may not reproduce, forward, modify, create derivative works based upon, transmit, distribute, disseminate, sell, publish, or sublicense the Content or in any way commingle the Content with other third-party content without McGraw Hill's consent.

Limited Warranty   The McGraw Hill Content is provided on an "as is" basis. Neither McGraw Hill nor its licensors make any guarantees or warranties of any kind, either express or implied, including, but not limited to, implied warranties of merchantability or fitness for a particular purpose or use as to any McGraw Hill Content or the information therein or any warranties as to the accuracy, completeness, correctness, or results to be obtained from, accessing or using the McGraw Hill Content, or any material referenced in such Content or any information entered into licensee's product by users or other persons and/or any material available on or that can be accessed through the licensee's product (including via any hyperlink or otherwise) or as to non-infringement of third-party rights. Any warranties of any kind, whether express or implied, are disclaimed. Any material or data obtained through use of the McGraw Hill Content is at your own discretion and risk and user understands that it will be solely responsible for any resulting damage to its computer system or loss of data.

Neither McGraw Hill nor its licensors shall be liable to any subscriber or to any user or anyone else for any inaccuracy, delay, interruption in service, error or omission, regardless of cause, or for any damage resulting therefrom.

In no event will McGraw Hill or its licensors be liable for any indirect, special or consequential damages, including but not limited to, lost time, lost money, lost profits or good will, whether in contract, tort, strict liability or otherwise, and whether or not such damages are foreseen or unforeseen with respect to any use of the McGraw Hill Content.

TotalTester Online

TotalTester Online provides you with sample questions from the CTS-I exam. The sample questions can be used in Practice Mode or Exam Mode. Practice Mode provides an assistance window with hints, references to the book, explanations of the correct and incorrect answers, and the option to check your answer as you take the test. Exam Mode provides a timed review of the questions. The number of questions, the types of questions, and the time allowed are intended to be an accurate representation of the exam environment. The option to customize your quiz allows you to create custom exams from selected duties (domains) or chapters, and you can further customize the number of questions and time allowed.

To take a test, follow the instructions provided in the previous section to register and activate your Total Seminars Training Hub account. When you register, you will be taken to the Total Seminars Training Hub. From the Training Hub Home page, select "CTS-I Certified Technical Specialist-Installation Exam Guide, 2e" from the Study drop-down menu at the top of the page, or from the list of Your Topics on the Home page. You can then launch the Total Tester and select the option to customize your quiz and begin testing yourself in Practice Mode or Exam Mode. All exams provide an overall grade and a grade broken down by domain.

Technical Support

For questions regarding the TotalTester or operation of the Training Hub, visit **www.totalsem.com** or e-mail **support@totalsem.com**.

For questions regarding book content, visit **www.mheducation.com/customerservice**.

For questions about the video or standards URLs or QR codes, please visit the AVIXA customer service page at **https://www.avixa.org/membership/contact** or e-mail AVIXA customer service at **membership@avixa.org**.

1/3 octave equalizer   A graphic equalizer that provides 30 or 31 slider adjustments corresponding to specific fixed frequencies with fixed bandwidths, with the frequencies centered at every one-third of an octave. The adjustment points shape the overall frequency response of the system.

1/4-inch phone connector   A connector typically used to transport unbalanced line-level audio signals from musical instruments and between processing devices.

1/8-inch phone connector   A connector typically used to transport unbalanced line-level audio signals from portable devices and computers.

3.5mm phone connector   A connector typically used to transport unbalanced line-level audio signals from portable devices and computers.

4K   An ultra-high-resolution video format with a minimum resolution of 3840×2160 pixels in a 16:9 (1.77:1) aspect ratio, approximately four times that of full HD (1920×1080 pixels). 4K formats include 3840×2160 pixels (UHDTV-1) and 4096×2160 pixels (digital cinema, DCI).

4K ecosystem   The video cameras, recorders, editors, processors, servers, distribution networks, and display technologies used for the production, distribution, and display of 4K ultra-high-resolution video.

6.5mm phone connector   A connector typically used to transport unbalanced line-level audio signals from musical instruments and between processing devices.

8K   An ultra-high-resolution video format with a minimum resolution of 7680×4320 pixels, which is 16 times the resolution of full HD (1920×1080 pixels). The related consumer television format is known as 4320p, 8K Ultra HD, and UHDTV-2.

8P8C connector   An eight-pin (eight position, eight conductor) modular connector typically used for the termination of multipair cables. Often used in Ethernet data networks. It is commonly incorrectly referred to as an RJ-45 connector, which is actually an 8P2C telephone connector.

acceptable viewing area   The viewing range for a screen; suggested as a 60-degree arc off the far vertical edge of the screen being viewed.

access point   A network device that allows other devices to connect to a network.

acoustic echo canceller   An echo cancelling device used in conferencing systems that attempts to remove environmental echoes that are created at the far site from sound reflected off hard surfaces and returned to conferencing microphones.

acoustics The properties or qualities of a room or building that determine how sound is transmitted and reflected. The study of the properties of sound.

AES Audio Engineering Society.

allied trade A business that collaborates with AV professionals to complete a project.

alternating current (AC) An electric current that reverses its direction periodically.

ambient light The sum of all lighting in an area.

ambient noise The sum of all sounds in an area.

amperage The amount of electric current flowing in a circuit. Current is measured in amperes (amps), abbreviated as A.

amplifier A device for increasing the strength of signals.

amplitude The height of the waveform of a signal.

analog (analogue) A continuously variable signal.

analog-to-digital/digital-to-analog (AD/DA) converters Devices that convert signals from analog to digital or from digital to analog.

angularly reflective screen A screen that reflects light at the same angle at which it arrived.

ANSI American National Standards Institute.

aperture The opening in an optical train that controls the amount of light passing through.

arc-fault circuit interrupter (AFCI) Also known as an arc-fault detection device (AFDD), a type of circuit breaker that triggers on detecting an electrical-arc fault in its load circuit.

arrayed loudspeaker system A loudspeaker arrangement that delivers sound from a single point in space.

artifact An element introduced into a signal during processing. Not always a good thing.

aspect ratio The ratio of image width to image height.

attack time The time taken for an action to complete its effect once triggered.

attenuate To reduce the amplitude of a signal.

Audio Coverage Uniformity measurement locations (ACUMLs) The test points within a venue that have been determined to carry out the measurements for the Audio Coverage Uniformity test.

Audio Coverage Uniformity Plan (ACUP)   A stand-alone document that identifies the Audio Coverage Uniformity measurement locations for a particular venue, using the AVIXA indication symbol.

audio processor   A device used to manipulate audio signals.

Audio Return Channel (ARC)   Introduced to the HDMI standard with version 1.4. ARC allows a display to send audio data upstream to a receiver or surround-sound controller, eliminating the need for a separate audio connection.

audio signal   An electrical representation of sound.

audio transduction   The process of converting acoustical energy into electrical energy or electrical energy back into acoustical energy.

Audio Video Bridging (AVB)   A standards-based audiovisual Data Link layer protocol defined under IEEE 802.1-AVB. It runs across a standard Ethernet network but requires AVB-enabled switches and network components that handle QoS prioritization of data. AVB has been renamed Time-Sensitive Networking to reflect the standard's applicability to communication among different types of devices, such as network sensors. See *Time-Sensitive Networking (TSN)*.

audiovisual infrastructure   The physical building components that make up the pathways, supports, and architectural elements required for audiovisual technical equipment installations.

audiovisual rack   A housing unit for electronic equipment. The inside of a typical AV industry rack is 19 inches (482.6mm) wide. Many of the technical specifications for a rack, including size and equipment height, are determined by standards that have been established by numerous standards-setting organizations. The outside width of the rack varies from approximately 530mm to 630mm (21in. to 25in.).

authority having jurisdiction (AHJ)   An organization, office, or individual responsible for enforcing the requirements of a code or standard or for approving equipment, materials, installation, or procedures. In some places the authority having jurisdiction may be known as the *regional regulatory authority*.

automatic gain control (AGC)   A circuit or process that maintains a constant output gain in response to input variables such as signal strength or ambient noise level.

Automatic Private IP Addressing (APIPA)   A Windows function that assigns locally routable addresses from the reserved network 169.254.0.0/16 to devices that do not have or cannot obtain an IP address. This allows devices to communicate with other devices on the same LAN. APIPA operates at the Network layer of the OSI Model and the Internet layer of the TCP/IP protocol stack.

balanced circuit   A two-conductor circuit in which both conductors and all circuits connected to them have the same impedance with respect to ground.

PART VII

balun   A contraction of BALanced-to-UNbalanced, a device used to connect a balanced circuit to an unbalanced circuit. For example, a transformer used to connect a 300-ohm television antenna cable (balanced) to a 75-ohm antenna cable (unbalanced).

band   A grouping or range of frequencies.

bandwidth   **1.** A range of frequencies. **2.** A measure of the amount of data or signal that can pass through a system during a given time interval.

bandwidth limiting   The process of limiting the bandwidth of a signal, usually to allow the signal to be transmitted over a lower-bandwidth path.

bandwidth (networking)   The available or consumed data communication resources of a communication path, measured in bits per second (bps). It is also called *data through-put* or *bit rate*.

baseband   A signal that has not been modulated onto a higher-frequency carrier.

benchmarking   The process of examining methods, techniques, and principles to establish a standard to which comparisons can be made.

bend radius   The radial measure of a curve in a cable, conductor, waveguide, or interconnect that defines the physical limit beyond which further bending has a measurable effect on the signal being transported.

bi-directional polar pattern   The shape of the region where some microphones will be most sensitive to sound from the front and rear, while rejecting sound from the top, bottom, and sides.

bill of materials (BOM)   A complete equipment list of components that must be procured in order to build the system as specified. The BOM also lists the costs associated with each aspect of designing and implementing the system.

bit   A contraction of the words *binary digit*. This is the smallest unit of binary digital information and may have the value 1 or 0.

bit depth   The number of bits used to specify a parameter.

bit error rate (BER)   The number of error bits present in a signal stream per unit of time.

bit rate   The measurement of the quantity of data transmitted over a digital signal stream. It is measured in bits per second (bps).

block diagram   A diagram of a system or device in which the principal parts are represented by suitably annotated geometrical figures to show both the functions of the parts and their functional relationships.

Bluetooth   A wireless technology for low-cost, short-range radio links between devices. It operates in the 2.402GHz to 2.480GHz industrial, scientific, and medical (ISM) frequency band.

BNC connector   A type of connector featuring a two-pin bayonet-type lock. The most common professional coaxial cable connector because of its reliability and ruggedness. It is used to terminate cables transporting signals such as SDI, RF, video, and time code.

bonding   Joining conductive material by a low-impedance connection, thus ensuring that they are at the same electrical potential.

boundary microphone   A microphone design where the diaphragm is placed close to a sonic "boundary" such as a wall, ceiling, or other flat surface. This arrangement prevents the acoustic reflections from the surface from mixing with the direct waveform and causing phase distortions. It is used in conference and telepresence systems.

branch circuit   The circuit conductors between the final overcurrent device protecting the circuit and the load connection.

breaker box   Another name for an electrical load distribution panel. See *Panelboard*.

broadcast domain   A set of devices that can send Data Link layer frames to each other directly, without passing through a Network layer device. Broadcast traffic sent by one device in the broadcast domain is received by all devices in the domain.

buffer amplifier   An electronic device that provides isolation and some load independence between circuit components.

building information modeling (BIM)   A data repository for building design, construction, and maintenance data shared by multiple disciplines on a single project.

bus   A wiring system that links multiple devices.

busbar   An electrically conductive path that serves as a common connection for two or more circuits.

buzz   A noise generated by the higher-order harmonics of the hum (50Hz or 60Hz) generated by the electrical mains.

byte   A data word containing 8 bits, also known as an *octet*. The symbol for byte is B.

cable   An assembly of more than one conductor (wire).

cable tray   A structure to provide rigid continuous support for cables.

campus area network (CAN)   A type of network linking multiple LANs in a limited geographical area such as a university campus or a cluster of buildings.

candela   The unit of luminous intensity. A candela is the luminous intensity emitted by a reference point light source in a given direction. The symbol for candela is cd. By sheer "coincidence," the luminous intensity of a common wax candle is approximately 1 candela.

capacitance   The ability of a nonconductive material to store an electrical charge. Capacitance measured in farads (F). The symbol for capacitance is C.

capacitive reactance   The opposition a capacitive device offers to alternating current flow. It is measured in ohms (Ω). The symbol for capacitive reactance is X_C. Capacitive reactance is inversely proportional to the frequency of the current in a circuit.

capacitor   A passive electrical component in which plates of conductive material are separated by a dielectric. For a given capacitance value, expressed in farads, a capacitor will have a greater opposition to AC current flow at lower frequencies than at higher frequencies.

captive screw connector   Also known as a Euroblock or Phoenix connector. A termination method where a stripped wire is inserted into the connector and secured by a set screw that pushes down a gate to form the electrical bond and clamp the wire in place.

cardioid polar pattern   A heart-shaped region where some microphones will be most sensitive to sound predominately from the front of the microphone diaphragm and reject sound coming from the sides and rear.

carrier   Modulated frequency that carries a communication signal.

Category 5 (Cat 5)   The designation for 100Ω unshielded twisted-pair (UTP) cables and associated connecting hardware whose characteristics are specified for data transmission up to 100Mbps (part of the EIA/TIA 568A standard).

Category 5e (Cat 5e)   An enhanced version of the Cat 5 cable standard that adds specifications to reduce far-end crosstalk for data transmission up to 1Gbps (part of the EIA/TIA 568A standard).

Category 6 (Cat 6)   A cable standard for Gigabit Ethernet and other interconnections that is backward-compatible with Category 5, Cat 5e, and Cat 3 cable (part of the EIA/TIA 568A standard). Cat 6 features more stringent specifications for crosstalk and system noise.

Category 6A (Cat 6A)   A cable standard for 10 Gigabit Ethernet and other interconnections that is backward-compatible with Category 5, Cat 5e, and Cat 6 cable. Constructed using a mechanism to physically separate the twisted pairs, Cat 6A features more stringent specifications for alien and near-end crosstalk.

Category 7 (Cat 7)   A cable specification for 10 Gigabit Ethernet and other interconnections that is backward-compatible with Cat 5e and Cat 6A cable. To further reduce crosstalk, in Cat 7 cables, each twisted pair is shielded and the entire cable is shielded.

Category 8 (Cat 8)   A cable standard for 40 Gigabit Ethernet and other interconnections that is backward-compatible with Category 5e, Cat 6, and Cat 6A cable (part of the EIA/TIA 568A standard). Cat 8 features a shield around each twisted pair and a further shield enclosing all twisted-pairs. It is only specified at 40Gbps for runs up to 30m (100ft).

cathode ray tube (CRT)   A high-vacuum glass tube containing an electron gun to produce the images seen on the phosphor-coated face of the tube. This video display technology was used in early-generation video monitors, television receivers, radar displays, and oscilloscopes.

CATV   Community antenna television. A system where broadcast television signals are received by a single central antenna and distributed to multiple end users.

center tap   A connection point located halfway along the track or winding of an electronic device such as an inductor or resistor.

central cluster   A single-source configuration of loudspeakers. In a central cluster, the sound is coming from one point in the room. The central cluster is normally located directly above (on the proscenium) and slightly in front of the primary microphone location.

central processing unit (CPU)   The portion of a computer system that reads and executes commands.

charged-coupled device (CCD)   A semiconductor light-sensitive device, commonly used in video and digital cameras, that converts optical images into electronic signals.

chassis   Also called a *cabinet* or *frame,* an enclosure that houses electronic equipment and is frequently electrically conductive (metal). The metal enclosure acts as a shield and is connected to the equipment-grounding conductor of the AC power cable, if so equipped, to provide protection against electric shock.

chassis ground   A 0V (zero volt) connection point of any electrically conductive chassis or enclosure surrounding an electronic device. This connection point may be extended to the earth/ground.

chroma   The saturation, or intensity, of a specific color. It is one of the three attributes that define a color; the other two are hue and value or luminance.

chrominance   The color part of a composite or S-video signal.

Classless Inter-Domain Routing (CIDR) notation   A compact representation of an IP address and its subnet mask using a slash and a decimal number to indicate how many leading 1 bits are in the network mask, with the remaining bits being the network host identifiers (e.g., 192.168.220.16/24). The subnet mask of 255.255.255.0 (24 mask bits) is represented as /24 in CIDR notation. CIDR replaces the class designations (A, B, and C) for IP address ranges.

cliff effect   The sudden loss of digital signal reception. When a digital signal is attenuated to the point where signal for a digital one is indistinguishable from the signal for a digital zero.

clipping   The deformation of a signal when a device's peak amplitude level is exceeded.

clock adjustment   The process used to align the timing of digital signals between transmitting and receiving devices.

coaxial (coax) cable   A cable consisting of a center conductor surrounded by insulating material, concentric outer conductor, and optional protective covering, all of circular cross section.

CobraNet   CobraNet is a proprietary digital audio Data Link layer protocol designed by Cirrus Logic. It uses standard Fast Ethernet cabling, switches, and other components. CobraNet signals are nonroutable.

codec   A contraction of the term *coder/decoder*. An electronic device or software process that encodes or decodes a data stream for transmission and reception over a communications medium.

collision domain   A set of devices on a carrier-sense multiple access local area network (LAN) whose packets may collide with one another if they transmit data at the same time. Only applies to non-switched Ethernet networks.

color difference signal   A signal that conveys color information such as hue and saturation in a composite format. Two such signals are needed. These color difference signals are R-Y and B-Y (sometimes referred to as Pr and Pb, or Cr and Cb).

color rendering index (CRI)   The effect a light source has on the perceived color of objects indexed against an incandescent source (CRI 100) of the same correlated color temperature.

color temperature   The quantification of the color of "white" light in reference to the light emitted by a standardized object at a specified temperature on the Kelvin scale. Measured in kelvins (K). Low color temperature light (~2,000K) has a warm (reddish) look, while light with a high color temperature (>4,000K) has a colder (bluish) appearance.

combiner   A device or process that combines signals of different frequencies together in a single medium. Used to combine multiple RF television signals into one cable for use in broadband cable television distribution.

common mode   **1.** Voltage fed in phase to both inputs of a differential amplifier. **2.** The signal voltage that appears equally and in phase from each current carrying conductor to ground.

common-mode rejection ratio (CMRR)   The ratio of the differential voltage gain to the common-mode voltage gain; expressed in decibels.

compander   An audio processing device that combines compression and expansion.

component video   Color video in which the brightness (luminance), color hue, and saturation (chrominance) are handled independently. The red, green, and blue signals—or more commonly, the Y, R-Y, and B-Y signals—are encoded onto three wires.

composite video signal   A single video signal that carries the complete color picture information and all synchronization signals.

compression   **1.** An increase in density and pressure in a compressible medium such as air. **2.** A process that reduces data size.

compression ratio   **1.** How much the volume on an audio compressor reduces depending on how far above the threshold the signal is. **2.** The ratio in size between the original signal and its compressed form.

compressor   A device that controls the overall amplitude of a signal by reducing the part of the signal that exceeds an adjustable threshold level set by the user. When the signal exceeds the threshold level, the overall amplitude is reduced by a ratio, also usually adjustable by the user.

compressor threshold   Sets the point at which the automatic volume reduction kicks in. When the input goes above the threshold, an audio compressor automatically reduces the volume to keep the signal from getting too loud.

condenser microphone   Also called a *capacitor microphone,* a microphone that transduces sound into electricity using capacitive principles.

conductor   In electronics, a material that easily conducts an electric current because some electrons in the material are easy to move.

conduit   A circular tube that houses cable.

cone   The most commonly used component in a loudspeaker system and found in all ranges of drivers.

constant voltage distribution   25V, 70V, 100V; a method of distributing signals to loudspeakers over a large area with lower losses than typical direct-coupled connections. Also known as a *high-impedance distribution system.*

Consumer Electronic Control (CEC)   A single-wire, bi-directional serial bus that uses AV link protocols to perform remote-control functions. It is an optional feature of the HDMI specification that allows for system-level automation when all devices in an AV system support it.

contact closure   A simple signaling system based on whether a contact or switch is open or closed. The conventional protocol is to interpret a closed contact (on) as a binary 1 and an open contact (off) as a binary 0.

content delivery network (CDN)   A distributed network of caching servers that can provide hosted unicast distribution of media for an organization. They are most often utilized by organizations whose content is in high demand.

contrast   The difference in luminance between dark and light elements of an image.

contrast ratio   Describes the dynamic video range of a display device as a numeric relationship between the brightest color (typically white) and the darkest color (typically black) that the system is capable of producing. Two methods are used to specify contrast ratio; the full on/full off method describes the dynamic contrast ratio, and the ANSI method measures the static contrast ratio.

control system   A system that controls subsystems such as audio, video, winches, drapes, mechanical devices, lighting, and atmospheric effects.

correlated color temperature (CCT)   The color appearance of a light source as compared with a standardized object heated to a temperature measured on the Kelvin scale. CCT is measured kelvins (K).

coverage pattern   The pattern of sound energy that a loudspeaker emits. The coverage pattern is dependent on the frequency of the sounds and the dimensions of the loudspeaker.

critical distance (d_c)   The point where the sound pressure level of the direct and reverberant sound fields is equal.

critical path schedule   Reveals the interdependence of activities and assesses resource and time requirements and trade-offs. It also determines the project's completion date and provides the capability to evaluate activity performance.

crossover   An electronic device that separates the frequency bands of an audio signal so that each driver in a multi-driver loudspeaker system is sent only those frequencies that it will transduce accurately.

crosspoint   A matrix-based switching device with multiple inputs and outputs wherein any input may be connected to any, many, or all outputs.

crosstalk   Any phenomenon by which a signal transmitted on one circuit or channel of a transmission system creates an undesired effect in another circuit or channel.

current   The quantity of electrical charge flowing in a circuit, measured in amperes (A). The symbol for current is C.

curvature of field   A blurry appearance around the edge of an otherwise in-focus object (or the reverse) when the angle incidence of light passing through the edges of a lens is different from the angle of incidence at the center of the surface. Curvature of field is due to the shape of the lens.

Dante   A proprietary digital audio Network layer protocol created by Audinate. Dante sends audio and video information as Internet Protocol (IP) packets. It is fully routable over IP networks using standard Ethernet switches, routers, and other components. Dante controller software manages data prioritization and signal routes.

dB SPL   A measure of sound pressure level with reference to 0dB SPL, a sound pressure of 20 microPascal (μPA).

decay time   The time taken for an effect to diminish.

decibel (dB)   A base-10 logarithmic representation of the ratio between two numbers. This ratio is used for quantifying differences in voltage, distance, and sound pressure as they relate to power.

delay   A signal-processing device or circuit used to delay the speed of transmission of a signal.

demodulator   A device or process that extracts the information from a modulated signal.

depth of field   The area in front of a lens that is in focus. The distance from the closest focused item to the item farthest away.

detail drawing   A detail drawing enlarges small items to show how they must be constructed or installed. They depict items too small to see at the project's typical drawing scale.

dielectric constant   Describes the ability of a material between two conductors to store an electrical charge. Dielectric strength is determined by the material's type and thickness and is the amount of voltage that insulation can stand before it breaks down.

differential mode   A signaling mode where two wires carry the same signal but one is in reverse polarity.

differentiated service (DiffServ)   A network quality-of-service strategy wherein data from specific applications or protocols is assigned a class of service. Flows assigned a high priority are given preferential treatment at the router, but delivery is not guaranteed.

diffusion   The scattering or random reflection of a wave from a surface.

digital media player   Devices or software applications that allow the playback of recorded or streamed audio and video content from media servers, the Internet, or local storage.

digital signage   Customized content shown on strategically located displays. Digital signage displays are placed in the same locations that would otherwise have been static signage: public spaces, museums, stadiums, corporate and educational buildings, retail stores, hotels, restaurants, etc. It is sometimes referred to as dynamic signage to differentiate it from large-format static signs.

digital signage media player   A specialized media player used to store and forward or play back digital signage content on display screens. Some commercial digital signage displays have integrated media players. Networked digital signage media players enable remote control and management of content.

digital signage template   Customizable layouts used for standardizing content across all displays on the network. They enable multiple messages or content from multiple sources to be displayed on a screen by presenting the information in zones.

digital signal processor (DSP)   A digital device designed to process signal streams such as audio, video, and RF data.

digital-to-analog converter   A device that converts digital signals into analog form.

direct coupled   A loudspeaker signal distribution system in which the amplifier is connected directly to the loudspeaker.

direct current (DC)   Electric current that does not reverse direction, unlike alternating current (AC). DC may vary in amplitude but not direction.

direct sound   Also known as *near-field.* Sound that is received directly from the source and not colored by the acoustics of its surroundings.

direct view display   A display such as an LCD, LED, OLED, vacuum fluorescent, or plasma screen where the viewer is looking directly at the image source, not at a projection screen.

directivity   The coverage pattern of loudspeakers.

dispersion   The separation of light into different frequencies/wavelengths, such as when a white light beam passes through a triangular prism. The different wavelengths of light refract at different angles, dispersing the light into its individual components.

Display Data Channel (DDC)   The data channel used between video sources and video displays to carry information about display capabilities and video formats. The channel used for exchanging EDID and HDCP data.

DisplayID   A standard developed by VESA outlining how video display data is structured to describe its performance and capabilities when communicating with other devices. By structuring data in a flexible, modular way, DisplayID enables devices to identify new display resolutions, refresh rates, audio standards, and other formats as they become available. For example, the standard can support a single image segmented across tiled displays using multiple video processors.

DisplayPort   A VESA-developed high-speed digital data transport protocol used to connect a video source to display devices. It can also carry audio, USB, and other data.

distributed sound system   A sound system using multiple distributed loudspeakers to provide sound coverage across an area at a constant sound pressure level.

distribution amplifier (DA)   An active device used to split one input signal into multiple isolated output signals at a constant level.

diversity receiver   An RF receiver that uses multiple antennas to receive a single RF transmission. The receiver calculates phase differences between the received signals to dynamically shift between antennas to avoid multipath signal cancellation.

DLP   Digital Light Processing. A projection technology based on the MEMS digital micromirror device (DMD) chip family. It uses a matrix of thousands of movable microscopic mirrors on a chip to display images on a screen.

Domain Name System (DNS) A hierarchical, distributed database that maps names to data such as IP addresses, name server addresses, and mail exchange addresses.

dome A type of loudspeaker driver construction. Fabric or woven materials are used to create a dome-shaped diaphragm. The voice coil is attached to the edge of the diaphragm.

dotted-decimal/dotted-quad notation A format commonly used for expressing 32-bit IPv4 addresses using four decimal numbers in the range 0 to 255, separated by decimal points (dots) (e.g., 192.168.12.254).

driver **1.** In audio, an individual loudspeaker unit. **2.** In electronics, a piece of software or firmware that takes a high-level command set and implements the commands in a specific format for the actual hardware or software present in the system.

dual channel In test equipment, refers to a test device with two independent input channels.

DVD Digital Video Disc or Digital Versatile Disc. An optical storage medium for data or video.

DVI Digital Visual Interface. A digital interface to connect an uncompressed video source to a display device. DVI has largely been replaced by HDMI, DisplayPort, and other formats.

Dynamic Host Configuration Protocol (DHCP) An IP address management process that automates the assignment of IP addresses and networking parameters to devices on an IP network.

dynamic microphone A microphone with a diaphragm attached to a moveable coil of wire located in a magnetic field. Sound pressure waves cause the diaphragm to move the coil in the magnetic field, inducing a small electric current in the coil.

dynamic range The difference between the loudest and quietest levels of a signal. Usually expressed in decibels (dB).

early reflected sound The sound waves that arrive at the listener's ear closely on the heels of the direct sound wave. These are the sound waves that are reflected (bounced) off surfaces between the source and the listener.

earthing conductor A conductor used to connect equipment or the grounded circuit of a wiring system to a grounding/earthing electrode (planet Earth).

echo A reflected or duplicated version of a signal that arrives at the listener with sufficient delay and separation from the original signal to allow the delayed signal to be perceived distinctly and later in time from the original signal.

echo cancellation A means of eliminating echo from a signal path.

electret microphone A type of condenser microphone using a prepolarized material, called an *electret,* which is applied to the microphone's diaphragm or backplate.

PART VII

electrical service   The conductors and equipment for delivering energy from the electricity supply system to the wiring system of the site served.

electromagnetic interference (EMI)   Interference signals produced by electromagnetic fields.

elevation drawing   A side view of an object or surface taken in the vertical plane from outside the object or surface.

emissive technology   Any display technology that emits light to create an image.

encoded   A signal that has been converted into another format.

energy management plan (EMP)   A document that details a systematic approach to implementing the most effective power consumption methods and procedures to achieve and maintain optimum energy usage.

equalizer   Electronic equipment that modifies the frequency characteristics of a signal.

equipment grounding   The connection to ground (earth) or to a conductive body that extends that ground connection of all normally noncurrent-carrying conductive materials enclosing electrical conductors or equipment or forming part of such equipment. The purpose is to limit any voltage potential between the equipment and earth/ground.

equipment grounding conductor (EGC)   The conductive path installed to connect normally noncurrent-carrying metal parts of equipment to the grounding/earthing electrode.

equipment rack   A centralized housing unit that protects and organizes electronic equipment.

equivalent acoustic distance (EAD)   The farthest distance one can go from the source without the need for sound amplification or reinforcement to maintain good speech intelligibility. It is a design parameter dependent on the level of the presenter and the noise level in the room.

ergonomics   Also known as human factors or human factors engineering. This is the scientific study of the way people interact with a system. It focuses on effectiveness, efficiency, reducing errors, increasing productivity, improving safety, reducing fatigue and stress, increasing comfort, increasing user acceptance, increasing job satisfaction, and improving quality of life.

Ethernet   A set of network cabling, signaling, and network access protocol standards. Before the introduction of switch-based networks Ethernet was based on carrier-sense multiple access technology with collision detection (CSMA/CD).

EtherSound   A proprietary digital audio Data Link layer protocol designed by Digigram. It uses standard 100Mbps or 1Gbps Ethernet cabling, switches, and other components. It requires a separate network with dedicated bandwidth. EtherSound signals are nonroutable.

expander   An audio processor that comes in two types: a downward expander and as part of a compander.

extended display identification data (EDID)   A data structure within a sink that is used to describe the sink's capabilities to a source. These capabilities include native resolution, color space information, and audio type (mono or stereo).

external configuration   Refers to the ability of one device to configure other devices and subsystems.

far field   The sound field distant enough from the sound source so the SPL decreases by 6dB for each doubling of the distance from the source.

farthest viewer   The viewer positioned at the farthest distance from the screen as defined by the viewing area.

feedback   **1.** In audio, unwanted noise caused by the loop of an audio system's output back to its input. **2.** In a control system, data supplied to give an indication of system performance or status.

feedback stability margin (FSM)   Extra margin added into the needed acoustic gain formula that represents extra gain that a sound system may need.

fiber-optic   A technology that uses total internal reflection of light along a transparent fiber to transmit information.

field   In interlaced video, one-half of a video frame containing every other line of information. Each standard interlaced video frame contains two fields.

filter   A device or process that blocks or passes certain frequencies from a signal.

firewall   Any technology, hardware, or software that regulates the traffic permitted to enter or exit a network. Firewalls control access across network boundaries.

firmware   A type of software that is permanently stored in a piece of hardware.

fixture   A luminaire that is mounted or fixed in place.

flex life   The number of times a cable can be bent before it breaks. A wire with more strands or twists per meter will have a greater flex life than one with a lower number of strands or fewer twists per meter.

focal length   The distance in millimeters between the center of a lens and the point where the image comes into focus. The shorter the focal length, the wider the angle of the image will be.

foot-candle   The U.S. customary unit of illuminance. The incident light measured when one lumen of light is spread over an area of one square foot. Its symbol is fc. 1 foot-candle = 10.76 lux.

footlambert   The footlambert is the U.S. customary unit for luminance. It is equal to $1/\pi$ candela per square foot. Its symbol is fl. 1 footlambert = 3.43 candela per square meter.

frame rate   The number of frames sent from a display source per second.

frequency   The number of complete cycles in a specified period of time. It is measured in hertz (Hz). 1Hz = 1 cycle per second.

frequency domain   A signal viewed as frequency versus amplitude is in the frequency domain. This allows you to view the amount of energy present at different frequencies.

frequency response   The amplitude response versus frequency for a given device.

Fresnel lens   A flat lens in which the curvature of a normal lens surface has been collapsed in such a way that concentric circles are impressed on the lens surface. A Fresnel lens is often used for the condenser lens in overhead projectors, in rear-screen projection systems, and in Fresnel spotlights.

front-screen projection   An image projection system where the image is projected from a source on the viewer's side of the screen.

full-duplex communication   A form of bi-directional data transmission in which messages may simultaneously travel in both directions.

full HD   A high-definition video mode with a resolution of 1920×1080 pixels.

fundamental frequency   Known as *pure tone,* the lowest frequency in a harmonic series.

gain   The increase in the amplitude of a signal.

gain control   A gain control is an electronic adjustment through which the gain of a signal can be increased or decreased.

gain-sharing automatic mixers   A gain-sharing automatic mixer is an audio mixer that automatically turns up microphone channels that are in use and turns down microphone channels that are not being used.

gate   An audio processor that allows signals to pass only above a certain setting or threshold.

gated automatic mixer   An audio mixer that turns microphone channels either "on" or "off" automatically.

gateway   A router device that connects a local network to an outside network, and all traffic must travel through it. A gateway will pass traffic to the routers below, and the routers below look to the gateway to find names (DNS addresses) that are not found on the local network.

gauge   The thickness or diameter of a wire or plate.

genlock   To lock the synchronization signals of multiple devices to a single synchronization source.

graphic equalizer   An equalizer with an interface that has a graph comparing amplitude on the vertical axis with frequency on the horizontal.

graphical user interface (GUI)   Often pronounced "gooey," provides a visual representation of a system's elements and functions.

graphics adapter   Commonly referred to as a *video card,* outputs computer video signals.

graphics processing unit (GPU)   A specialized circuit designed for processing display functions. The processor is optimized to render and manipulate images in a video frame buffer.

grayscale   The luminance (brightness and darkness) of a color. It is sometimes called *value.* It is one of the three attributes of color; the other two are hue and chroma.

grayscale test pattern   A pattern that displays a range of known gray values between black and white.

ground   **1.** The earth. **2.** In the context of an electrical circuit, the earth or some conductive body that extends the ground (earth) connection. **3.** In the context of electronics, the 0V (zero volt) circuit reference point. This electronic circuit reference point may or may not have a connection to the earth.

ground fault   **1.** An unintentional, electrically conducting connection between an ungrounded conductor of an electrical circuit and a normally noncurrent-carrying conductor, metallic enclosure, metallic raceway, metallic equipment, or earth. **2.** The electrical connection between any ungrounded conductor of the electrical system and any noncurrent-carrying metal object.

ground loop   An electrically conductive loop that has two or more ground reference connections. The loop can be detrimental when the reference connections are at different potentials, which causes current flow within the loop.

ground plane   A continuous conductive area. The fundamental property of a ground plane is that every point on its surface is at the same potential (low impedance) at all frequencies of concern.

ground potential   A point of zero potential in a circuit.

ground reference   The 0V (zero volt) reference point for a circuit.

grounded conductor   A system or circuit conductor that is intentionally grounded.

ground-fault circuit interrupter (GFCI)   A safety device that de-energizes a circuit within an established period of time when a current to ground exceeds a specified level. Similar in function to an earth leakage circuit breaker (ELCB), a residual current device (RCD), or a core balance relay (CBR).

ground-fault current path An electrically conductive path from the point of a ground fault on a wiring system through normally noncurrent-carrying conductors, equipment, or earth to the electrical supply source.

grounding Connecting to ground or to a conductive body that extends the ground connection. The connected connection is referred to as *grounded*.

grounding conductor A conductor used to connect equipment or the grounding circuit of a wiring system to a grounding/earth electrode or electrodes.

grounding electrode A conducting object through which a direct connection to earth (planet Earth) is established.

grounding electrode conductor The conductor used to connect the system grounded conductor or the equipment to a grounding electrode or to a point on the grounding electrode system.

group management protocol (GMP) Allows a host to inform its neighboring routers of its desire to start or stop receiving multicast transmissions. Without a GMP, multicast traffic is broadcast to every client device on the network segment, impeding other network traffic and overtaxing device CPUs.

group of pictures (GoP) A set of successive frames that are required to display a complete series in a digital video signal. It includes the visible picture, timing/sync information, and compression frames.

half-duplex communication A form of bi-directional data transmission in which messages can only travel in one direction at any time.

harmonic distortion Signal distortion that arises when harmonics of an input signal are produced during processing and appear in the output together with the processed input signal.

harmonics Higher-frequency waves that are a multiple of the fundamental frequency.

HD-15 connector HD-15 connector, sometimes called a VGA connector, is a video connector that is typically associated with the output of an analog computer graphics card. It has three rows of five pins, which carry analog red, green, blue, and sync signals along with display data channel information.

HDBaseT A connectivity standard for the transmission of high-definition video, audio, DC power, Ethernet, and serial signaling, including USB and other protocols, over standard twisted-pair data cables such as Cat 5e and above.

HDCP key A long number that a program uses to verify authenticity and encode/decode content. HDCP processes use multiple types of keys. These keys are strongly protected by Digital Content Protection, LLC.

HDCP receiver A device that can receive and decode HDCP signals. A television is an example of a receiver.

HDCP repeater   A device that can receive HDCP signals and transmit them to another device, such as a switcher or distribution amplifier.

HDCP sink   A device that receives and decodes the HDCP signals.

HDCP source   A device that sends HDCP-encoded signals and content.

HDCP transmitter   A device that can send HDCP-encoded signals and content. A Blu-ray Disc player is an example of an HDCP transmitter.

HDMI Ethernet Audio Control (HEAC)   In HDMI, the combining of HDMI Ethernet Channel (HEC) and Audio Return Channel (ARC) into one port or cable. See *HDMI Ethernet Channel (HEC)* and *Audio Return Channel (ARC)*.

HDMI Ethernet Channel (HEC)   Consolidates video, audio, and data streams into a single HDMI cable. A dedicated data channel enables high-speed, bi-directional networking to support future IP solutions and allow multiple devices to share an Internet connection.

HDTV   High-definition television. Generally includes image resolutions above 1280×720.

headend   The equipment located at the start of a cable distribution system where the signals are processed and combined prior to distribution.

headroom   The difference in dB between peak- and average-level performance of a system.

heat load   Heat load is the heat that is generated and released by a device. It is measured in joules (or British thermal units).

heat sink   A device that absorbs and dissipates the heat produced by a component.

hemispheric polar pattern   The dome shape of the region where some microphones will be most sensitive to sound. This pattern is used for boundary microphones.

hertz (Hz)   The unit of frequency. 1Hz = 1 cycle per second.

hextet   A group of 16 bits, usually written as four lowercase hexadecimal digits (e.g., 0fe8). The 128-bit addresses used in IPv6 are written as eight hextets separated by colons (e.g., 2006:0fe8:85a3:0000:0002:8a2e:0a77:c082).

High-Bandwidth Digital Content Protection (HDCP)   A form of encryption developed by Intel to control digital audio and video content as it travels across Digital Video Interface (DVI) or High-Definition Multimedia Interface (HDMI) connections. It prevents transmission or interception of unencrypted HD content.

High Definition Multimedia Interface (HDMI)   A point-to-point connection protocol between video devices for digital video and audio. HDMI signals include audio, control, and digital asset rights management information. It is fully compatible with DVI.

PART VII

high dynamic range (HDR)   Digital images having a bit depth of at least 10 bits (1,024 levels) per channel. Standard dynamic range images have a bit depth of 8 bits (256 levels) per channel.

high-pass filter   A circuit that allows signals above a specified frequency to pass unaltered while simultaneously attenuating frequencies below the specified limit.

hiss   Broadband higher-frequency noise generated by the amplification stages of an audio system. Typically associated with poor audio system gain structure.

horn   A loudspeaker that reproduces mid to high frequencies.

hot plug   **1.** A low-level signal sent by an EDID source that indicates whether a sink or display is connected. **2.** A system that can detect and respond to devices being connected or disconnected during normal operations (e.g., USB).

hue   The attribute of a color that represents its position in a defined color space or on the visible spectrum. Hue is usually described with a color name such as "red," "blue," "yellow," or "purple." It is one of the three attributes that define color; the other two are luminance and chroma.

hum   Undesirable (usually 50 to 60Hz-plus harmonics) noise emanating from an audio device or evidenced by a rolling hum bar on a display.

IEEE   The Institute of Electrical and Electronics Engineers.

illuminance   The amount of light falling on a surface, measured in lux (lx) or foot-candle (fc). 1 foot-candle = 10.76 lux.

image constraint token (ICT)   A digital flag signal incorporated into some high-definition digital video streams. If present, the video stream can be decoded at full resolution only on HDCP-enabled devices.

image resolution   Image resolution is the total number of pixels in the image. It is normally expressed as the number of horizontal pixels multiplied by the number of vertical pixels.

imager   A light-sensitive electronic device behind a camera's lens, usually made up of thousands of sensors that convert the light input into an electrical output.

impedance   The total opposition that a circuit presents to an alternating current. It includes resistance (R), inductive reactance (X_L), and capacitive reactance (X_C). Impedance is measured in ohms (Ω). Its symbol is Z.

impedance matching   Having an impedance value on an input that an output is expecting. It does not necessarily mean having comparable impedances on an input and an output.

impedance meter   Device used to measure the impedance of an electrical circuit.

inductance   The magnetic property of a circuit that opposes any change in current, represented by the symbol L and measured in henrys (H).

induction   The influence exerted on a conductor by a changing magnetic field.

inductive reactance (X_L)   Opposition to the current flow offered by the inductance of a circuit. Inductive reactance is measured in ohms (Ω). Its symbol is X_L. The inductive reactance in a circuit is directly proportional to the frequency of the current.

InfoFrames   Structured packets of data that carry information regarding aspects of audio and video transmission, as defined by the EIA/CEA-861 standard. Using this structure, a frame-by-frame summary is sent to the display, permitting selection of appropriate display modes automatically. InfoFrames typically include auxiliary video information, generic vendor-specific source product description, MPEG, and audio information.

infrared (IR)   The range of non-visible light frequencies below the red end of the visible spectrum. IR signal transmission requires an uninterrupted line of sight between transmitter and receiver.

input   A connection point that receives information from another piece of equipment.

I/O port   Input and/or output port. A connection port on a device for handling input and/or output signals.

insulation   A material of high dielectric strength used to isolate the flow of electric current between conductors.

Integrated Services Digital Network (ISDN)   A communications standard for transmitting data over digital telephone lines.

intellectual property (IP)   A type of property that includes the creations of the human intellect. It includes ideas; written material such as books, articles, poems, and essays; designs including plans, specifications, diagrams, sketches, and program code; images such as photographs, graphic arts, and other artworks; music and songs; and movie and video contents. Intellectual property rights are usually asserted through the use of patents, trademarks, copyrights, and trade secrets. All IP used in the design, execution, and operation of an AV installation must be correctly licensed from each of the owners of that IP.

intelligibility   A sound system's ability to produce an accurate reproduction of sound, allowing listeners to identify words and sentence structure.

Internet Corporation for Assigned Names and Numbers (ICANN)   A nonprofit organization chartered to oversee several Internet-related tasks. ICANN manages Domain Name System (DNS) policy, including the top-level domain space for the Internet.

Internet Group Management Protocol (IGMP)   The group management protocol for IPv4. IGMPv1 allowed individual clients to subscribe to a multicast channel. IGMPv2 and IGMPv3 added the ability to unsubscribe from a multicast channel.

interlaced scanning   The scanning process that alternately displays the odd and even lines of a video frame to construct a full frame of video signal.

internal configuration   Refers to the setup and customization of management or control of a device.

Internet of Things (IoT)   Refers to situation where network connectivity and computing capability has been extended to include objects, sensors, and other devices, allowing these devices to exchange data.

Internet Protocol (IP)   A TCP/IP protocol defined in the IETF standard RFC 791. IP defines rules for addressing, packaging, fragmenting, and routing data sent across an IP network. IP falls under the Internet layer of the TCP/IP protocol stack and the Network layer of the OSI Model.

Internet Protocol Television (IPTV)   A system that delivers television services over an IP network such as a LAN or the Internet.

inverse square law   The law of physics stating that a physical quantity or strength is inversely proportional to the square of the distance from the source of that physical quantity.

IoT   See *Internet of Things (IoT)*.

IP   See *Internet Protocol (IP)* and *intellectual property (IP)*.

IR   See *Infrared (IR)*.

isolated ground (IG)   An equipment grounding method permitted by the North American NEC for reducing electrical noise (electromagnetic interference) on the grounding circuit. The isolation between IG receptacles and circuits and the normal equipment grounding is maintained up to the point of the service entrance (or a separately derived system) where the grounded (neutral) conductor, equipment grounding, and isolated equipment grounding conductor are bonded together and to earth/ground (planet Earth).

isolated grounding circuit   A circuit that allows an equipment enclosure to be isolated from the raceway containing circuits. The equipment on the circuit is grounded via an insulated earthing/grounding conductor.

isolated receptacle   A power receptacle or mains power outlet in which the grounding terminal is purposely insulated from the receptacle-mounting means. In North America isolated receptacles are identified by a triangle engraved on the face. The receptacle (and so the equipment plugged into the receptacle) is grounded via an insulated earthing/ grounding conductor.

jacket   Outside covering used to protect cable wires and their shielding.

junction box   A metal or plastic enclosure for enclosing the junction of electrical wires and cables. A junction box can be used as a termination point with a custom connector plate or interface plate. A junction box can also be installed and used as a pull box for longer cable runs.

keystone error   The trapezoidal distortion of an image due to the projection device being at an angle to the plane of the screen.

lamp   The light source in some luminaires or projectors.

latency   Response time of a system. The delay between an input being received by a system and the corresponding output being generated. Measured in seconds.

lavalier   A small microphone usually worn either around the neck or attached to apparel. Derived from the name of an item of jewelry worn as a pendant.

Law of Conservation of Energy   States that energy cannot be created or destroyed. Energy can be transformed from one form to another and transferred from one body to another, but the total amount of energy remains constant.

lenticular screen   A screen surface characterized by embossing designed to transmit or reflect maximum light over wide horizontal, but narrow vertical, angles.

lighting fixture   A lighting instrument or luminaire.

limiter   An audio signal processor that functions like a compressor except that signals exceeding the threshold level are reduced at ratios of 10:1 or greater.

limiter ratio   Defines how much the limiter will compress signals that exceed its threshold. The limiter compresses only the portion of the signal that exceeds its threshold.

limiter threshold   Defines which portions of the signal the limiter will affect. All decibel levels below or equal to the threshold will pass through the limiter unchanged. All signals above the threshold will be compressed.

line conditioner   See *Power conditioner*.

line driver   An amplifier used to compensate for signal attenuation created by cable impedance.

line level   The specified strength of an audio signal used to transmit analog audio between the elements of an audio system. This is generally considered to be approximately 1V at 1kHz into a 600Ω impedance.

liquid crystal display (LCD)   A video display technology that uses light transmission through polarizing liquid crystals to display an image.

liquid crystal on silicon (LCoS)   A reflective liquid crystal imaging technology. A liquid crystal layer is applied to a reflective complementary metal-oxide semiconductor (CMOS) mirror substrate. The embedded CMOS circuitry controls the reflectivity of the liquid crystal pixels.

listed   Equipment, materials, or services included in a list published by a recognized testing laboratory. The term is usually applied to products or processes tested by the U.S. Underwriters Laboratories (UL).

load center   An electrical industry term used to identify a lighting and appliance electrical distribution board in residential and light-commercial applications.

local area network (LAN)   A computer network connecting devices within a confined geographical area, such as a building or living complex.

local monitor   A local device used to monitor the output signal from a system.

logarithm   The exponent of 10 that equals the value of a number.

logic network diagram   A project management tool that aids in sequencing and ultimately scheduling a project's activities and milestones. It represents a project's critical path as well as the scope for the project.

lossless compression   A process that compresses data without losing any information.

lossy compression   A form of compression that produces an approximation of the original data by eliminating noncritical or redundant information.

loudspeaker   A transducer that converts electrical energy into acoustical energy.

loudspeaker circuit   A group of wired loudspeakers.

low-pass filter   A circuit that allows signals below a specified frequency to pass unaltered while simultaneously attenuating frequencies above the one specified.

low smoke zero halogen (LSZH) cable   A type of cable with insulation and sheathing that produces only low levels of smoke and no halogen products on combustion. Suitable for use in ventilation plenum spaces.

low voltage   An ambiguous term. It may mean less than 70V AC to an AV technician, while an electrician may use the same term to describe circuits less than 600V or 1kV AC. The meaning of the term may also be determined by the authority having jurisdiction (AHJ).

lumen   The unit of luminous flux. A measure of the total quantity of visible light emitted from light source per unit of time. Its symbol is lm.

luminaire   A lighting instrument. Consists of a light source, optical system, housing, and mounting mechanism.

luminance (Y)   Also called *luma,* the brightness component of a combined video signal that includes synchronization, color, and brightness information. Its symbol is Y.

lux   The international unit of illuminance. The incident light measured when one lumen of light is spread over an area of one square meter. Its symbol is lx. 10.76 lux = 1 foot-candle.

MAC address   The unique 48-bit hardware address of a network device, usually written as six groups of two hexadecimal numbers, separated by a hyphen or colon (e.g., 01-23-45-67-89-ab or 01:23:45:67:89:ab).

mains buzz   A mixture of higher-order harmonics of the 50Hz or 60Hz noise (hum) originating from the AC mains power system and audible in a sound system.

matrix decoder   A video decoder that extracts red, green, and blue signals from either composite or Y, R-Y, and B-Y signals.

matrix switcher   An electronic switching device with multiple inputs and outputs. The matrix allows any input to be connected to any one or more of the outputs.

matte-white screen   A screen that uniformly disperses light, both horizontally and vertically, creating a wide viewing cone and wide viewing angle.

maximum transmission unit (MTU)   The size in bytes of the largest frame that can pass over a Data Link layer connection. Header information must be included within the MTU.

mechanical switcher   A switch that mechanically opens and closes to connect circuit elements. It functions like a wall switch, meaning there is a mechanical connection or disconnection.

MEMS   See *Microelectromechanical systems (MEMS)*.

MEMS microphone   A microphone built using MEMS technology. Generally, MEMS implementations of either condenser or piezo-electric microphones.

mesh topology   A network where each node is connected via bridges, switches, or routers to at least one other node.

metropolitan area network (MAN)   A communications network that covers a single geographic area, such as a suburb or city.

mic level   The very low-level signal from a microphone. Typically only a few millivolts.

microphone sensitivity   A specification that indicates the electrical output of a microphone when it is subjected to a known sound pressure level. Usually measured in dBV/Pa.

microelectromechanical systems (MEMS)   Mechanical devices built directly onto silicon chips using the same fabrication processes as microprocessors and memory systems. Best known as the digital micromirror devices (DMD) used for light switching in DLP projectors.

middleware   Software that provides services to applications that aren't available from the operating system. In a streaming system, for example, middleware may perform transcoding, compression, or remote access authentication.

midrange   A loudspeaker that reproduces midrange frequencies, typically 300Hz to 8kHz.

milestone   A significant or key event in the project, usually the completion of a major deliverable or the occurrence of an important event. It can often be associated with payment milestones and client approvals.

millwork   Carpentry work produced in a mill. Usually refers to finished woodwork such as doors, molding, trim, flooring, cupboards, and wall paneling.

mixer   A device for blending multiple signal sources.

mix-minus system   A type of speech reinforcement system that allows both meeting presenters and participants to be heard. Each loudspeaker is given a separate subsystem, which mixes the microphone signals, minus the closest microphone.

Mobile High-Definition Link (MHL)   A standard audio/video interface for connecting mobile electronics to high-definition televisions and audio receivers. SuperMHL is capable of 36Gbps with HDR and WCG video up to 8K at 120fps.

modular connector   A latching connector used with four, six, or eight pins. Common modular connectors are RJ-11 and RJ-45 (8P8C).

modulator   A device that varies one or more properties of a carrier signal (frequency, amplitude, phase) with information from another signal.

Multicast Listener Discovery (MLD)   The IPv6 group management protocol. Multicast is natively supported by IPv6; any IPv6 router will support MLD. MLDv1 performs roughly the same functions as IGMPv2, and MLDv2 supports roughly the same functions as IGMPv3.

multicast streaming   A one-to-many transmission, meaning one server sends out a single stream that can then be accessed by multiple clients. In IPv4, class D IP addresses (224.0.0.0 to 239.255.255.255) are reserved for multicast transmissions. In IPv6, multicast addresses have the prefix ff00::/8.

multimeter   A test instrument with multiple ranges for measuring current, voltage, and resistance. Many instruments also include a simple continuity test capability.

multiplexing   The sharing of a single communications channel for multiple signals.

multipoint   Also called *continuous presence,* videoconferencing that links many sites to a common gateway service, allowing all sites to see, hear, and interact at the same time. Multipoint requires a bridge or bridging service.

Multi-Protocol Label Switching (MPLS)   A networking protocol that allows any combination of Data Link layer protocols to be transported over any type of Network layer. MPLS routes data by examining each packet's MPLS label without examining packet contents. Implementing MPLS improves interoperability and routing speed.

Murphy's law   Also known as Sod's law. *If anything can go wrong, it will go wrong.* A reminder that no assumptions should ever be made about the cause of a fault in a system. If you saw the identical fault last week, it will have an entirely different cause this week.

native resolution   The number of rows of horizontal and vertical pixels that create the picture. The native resolution describes the actual resolution of the imaging device and not the resolution of the delivery signal.

near-field   The sound field close to the sound source that has not been colored by room reflections. This is also known as *direct sound*.

needed acoustic gain (NAG)   The gain the sound system requires to achieve an equivalent acoustic level at the farthest listener equal to what the nearest listener would hear without sound reinforcement.

needs analysis   A needs analysis identifies the activities that the end users need to perform, then develops the functional descriptions of the systems that support those needs.

network address translation (NAT)   Any method of altering IP address information in IP packet headers as the packet traverses a routing device. NAT is typically implemented as part of a firewall strategy. The most common form of NAT is port address translation (PAT).

network bridge   A network device that connects between two networks. It may translate between different protocols on the bridged networks.

network interface card (NIC)   An interface that allows a device to be connected to a network.

network segment   A network segment is any single section of a network that is physically separated from the rest of the network by a networking device such as a switch or router. A segment may contain one or more hosts.

network switch   A networking device that connects multiple network segments by storing the data packets sent from the transmitting segment and forwarding them to the receiving segment. Hosts on the segments are identified by their MAC addresses.

network topology   The physical arrangement of the elements connected to a network.

neutral conductor   The conductor in an electrical supply system that is connected to the earth/ground for current return. This is not part of the protective earth/grounding system. Also see *Grounded conductor* for U.S. usage.

nibble/nybble   A group of four bits. Half a byte or half an octet. Usually written as a single hexadecimal digit (values 0 to f).

nine-pin connector   The DB 9 is the most common type of connector used in RS-232 control systems.

nit   The unit of luminance (cd/m^2). Used to measure screen or surface brightness.

noise   Any signal present in a system other than the desired signal.

noise criterion (NC) rating   Developed to establish satisfactory conditions for speech intelligibility and general living environments. Measurements are taken at eight center octave frequencies from 63Hz to 8kHz and plotted against a standardized curve.

noise-masking system   Also known as a sound-masking or speech privacy system. A sound system that deliberately introduces background noise into a space to raise the threshold of hearing and thus increase privacy between occupants in a shared space.

noisy ground   An electrical connection to a ground point that produces or injects spurious voltages into the computer system through the connection to ground (IEEE Std. 142-1991).

nominal impedance   An estimate of the typical impedance of a device or system.

notch filter   A filter that notches out, or eliminates, a specific band of frequencies.

number of open microphones (NOM)   Takes into account the increased possibility of feedback by adding more live microphones in a space. Each time the number of open microphones is doubled, you lose 3dB of gain before feedback.

Nyquist-Shannon Sampling Theorem   States that an analog signal can be reconstructed if it is encoded using a sampling rate that is greater than twice the highest frequency sampled. For example, since the range of human hearing extends to 20kHz, the minimum sampling rate for digital audio should be greater than 40kHz. The higher the sampling rate above this minimum, the more accurate the digital sample.

octave   A band, or group, of frequencies in which the highest frequency in the band is double the lowest frequency. For example, 200Hz to 400Hz is an octave, 6kHz to 12kHz is an octave, and so on.

octet   A group of eight bits, often called a byte.

Ohm's law   A law that defines the relationship between current, voltage, and resistance in an electrical circuit. The current is proportional to the applied voltage and inversely proportional to the resistance, giving the formula $I = V \div R$, where I is the current (in amps), V is the voltage (in volts), and R is the resistance (in ohms).

omnidirectional   Receiving signals from or transmitting in all directions. Used to describe the sensitivity or radiation pattern for devices that operate equally in nearly all directions.

on-axis   Along the axis of a symmetrical pattern. In projection, the center line of a screen, perpendicular to the viewing area for a displayed image. In audio, along the central axis of a microphone's pick-up pattern or along the central axis of a loudspeaker's dispersion pattern.

organic light-emitting diode (OLED)   A semiconductor light-emitting diode constructed from organic compounds. Displays built from OLEDs generally use separate layers for emitting the red, green, and blue components of an image.

oscilloscope   A device that allows the viewing and measurement of electronic signals on a visual display.

OSI Model   Open Systems Interconnection Model. This is a reference model developed by ISO in 1984 as a conceptual framework of standards for communication in the data network across different equipment and applications by different vendors. Network communication protocols fall into seven categories, or layers.

overcurrent   Any current in excess of the rated current for equipment or a conductor. It may result from overload, a short circuit, or a ground fault.

overcurrent protection device   A safety device designed to disconnect a circuit if the current exceeds a predetermined value. Examples are circuit breakers and fuses.

packet filtering   A network data filtering process that uses rules to determine whether a data packet will be allowed to pass through a router or gateway. The filtering rules are based on the contents of the protocol header of each packet.

panelboard   The North American name for an electrical load distribution board.

parallel circuit   A circuit in which the voltage is the same across each load, but the current divides and takes all the available paths and returns to the source.

parametric equalizer   An equalizer that allows discrete selection of a center frequency and adjustment of the width of the frequency range that will be affected. This can allow for precise manipulation with minimal impact of adjacent frequencies.

peak   The highest level of signal strength, determined by the height of the signal's waveform.

peaking   An adjustment method that allows compensation for high-frequency loss in cables.

peaking control   Electronic adjustments within a video component that can be used to compensate for system losses.

permissible area   The maximum amount of space that cables should occupy inside an electrical conduit.

personal area network (PAN)   A limited-range, usually wireless, network that serves a single person or small workgroup.

phantom power   A DC power supply delivered as an "invisible" overlay on the signal wires of a system. In audio phantom power systems, a voltage is overlaid on microphone signal lines to enable the remote powering of devices such as condenser microphones and active direct input boxes. In power over Ethernet (PoE) systems, the voltage is overlaid across both wires in a twisted pair.

phase   A particular value of time for any periodic function. For a point on a sine wave, it is a measure of that point's distance from the most recent positive-going zero crossing of the waveform. It is measured in degrees; 0 to 360 degrees is a complete cycle.

phono connector   The international name for the type of coaxial connector known as RCA in North America.

phosphor   A substance that produces light when stimulated by radiation. Phosphors are used to produce some colors of visible light in fluorescent lamps, LED sources, some laser light sources, and CRT and plasma displays.

pink noise   A signal with a broad spectrum of random frequencies that has equal energy in each octave band.

pink noise generator (PNG)   A device to generate pink noise. An audio PNG is commonly used in conjunction with an audio spectrum analyzer to evaluate and align a sound system in an environment.

pixel   A contraction of the words *picture* and *element*. The smallest element used to build an image.

plan view   A drawing of a space from the "top view" taken directly from above. Examples include a floor plan and site plan.

plane of screen   Identification of image position on a plan or drawing relative to other plotted locations. It is a notional line, whether in plan view or elevation, that aligns with the front surface of the screen (that is, image position) used as a datum to define viewers' relative positions.

plasma display panel (PDP)   A direct-view display technology consisting of an array of pixels, which are composed of three subpixels, corresponding to the colors red, green, and blue. Gas in the plasma state is used to react with phosphors in each subpixel to produce colored light (red, green, or blue) from a phosphor in each subpixel.

playback system   A system designed specifically to play back recorded material.

plenum space   The plenum space is also called *environmental air space*. It is an area connected to air ducts that forms part of the air distribution system.

point source   A sound system that has an apparent central location for the loudspeakers. This type of sound system is typically used in a performance venue or a large house of worship.

point-to-point   Conferencing where sites are directly linked.

polar pattern   Also known as *pickup pattern* or *transmission pattern*. The shape of the pattern covered by the pickup or transmission device.

polar plot   A polar plot is a graphical representation of the relationship between a device's directionality and its input or output.

port   **1.** An input and/or output socket on an electronic device. **2.** In a TCP/IP network, a 16-bit number included in the TCP or UDP Transport layer header. The port number typically indicates the Application layer protocol that generated a data packet. A port may also be called by its associated service (e.g., port 80 may be called HTTP, or port 23 may be called Telnet). **3.** To move an application or function to a new platform.

port address translation (PAT)   A method of network address translation (NAT) whereby devices with private, unregistered IP addresses can access the Internet through a device with a registered IP address. Unregistered clients send datagrams to a NAT server with a globally routable address (typically a firewall, a gateway, or a router). The NAT server forwards the data to its destination and relays responses to the original client.

post tension type construction   A type of structure that uses metal cables embedded within the concrete slab to support the structure. The cables act as a suspension support system that allows for wider spacing of support structures within a building.

potential acoustic gain (PAG)   The potential gain that can be delivered by the sound system without ringing and before feedback occurs. It is based upon the number of open microphones and the distances between sources (like a presenter) and microphones, microphones and loudspeakers, and listeners and loudspeakers.

power   The rate at which work is done. Measured in watts (W); 1 watt = 1 joule/second. The symbol for power is P.

power amplifier   Amplifies an audio signal to a level sufficient to drive loudspeakers.

power conditioner   Also known as a line conditioner or power line conditioner. A device that conditions the quality of power being fed to equipment by regulating the voltage and eliminating line noise.

power distribution unit (PDU)   An electrical device that distributes mains power to multiple electrical devices. A PDU may contain switches, overcurrent protection, voltage and/or current monitoring, remote circuit-controllers, and power receptacles.

Power over Ethernet (PoE)   A DC power supply delivered as an "invisible" overlay on the data pairs of an Ethernet network system. The voltage is overlaid equally across both wires in a twisted pair to eliminate any effect on the data. PoE is used to power a wide range of Ethernet-connected devices.

power sequencing   The act of powering on and off equipment that requires a progressive startup or shutdown sequence for safe or convenient operations. Sequencing may help prevent tripping circuit breakers by limiting surge or inrush currents when devices are powered up.

preamplifier   An amplifier that boosts a low-level electronic signal before it is sent to other processing equipment.

primary optic   Also known as the primary lens. The major lens in an optical system. In a projector, the primary lens usually controls the focus of the image on a screen.

program report   A document that describes the client's specific needs, the system purpose and functionality, and the designer's best estimate of probable cost in a nontechnical format for review and approval by the owner. This is also known as the AV narrative, the discovery phase report, the return brief, or the concept design report.

progressive scanning   Scanning that traces the image's scan lines sequentially.

Protocol Independent Multicast (PIM)   Allows multicast routing over LANs, WANs, or the open Internet. Rather than routing information on their own, PIM protocols use the routing information supplied by whatever routing protocol the network is already using.

pulling tension   The maximum amount of tension that can be applied to a cable or conductor before it is damaged.

pure tone   See *Fundamental frequency.*

Q factor   The quality factor of an audio filter is the ratio of the height of the center frequency of the filter compared with the bandwidth of the filter at the 3dB point.

quality of service (QoS)   Any method of managing network data traffic to preserve system functionality and provide the required bandwidth for critical applications. Typically, QoS involves some combination of bandwidth allocation and data prioritization.

quiet ground   A point on a ground system that does not inject spurious voltages into the computer system. There are no standards to measure how quiet a quiet ground is.

raceway   An enclosed channel of metal or nonmetallic materials designed for carrying wires, cables, or busbars.

rack   See *Equipment rack.*

rack elevation diagram   A rack elevation diagram is a pictorial representation of the front of a rack and the location of each piece of equipment within that rack, typically labeling the number of RUs used for each piece of gear.

rack unit (RU)   A unit of measure of the vertical space in a rack. One RU equals 1.75 inches (44.5mm).

radio frequency (RF)   The portion of the electromagnetic spectrum that is suitable for radio communications. Generally, this is considered to be from 10kHz up to 300MHz. This range extends to 300GHz if the microwave portion of the spectrum is included.

radio frequency interference (RFI)   Radiated electromagnetic energy that interferes with or disturbs an electrical circuit.

rarefaction   A decrease of density and pressure in a compressible medium such as air.

ratio   A mathematical expression that represents the relationship between the quantities of numbers of the same kind. A ratio is typically written as X:Y or X/Y.

RCA connector   The North American name for the *phono* connector, a coaxial connector most often used with line-level audio signals and consumer composite video signals.

reactance   Opposition to current flow in a circuit resulting from the reaction of the capacitance and inductance in the circuit. Measured in ohms (Ω). The symbol for reactance is X.

rear-screen projection   A system in which the image is projected toward the audience through a translucent screen material for viewing from the opposite side.

reference level   In the context of decibel measurements, the reference level is the established starting point represented by 0dB. The reference level varies according to unit and application.

reference point   The point of zero potential used as the voltage reference for a circuit.

reflected ceiling plan   A plan used to illustrate elements in the ceiling with respect to the floor. It should be interpreted as though the floor is a mirrored surface, reflecting the features within the ceiling.

reflecting server   A caching server in a content delivery network (CDN). Takes in a unicast stream and broadcasts out a multicast stream.

reflection   Light or sound energy that has been redirected by a surface.

reflective technology   A display device that reflects light to create an image.

refraction   The bending or changing of the direction of a light ray when passing between transparent mediums such as air, water, glass, or a vacuum. The *refractive index* of a material is a measure of the speed that light travels through the medium in comparison to its speed through a vacuum.

refresh rate   The number of times per second a display device will update the display of a received image.

release time   The release time of an audio compressor determines how quickly the volume increases when an audio signal returns below the threshold.

relocatable power tap (power strip)   A North American cord-connected product rated 250V AC or less and 20A or less with multiple receptacles. This tap is intended only for indoor use and plugged directly into a branch circuit. It is not intended to be connected to another relocatable power tap.

reserve DHCP   A hybrid approach to IP address allocation. Using reserve DHCP, a block of addresses is reserved for devices requiring a static IP address. The remaining IP addresses in the subnet pool are assigned dynamically using DHCP.

residual current device (RCD)   A safety device that de-energizes a circuit (or a portion of that circuit) within an established period of time when a current to ground exceeds a specified level. Similar in function to an earth leakage circuit breaker (ELCB), a ground-fault circuit interrupter (GFCI), or a core balance relay (CBR).

resistance   The property of a material to impede the flow of electrical current. Measured in ohms (Ω). The symbol for resistance is R.

resistor   A passive electrical device that produces opposition to current flow. Current passes through a resistor in direct proportion to voltage, independent of frequency. The relationship between voltage across and current through a resistor is defined in Ohm's law.

resolution   **1.** The amount of detail in an image. **2.** The number of picture elements (pixels) in a display.

reverberant sound   Sound waves that bounce off multiple surfaces before reaching the listener but arrive at the listener's ears quite a bit later than early reflected sound.

reverberation   Numerous, persistent reflections of sound energy.

RF   See *Radio frequency (RF)*.

RF control   A method of control employing RF wireless signaling. RF control systems vary in complexity from simple one-way on/off signals to high-bandwidth, multi-channel, bi-directional systems with complex user interfaces and rich feedback. Wireless control may use many formats, including Wi-Fi, DECT, ISM, Bluetooth, UWB, or LTE frequencies and signaling protocols.

RF distribution system   A closed-circuit television distribution system with each of the composite video and audio program signals modulated onto a radio frequency carrier signal for distribution via RF cables. Receiving devices must have a demodulator capable of extracting the separate program channels.

RGBHV signal   A high-bandwidth video signal with separate conductors for the red signal, green signal, blue signal, horizontal sync, and vertical sync.

RGBS signal   A four-component video signal composed of a red signal, a green signal, a blue signal, and a composite sync signal.

rigid metal conduit   Rigid metal conduit, called *rigid* in North America, is the heaviest conduit and offers the best physical and EMI protection.

rigid nonmetallic tubing   Rigid nonmetallic tubing is very stiff with a thick wall but lightweight. It is similar to plumbing tubing. Because it is not flexible, it is available in preformed pieces at various angles.

ring   A network topology that connects terminals, computers, or nodes in a continuous loop.

room criteria (RC) rating   With measurements taken at eight center-octave frequencies from 31.5Hz to 8kHz, the average of the measurements taken from 500, 1kHz, and 2kHz. This includes additional steps to rate the background noise as (N) for neutral, (R) for rumbly, or (H) for hissy.

room mode   An acoustic wave-interference phenomenon that occurs between the parallel surfaces of an enclosure where the dimension between those parallel surfaces equals one-half wavelength (and the harmonics thereof) of a frequency. The wave is thus reflected on itself out of polarity creating location-specific areas of maximum and minimum pressure.

router   A network device that forwards data packets between computer networks. It operates on the OSI Network layer (layer 3).

RS-232   A point-to-point serial data protocol. The interface between data terminal equipment and data circuit-terminating equipment employing serial binary data interchange. It supports a single-ended mode of operation with one driver and one receiver. At a cable length of 15m (50ft) RS-232 supports a data rate of up to 19.2kbps. At its maximum cable length of 900m (3,000ft), it supports a data rate of 2.4kbps.

RS-422   A multidrop serial data protocol. Provides the electrical characteristics of balanced-voltage digital interface circuits. It is a balanced signal with one driver and 10 receivers with multidrop capability. The maximum cable length for RS-422 is 1,220m (4,000ft) with a data rate of 10Mbps.

RS-485   A transmission line serial data protocol. Supports a differential mode of operation with 32 tri-state drivers and 32 tri-state receivers and multidrop capability. The maximum cable length for RS-485 is 1.2km (4,000ft) with a data rate of 10Mbps.

RsGsBs   A video transmission system using red, green, and blue signals with composite sync added to each color channel. This requires three cables to carry the entire signal. It is often referred to as *RGB sync on all three*.

RT$_{60}$ The time taken for the energy in an initially steady reverberant sound field to decay by 60dB after the source of the sound ceases.

RU   **1.** See *Rack unit (RU)*. **2.** A CTS renewal unit. The completion of renewal unit–accredited training is required by CTS, CTS-I, and CTS-D certification holders to maintain the currency of their certification.

sampling rate   The number of samples taken per unit of time when converting a continuous analog signal to a discrete digital signal.

scale   The representation of a number by another number differing by a fixed ratio.

scale drawing   A drawing that shows objects in accurate proportion, with all dimensions enlarged or reduced in a fixed ratio.

scaler   A processor that changes the resolution of an image without changing its apparent content. Scaling may be required when the image resolution does not match the resolution of the display device.

scan rate   The rate at which a raster-scan image is displayed. Horizontal scan rate is the rate at which a single horizontal line is displayed. Vertical scan rate (refresh rate) is the rate at which an entire screen image is displayed.

scattering   When a wave hits a textured surface, the incoming waves are reflected in multiple directions because the surface is uneven.

scene   A recallable preset of lighting levels for one or more zones.

scope statement   A written agreement between the client, the project sponsor, the key stakeholders, and the project team that defines the boundaries of the project.

screen gain   The ability of a projection screen to concentrate the light reflected from it. Gain is compared to the reflection of a matte-white screen, which reflects light uniformly in all directions (a gain of 1).

SDTV   Standard-definition television. It has a 4:3 aspect ratio and a resolution of 576i for PAL and SECAM, and 480i for NTSC.

section drawing   A section drawing is a view of the interior of a building in the vertical plane. Section drawings show a bisected wall, which allows you to view what is behind it.

sensitivity specification   The measure of a device's capacity to convert one form of energy into another form of energy. Usually stated as a ratio of input units to output units, such as mV/Pa for microphones.

serial digital interface (SDI)   A set of serial data standards to transport digital video data over BNC-terminated 75Ω coaxial cable. Variants include HD-SDI for HD video, 6G-SDI (6Gbps) for 4kp30, 12G-SDI (12Gbps) for 4kp60, and 24G-SDI (24Gbps) for 8kp30 video. SDI can also be transported over optical fiber.

series circuit   A circuit arrangement where all the current supplied by the source will flow through the entire circuit. While all the current flows through all the circuit elements, the voltage is divided between the loads and the wires that connect them.

series/parallel loudspeaker circuit   When groups of loudspeakers in circuit branches are wired together in series. Typically, loudspeakers in the same branch have the same impedance. Each branch is connected to the positive and negative lines of the amplifier in parallel.

server   A computer system that shares resources and services with other connected devices.

service-level agreement (SLA)   A document used to record an agreement between a service provider and a customer. It describes the services to be provided, documents service-level targets, and specifies the roles and responsibilities of the service provider(s) and the customer(s).

shear force   The force exerted on an object in the direction of the object's cross section. In the case of a wall-mounting bolt, the shear force across the bolt caused by the gravitational load of the object it is supporting.

shield or screen   A grounded conductive partition placed between two regions of space to control the propagation of electric and magnetic fields between them. The screen or shield acts as a Faraday cage. It can be the chassis (metallic box) that houses an electronic device or the conductive enclosure (aluminum foil, conductive polymer, or copper braid) that surrounds a screened/shielded wire or cable.

shotgun microphone   A long, cylindrical, highly sensitive microphone with a very narrow pickup pattern.

sightline   The line of sight between a viewer and an object that needs to be seen.

signal flow   The traceable path of signals through a system.

signal generator   A test instrument that produces calibrated electronic signals for the testing or alignment of electronic circuits or systems.

signal ground   **1.** A 0V (zero volt) point of no potential that serves as the circuit reference. **2.** A low-impedance path for the current to return to the source.

single-phase power   Alternating current electrical power supplied by two current-carrying conductors. This type of supply is used for residential and many light-commercial applications.

single-point ground (SPG)   In the context of IEEE Std. 1100, refers to implementation of an isolated equipment grounding configuration for the purposes of minimizing problems caused by circulating current in ground loops.

signal-to-noise (S/N) ratio   The ratio, measured in decibels, between an information signal and the accompanying noninformation noise. The higher this ratio, the cleaner the signal.

Simple Network Management Protocol (SNMP)   A set of Internet Engineering Task Force (IETF) standards for network management.

socket   In a TCP/IP network, the combined port number, Transport Layer protocol identifier, and IP addresses of communicating end systems. A socket uniquely identifies a session of a given transport protocol.

sound pressure level (SPL)   The effective pressure level of a sound, usually stated in relation to a reference pressure such as 20µPa (the threshold of human hearing). In the context of AVIXA Standard 1M, expressed in unweighted decibels.

sound reinforcement system   The combination of microphones, audio mixers, signal processors, power amplifiers, and loudspeakers that are used to electronically amplify and distribute sound.

source-specific multicast (SSM)   In data streaming SSM allows clients to specify the sources from which they will accept content. This has the dual benefits of reducing demands on the network while improving network security.

specification   A written, precise description of the design criteria for a piece of work. Specifications define the level of qualitative and/or quantitative parameters to be met and the criteria for their acceptance. All specifications must be formulated in terms that are specific, measurable, and verifiable and unambiguous.

specular reflection   The mirror-like reflection of electromagnetic radiation, in which most of the radiation is reflected in a single direction.

speech privacy system   A sound system that adds background noise to an environment to raise the hearing threshold to make it more difficult to hear low-level sounds such as traffic noise, machinery, and distant human speech. Used to assist with speech privacy in open environments.

speech-reinforcement system   An audio system that reinforces or amplifies a presenter's voice.

splitter   An electronic device that splits a signal to route it to different devices.

spot photometer   Also known as a spot meter. A type of photometer used to measure illuminance over a narrow angle.

standing wave   Occurs between parallel reflecting surfaces (e.g., walls in a room or a loudspeaker cabinet) where the distance between those parallel surfaces equals one-half wavelength of a wave (and the harmonics thereof). The wave is thus reflected on itself out of phase, creating an interference pattern with location-specific areas of maximum and minimum amplitude.

star topology   A network topology where all network devices are connected to a central network device, usually a switch.

static IP address   A permanently assigned IP address.

stereophonic   An audio reproduction system where multiple outputs are designed to create the illusion of sound perspective.

streaming video/streaming audio   Sequence of moving images or sounds sent in a continuous stream over the Internet and displayed by the viewer as they arrive. With streaming video or audio, a web user does not need to wait to download a large file before seeing the video or hearing the sound.

subnet   A logical group of hosts within a local area network (LAN). A LAN may consist of a single subnet, or it may be divided into several subnets. Additional subnets may be created by modifying the subnet mask on the network devices and hosts.

subnet mask   A number that identifies which part of an IP address corresponds to the network address and which corresponds to a host address on that network. In CIDR notation the mask is represented by a slash (/), followed by the number of bits as a decimal number (e.g., /24). In dotted-decimal notation the mask is represented by a dotted-decimal representation of the bit pattern for network and host address bits, where bits equal to 1 indicate that the corresponding bits in the IP address identify the network address, and bits equal to 0 identify the host address (e.g., 255.255.255.0, which is equivalent to /24). IP addresses with the same network identifier bits are on the same subnet.

subwoofer   A loudspeaker that reproduces lower frequencies, typically 20Hz to 200Hz.

supercardioid polar pattern   The exaggerated heart shape of the area where a highly directional microphone is most sensitive to sound.

surface-mount microphone   Also called a *boundary microphone,* a microphone designed to be mounted directly against a hard boundary or surface, such as a conference table, a stage floor, or a wall, to pick up sound.

surround-sound system   An audio system that uses multiple channels to produce an acoustic experience where the sound appears to surround listeners.

switcher   A device used to select one of several available signals.

switch-mode power supply   A type of DC power supply that uses a switching regulator to control the output voltage.

sync   Synchronization. The timing information used to coordinate signals and events.

system   In the AV industry, a compilation of multiple individual AV components and subsystems interconnected to achieve a communication goal.

system black   The lowest level of luminance a video system is capable of producing for its operating conditions.

system ground   The point at which the safety earth/ground for an electrical system is connected to the earth, usually through a highly conductive earthing/grounding spike driven into the ground (planet Earth).

tap   A connection to a transformer winding that allows you to select a different power level from the transformer.

task lighting   Lighting directed to a specific surface or area that provides illumination for visual tasks.

tensile strength   The maximum force that a material can withstand before deforming or stretching.

three-phase power   Alternating current electrical power supplied by three current-carrying conductors, each carrying an AC voltage with a phase offset of 120 degrees from one another. A fourth conductor, a neutral, is used as the return conductor. This type of supply is used for commercial and industrial applications.

threshold   The point at which a function or effect becomes active.

throw distance   The distance between a light source, such as a projector or luminaire, and a focusing surface, such as a stage or a screen.

Thunderbolt   Interface technology that transfers audio, video, power, and data over one cable in two directions. Thunderbolt versions 1 and 2 use the same connector as Mini DisplayPort (MDP), while Thunderbolt 3 uses USB Type-C.

PART VII

time code   A sequence of numeric codes generated at fixed intervals to provide a time synchronization signal. The Society of Motion Picture and Television Engineers (SMPTE) time code used throughout the AV and production industries uses an eight-digit address scheme, representing the hour, minute, second, and frame number for each frame of a video sequence. The SMPTE time code is encoded in a wide variety formats, including being embedded in audio, video, and data streams.

time domain   A view of a signal as amplitude versus time. The display on a time-based oscilloscope shows the input signal in the time domain.

Time-Sensitive Networking (TSN)   The IEEE working group overseeing the Audio Video Bridging standard has been renamed Time-Sensitive Networking to reflect the standard's applicability to communication among different types of devices, such as network sensors. See *Audio Video Bridging (AVB)*.

transduction   The process by which one form of energy is changed into another.

transformer   A passive electrical device that electromagnetically transfers energy between two AC circuits. Commonly constructed of at least two electrically isolated induction coils sharing a common core.

transient disturbance   A momentary variation in a signal, such as a surge, spike, sag, dropout, or spurious noise.

transition-minimized differential signaling (TMDS)   A technology for transmitting high-speed serial data. The signaling method used in the HDMI and DVI interfaces.

transmission   The passage of a wave through a medium. Examples include the transmission of soundwaves through air and the transmission of electromagnetic waves through space.

Transmission Control Protocol (TCP)   A connection-oriented, reliable Transport layer protocol. TCP transport uses two-way communication to provide guaranteed delivery of information to a remote host. It is connection-oriented, meaning it creates and verifies a connection with the remote host before sending it any data. It is reliable because it tracks each packet and ensures that it arrives intact. TCP is the most common transport protocol for sending data across the Internet.

transmission loss   The attenuation that occurs as a signal moves through a medium. Usually expressed in decibels.

transmissive technology   Any display device that creates images by allowing or preventing light to pass.

tweeter   A loudspeaker that is specifically designed to reproduce frequencies above 3kHz.

twisted-pair   A pair of wires that are twisted around each other to facilitate common-mode noise rejection.

two-way/three-way loudspeaker   A loudspeaker enclosure containing two or three separate loudspeakers, each designed to optimally reproduce a portion of the audio spectrum. The multispeaker enclosure is intended to cover the entire spectrum from a single cabinet. Each loudspeaker may be fed from a separate amplifier (bi-amplification or tri-amplification), or the entire enclosure may be fed from a single amplifier with an internal crossover filter network used to send the optimal frequency band to each of the loudspeakers.

UHD ecosystem   The video cameras, recorders, servers, media players, displays, distribution, processing, and networking technologies used for recording, editing, producing, delivering, and displaying ultra-high-definition video.

ultra-high-definition (UHD or ultra HD)   A term used to describe video formats with a minimum resolution of 3840×2160 pixels in a 16×9 or wider aspect ratio.

ultra-wideband (UWB)   UWB is a low-power communications protocol that transmits an extremely wide bandwidth (500+Mhz) signal using time modulation to encode the data. Its signals are highly immune to interference from other RF systems, yet produce very little interference with them. UWB can detect the distance between the transmitter and receiver, enabling proximity detection and triggering.

unbalanced circuit   A circuit in which one conductor carries the signal and another conductor carries the return. The return conductor is often the cable shield or drain wire and is a low-impedance connection connected to the signal ground. As the impedances of the two conductors are quite different, they are unbalanced with respect to one another.

unicast streaming   A one-to-one connection between the streaming server sending out the data and client devices listening to the stream. Each client has a direct relationship with the server. Since the server is sending out a separate stream for each client, unicast streaming of media to three clients at 100Kbps actually uses 300Kbps of bandwidth. IP unicast streams may use either UDP or TCP transport, although TCP transport will inevitably require some buffering.

unity gain   Derived from the number 1, refers to no change in gain.

Universal Serial Bus (USB)   A standard for connecting, communicating, and supplying power between electronic devices. Version 3.2 of USB is capable of communicating at 20Gbps and can utilize a USB Type-C (USB-C) connector, which supports DisplayPort, HDMI, power, all USB generations, and VGA. USB 4 communicates at up to 40Gbps and includes handling the Thunderbolt protocol.

unshielded twisted-pair (UTP) cable   Typically used for data transfer, UTP cable contains multiple two-conductor pairs twisted at regular intervals, employing no external shielding.

vectorscope   A specialized oscilloscope-type display used in video systems to display and measure chrominance accuracy and levels. A vectorscope mode may be included in video waveform monitoring systems.

PART VII

vertical scan rate   The number of complete fields a device draws in a second. This may also be called the *frame rate, vertical sync rate,* or *refresh rate.* The vertical scan rate is measured in hertz (Hz).

video wall   A video display composed of a matrix of smaller video displays linked to display a contiguous image.

viewing angle   The angle at which a viewer is located in reference to the center axis of a display.

viewing area plan   A plan-view drawing of the viewing environment that identifies five viewing locations as defined in the requirements section of the ANSI/AVIXA V201.01 standard, Image System Contrast Ratio.

viewing cone   The volume of space containing the audience viewing a display. The term *cone* is used because there is width, height, and depth to the viewing space, which emanates from the center of the display.

virtual local area network (VLAN)   A network created when network devices on separate LAN segments are linked to form a logical group.

virtual private network (VPN)   A virtual point-to-point private connection established across a network via an encrypted tunneling protocol. VPNs are used for secure remote access, monitoring, troubleshooting, and control.

visual acuity   The eye's ability to discern fine details. There are several different kinds of acuity, including resolution acuity, which is the ability to detect that there are two stimuli, rather than one, in a visual field, and recognition acuity, which is the ability to identify correctly a visual target, such as differentiating between a *G* and a *C*.

visual field   The volume of space that can be seen when a person's head and eyes are absolutely still. It is specified as an angle, usually in degrees. The visual field of a single eye is termed *monocular vision,* and the combined visual field where the perceived image from both eyes overlap is called *binocular vision.*

Voice over Internet Protocol (VoIP)   A suite of technologies and protocols that allow the transmission of telephone calls and multimedia over Internet Protocol (IP) networks.

volt   The basic international unit of potential difference or electromotive force. It is represented by the symbol V.

voltage   The electrical potential difference across a circuit.

watt   The international unit of power. It is represented by the symbol w.

waveform monitor   An oscilloscope-type monitor used to display the waveforms of signals. A video waveform monitor is a specialized monitor used to display and analyze a video signal's sync, luminance, and chroma. Some waveform monitors include a *vectorscope* mode.

wavelength   The distance between the corresponding points on two consecutive cycles of a wave. Measured in meters.

wayfinding   The use of audiovisual guides or signage to assist with navigation to a destination.

webcasting   Allows the broadcast of digital media such as audio or video over the World Wide Web, which audience members can stream live or access on demand. Essential equipment for webcasting includes computers, streaming servers, production software, recording gear, appliances, and more.

white noise   A signal with a broad spectrum of random frequencies at the same energy level.

wide area network (WAN)   A data communications network that links local area networks (LANs) that are distributed over large geographic areas, such as cities, states, countries, and regions.

wire   A single conductive element intended to carry a current.

wireless access point   A network device that allows other devices to access a wireless network.

wireless local area network (WLAN)   A network that shares information by radio frequency (RF) wireless transmission.

woofer   A loudspeaker that produces low frequencies, typically 20Hz to 200Hz.

work breakdown structure (WBS)   A deliverable-oriented grouping of project elements that will ultimately organize and define the total scope of the project. Each descending level represents an increasingly detailed definition of a project component.

XLR connector   A secure, low-voltage connector used in professional audiovisual systems. The three-pin version is the standard audio signal cable for the production and AV industries. The four-pin version is widely used for communication headsets, and the five-pin version is the standard connector for the DMX512 digital lighting protocol. Also known as a *Cannon connector*.

zone   **1.** A defined area within a system. **2.** In lighting, a zone is a grouping of luminaires (lighting fixtures) that are focused on the same area. **3.** In digital signage, a zone is an area where specific content may be placed. **4.** In audio, a zone is an area where the same program is delivered.

INDEX